Stochastic Processes and Models

Stochastic Processes and Models

David Stirzaker
St John's College, Oxford

OXFORD
UNIVERSITY PRESS

Great Clarendon Street, Oxford OX2 6DP

Oxford University Press is a department of the University of Oxford.
It furthers the University's objective of excellence in research, scholarship,
and education by publishing worldwide in

Oxford New York

Auckland Cape Town Dar es Salaam Hong Kong Karachi
Kuala Lumpur Madrid Melbourne Mexico City Nairobi
New Delhi Shanghai Taipei Toronto

With offices in

Argentina Austria Brazil Chile Czech Republic France Greece
Guatemala Hungary Italy Japan Poland Portugal Singapore
South Korea Switzerland Thailand Turkey Ukraine Vietnam

Oxford is a registered trade mark of Oxford University Press
in the UK and in certain other countries

Published in the United States
by Oxford University Press Inc., New York

© Oxford University Press, 2005

The moral rights of the author have been asserted
Database right Oxford University Press (maker)

First published 2005

All rights reserved. No part of this publication may be reproduced,
stored in a retrieval system, or transmitted, in any form or by any means,
without the prior permission in writing of Oxford University Press,
or as expressly permitted by law, or under terms agreed with the appropriate
reprographics rights organization. Enquiries concerning reproduction
outside the scope of the above should be sent to the Rights Department,
Oxford University Press, at the address above

You must not circulate this book in any other binding or cover
and you must impose the same condition on any acquirer

British Library Cataloguing in Publication Data

Data available

Library of Congress Cataloging in Publication Data

Data available

Typeset by Newgen Imaging Systems (P) Ltd., Chennai, India
Printed in Great Britain
on acid-free paper by
Biddles Ltd., King's Lynn, Norfolk

ISBN 0–19–856813–4   978–0–19–856813–1
ISBN 0–19–856814–2(Pbk.)   978–0–19–856814–8(Pbk.)

3 5 7 9 10 8 6 4 2

Contents

Preface		ix
1	**Probability and random variables**	**1**
1.1	Probability	1
1.2	Conditional probability and independence	4
1.3	Random variables	6
1.4	Random vectors	12
1.5	Transformations of random variables	16
1.6	Expectation and moments	22
1.7	Conditioning	27
1.8	Generating functions	33
1.9	Multivariate normal	37
2	**Introduction to stochastic processes**	**45**
2.1	Preamble	45
2.2	Essential examples; random walks	49
2.3	The long run	56
2.4	Martingales	63
2.5	Poisson processes	71
2.6	Renewals	76
2.7	Branching processes	87
2.8	Miscellaneous models	94
2.9	Some technical details	101
3	**Markov chains**	**107**
3.1	The Markov property; examples	107
3.2	Structure and n-step probabilities	116
3.3	First-step analysis and hitting times	121
3.4	The Markov property revisited	127
3.5	Classes and decomposition	132
3.6	Stationary distribution: the long run	135
3.7	Reversible chains	147

3.8	Simulation and Monte Carlo	151
3.9	Applications	157

4 Markov chains in continuous time 169

4.1	Introduction and examples	169
4.2	Forward and backward equations	176
4.3	Birth processes: explosions and minimality	186
4.4	Recurrence and transience	191
4.5	Hitting and visiting	194
4.6	Stationary distributions and the long run	196
4.7	Reversibility	202
4.8	Queues	205
4.9	Miscellaneous models	209

5 Diffusions 219

5.1	Introduction: Brownian motion	219
5.2	The Wiener process	228
5.3	Reflection principle; first-passage times	235
5.4	Functions of diffusions	242
5.5	Martingale methods	250
5.6	Stochastic calculus: introduction	256
5.7	The stochastic integral	261
5.8	Itô's formula	267
5.9	Processes in space	273

6 Hints and solutions for starred exercises and problems 297

Further reading 323

Index 325

... and chance, though restrained in its play within the right lines of necessity, and sideways in its motions directed by free will, though thus prescribed to by both, chance by turns rules either, and has the last featuring blow at events.

Herman Melville, *Moby Dick*

In such a game as this there is no real action until a guy is out on a point, and then the guys around commence to bet he makes this point, or that he does not make this point, and the odds in any country in the world that a guy does not make a ten with a pair of dice before he rolls seven, is 2 to 1.

Damon Runyon, *Blood Pressure*

Preface

This book provides a concise introduction to simple stochastic processes and models, for readers who have a basic familiarity with the ideas of elementary probability. The first chapter gives a brief synopsis of the essential facts usually covered in a first course on probability, for convenience and ease of reference. The rest of the text concentrates on stochastic processes, developing the key concepts and tools used in mainstream applications and stochastic models.

In particular we discuss: random walks, renewals, Markov chains, martingales, the Wiener process model for Brownian motion, and diffusion processes. The book includes examples and exercises drawn from many branches of applied probability.

The principal purpose here is to introduce the main ideas, applications and methods of stochastic modelling and problem-solving as simply and compactly as possible. For this reason, we do not always display the full proof, or most powerful version, of the underlying theorems that we use. Frequently, a simpler version is substituted if it illustrates the essential ideas clearly. Furthermore, the book is elementary, in the sense that we do not give any rigorous development of the theory of stochastic processes based on measure theory.

Each section ends with a small number of exercises that should be tackled as a matter of routine. The problems at the end of any chapter have the potential to require more effort in their solution.

The ends of proofs, examples, and definitions are denoted by the symbols □, ○, and △, respectively. And all these, with key equations, are numbered sequentially by a triple of the form (1.2.3); this example refers to the third result in the second section of Chapter 1. Within sections a single number is used to refer to an equation or proof etc. in the same section. Exercises are lettered (a), (b), (c), and so on, whereas Problems are numbered; solutions to selected Exercises and Problems appear at the end of the book, and these are starred in the text.

1 Probability and random variables

> To many persons the mention of Probability suggests little else than the notion of a set of rules, very ingenious and profound rules no doubt, with which mathematicians amuse themselves by setting and solving puzzles.
>
> *John Venn,* The Logic of Chance

> "It says here he undoubtedly dies of heart disease. He is 63 years old and at this age the price is logically thirty to one that a doctor will die of heart disease. Of course," Ambrose says, "if it is known that a doctor of 63 is engaging in the rumba, the price is one hundred to one."
>
> *Damon Runyon,* Broadway Incident

This chapter records the important and useful results that are usually included in introductory courses on probability. The material is presented compactly for purposes of reference, with long proofs and elaborations mostly omitted. You may either skim the chapter briskly to refresh your memory, or skip it until you need to refer back to something.

1.1 Probability

Randomness and probability are not easy to define precisely, but we certainly recognize random events when we meet them. For example, randomness is in effect when we flip a coin, buy a lottery ticket, run a horse race, buy stocks, join a queue, elect a leader, or make a decision. It is natural to ask for the probability that your horse or candidate wins, or your stocks rise, or your queue quickly shrinks, and so on.

In almost all of these examples the eventualities of particular importance to you form a proportion of the generality; thus: a proportion of all stocks rise, a proportion of pockets in the roulette wheel let your bet win, a proportion of the dartboard yields a high score, a proportion of voters share your views, and so forth.

In building a model for probability therefore, we have it in mind as an extension of the idea of proportion, formulated in a more general framework. After some reflection, the following definitions and rules are naturally appealing.

Note that, here and elsewhere, a word or phrase in italics is being defined.

(1) **Experiment.** Any activity or procedure that may give rise to a well-defined set of outcomes is called an *experiment*. The term well-defined means simply that you can describe in advance everything that might happen. △

(2) **Sample space.** The set of all possible outcomes of an experiment is called the *sample space*, and is denoted by Ω. A particular but unspecified outcome in Ω may be denoted by ω. △

The main purpose of a theory of probability is to discuss how likely various outcomes of interest may be, both individually and collectively, so we define:

(3) **Events.** An *event* is a subset of the sample space Ω. In particular, Ω is called the *certain event*, and its complement Ω^c is an event called the *impossible event*, which we denote by $\Omega^c = \emptyset$, the empty set. △

If two events A and B satisfy $A \cap B = \emptyset$, then they are said to be *disjoint*. Note that while all events are subsets of Ω, not all subsets of Ω are necessarily events. To make it easy to keep track of events, we require that the family of events is closed under intersections and unions. Formally we make this definition:

(4) **Event space.** A collection of events is called an *event space*, and denoted by $\mathcal{F}$, if it satisfies

(a) $\emptyset$ is in $\mathcal{F}$;
(b) if A is in $\mathcal{F}$, then A^c is in $\mathcal{F}$;
(c) if A_n is in $\mathcal{F}$ for $n = 1, 2, \ldots$, then $\bigcup_{n=1}^{\infty} A_n$ is in $\mathcal{F}$.

(The event space may also be called an *algebra* of events or a *sigma field*.)△

Naturally, we seek to assign probabilities to the events in $\mathcal{F}$, and it follows that 'probability' is a function defined on the events in $\mathcal{F}$. We denote this function by P($\cdot$). Because we think of probability as related to the idea of proportion, it is consistent, convenient, and conventional to require that for any event A its probability P(A) satisfies $0 \leq P(A) \leq 1$.

Our experience of proportions and probabilities in transparent and trivial cases leads us to make the following general definition of the properties of the function P($\cdot$):

(5) **Probability function.** The function P($\cdot$), defined on the events in a space $\mathcal{F}$, is a *probability* if:

(a) $P(\Omega) = 1$;
(b) $0 \leq P(A) \leq 1$ for any event A;

(c) if $A_j \cap A_k = \emptyset$ for $j \neq k$, then

$$P\left(\bigcup_{n \in I} A_n\right) = \sum_{n \in I} P(A_n),$$

where $(A_n; n \in I)$ may be a finite or countably infinite collection of events.

Note that $P(\cdot)$ may be called a probability *measure*, or a probability *distribution*. The entire structure comprising Ω, with its event space $\mathcal{F}$, and probability function $P(\cdot)$, is called a *probability space*. Property (c) is often called the axiom of *countable additivity*. △

From these simple basic rules, we may immediately deduce a number of extremely useful and important relationships between the probabilities of events of interest. Note that henceforth symbols like A, B, A_n, B^c, A_3, and so on, will always denote events in $\mathcal{F}$.

(6) **Basic properties**. Using (5) we find that, for example,

(7) $$P(A^c) = 1 - P(A),$$

(8) $$P(\emptyset) = 0,$$

(9) $$P(A \cup B) = P(A) + P(B) - P(A \cap B),$$

(10) $$P(A) \leq P(B), \quad \text{if } A \subseteq B,$$

(11)
$$P\left(\bigcup_{r=1}^n A_r\right) = \sum_{r=1}^n P(A_r) - \sum_{r<s} P(A_r \cap A_s) + \cdots + (-1)^{n+1} P\left(\bigcap_{r=1}^n A_r\right),$$

(12) if $\lim_{n \to \infty} A_n = A$, then $\lim_{n \to \infty} P(A_n) = P(A)$.

Identity (11) is sometimes called the *inclusion–exclusion* identity, and the result in (12) is sometimes called the *continuity property* of $P(\cdot)$.

> **Exercises**
> (a)* Prove all the results (7)–(12) above.
> (b) Show that for any events A and B, $P(A) - P(B) \leq P(A \cap B^c)$.

(c)* **Boole's inequality.** For any events $A_1, A_2, \ldots, A_n$, show that

$$P\left(\bigcup_{r=1}^{n} A_r\right) \leq \sum_{r=1}^{n} P(A_r).$$

1.2 Conditional probability and independence

Frequently we may know that some event B in Ω has in fact occurred; or we may wish to investigate the consequences if it were supposed that it had. The sample space Ω is effectively reduced to B, and considerations of proportion lead us to define the concept of conditional probability thus:

(1) **Conditional probability.** If $P(B) > 0$, then the *conditional probability* that A occurs given that B occurs is denoted by $P(A \mid B)$, and defined by

(2) $$P(A \mid B) = P(A \cap B)/P(B).$$

More generally, whether $P(B) > 0$ or not, we agree that

(3) $$P(A \cap B) = P(A \mid B)P(B).$$

This is sometimes called the basic *multiplication rule*. △

It is most important to remember that $P(A \mid B)$ is a probability function as we defined above, so that if we define, for each A in $\mathcal{F}$, the function

$$Q(A) = P(A \mid B),$$

this satisfies all the basic properties in (1.1.6), and consequent results. For example (1.1.7) becomes

(4) $$P(A^c \mid B) = 1 - P(A \mid B)$$

and so on. Furthermore, the definition of conditional probability, together with the basic rules for probability, supply several further important and useful identities. For example,

(5) **Partition Rule.** In its simplest form this says

$$P(A) = P(A \mid B)P(B) + P(A \mid B^c)P(B^c).$$

More generally, if $B_j \cap B_k = \emptyset$ for $j \neq k$, and $A \subseteq \bigcup_r B_r$, then

$$P(A) = \sum_r P(A \mid B_r)P(B_r).$$

Both identities follow immediately from the definition of conditional probability and (1.1.11).

(6) **Multiplication Rule.** The elementary form $P(A \cap B) = P(A \mid B)P(B)$ in (3) is easily extended to the more general form

$$P\left(\bigcap_{r=1}^{n+1} A_r\right) = P(A_1)P(A_2 \mid A_1)P(A_3 \mid A_1 \cap A_2) \cdots P\left(A_{n+1} \mid \bigcap_{r=1}^{n} A_r\right).$$

This is a simple exercise for you.

Finally, if we combine the extended Partition Rule (5), with definition (2), we obtain the celebrated

(7) **Bayes' Rule.** If $A_j \cap A_k = \emptyset$ for $j \neq k$, and $B \subseteq \bigcup_{r=1}^{n} A_r$, and $P(B) > 0$, then

$$P(A_r \mid B) = \frac{P(B \mid A_r)P(A_r)}{\sum_{r=1}^{n} P(B \mid A_r)P(A_r)}.$$

A popular application of this result is to the interpretation of test results; see the exercises. ○

The idea of conditioning leads us to another key concept. It may happen that $P(A \mid B) = P(A)$, and this is sufficiently important to deserve a definition:

(8) **Independence**

(a) Events A and B are *independent* if

$$P(A \cap B) = P(A)P(B).$$

(b) A family $(A_i, 1 \leq i \leq n)$ is *independent* if

$$P(A_{i_1} \cap A_{i_2} \cap \cdots \cap A_{i_r}) = P(A_{i_1}) \cdots P(A_{i_r})$$

for any selection $1 \leq i_1 < i_2 < \cdots < i_r \leq n$.

(c) A family $(A_i, 1 \leq i \leq n)$ is *pairwise independent* if $P(A_i \cap A_j) = P(A_i)P(A_j)$, $i \neq j$. This is implied by independence, but does not imply it.

(d) Events A and B are *conditionally independent* given C if

$$P(A \cap B \mid C) = P(A \mid B)(P(A \mid C).$$

This does not imply the independence of A and B, nor is it implied by the independence of A and B, (except in trivial special cases). △

Exercises

(a)* Three dice are rolled. Define the events

$A \equiv$ The first and second show the same.
$B \equiv$ The second and third show the same.
$C \equiv$ The first and third show the same.

Show that the events A, B, and C are pairwise independent but not independent.

(b) A total of n coins are flipped, of which exactly one is biased, and the rest are fair. Let A be the event that the n coins together show exactly one head or exactly one tail; let B be the event that the biased coin shows a head. Prove that A and B are independent. Now let C be the event that one of the fair coins shows a head. Is A independent of C? Is A independent of $B \cap C$?

(c) **Testing.** A test for a disease gives a positive result with probability $\frac{99}{100}$ when given to an ill person, and with probability $\frac{1}{100}$ when given to a well person. What is the probability that you have the disease if the test shows positive and

(i) the disease affects one person in 100?

(ii) the disease affects one person in 10^5?

Discuss the implications of your results.

1.3 Random variables

In real life, we are often not so much interested in the actual outcomes ω in the sample space Ω, as in numerical results that depend on ω. (For example, in any bet, one is commonly more interested in the resulting gain or loss than in the outcome of the randomizer that decides the issue.) Also, in many cases the sample space Ω actually is the real line $\mathbb{R}$, or some subset of it. And finally, even if neither of these is true, we may just assign a number $X(\omega)$ to any outcome ω in Ω. This effectively transfers our interest from Ω to the possible values in $\mathbb{R}$ taken by the function $X(\omega)$. Then, naturally, we are interested in the probability of $X(\omega)$ lying in various sets of interest in $\mathbb{R}$.

This preamble makes all the following definitions natural and inevitable. Note that we always denote a random variable by a capital letter (X, Y, Z, S, T, etc.), but the possible values of a random variable are always denoted by lowercase letters ($x, y, z, s, t, i, j, k, n$, etc.). (However, not all uppercase letters denote random variables.)

(1) **Random variable.** Given Ω and a probability distribution P, a *random variable* is a function $X(\omega)$, defined on Ω, taking values in $\mathbb{R}$. △

(2) Distribution function. The *distribution function* $F_X(x)$ of X is denoted by $F_X(x) = P(X \leq x)$, and defined by

$$F_X(x) = P(B_x),$$

where
$$B_x = \{\omega : X(\omega) \leq x\}.$$

(Note that this definition entails the condition on any random variable $X(\cdot)$ that $B_x \in \mathcal{F}$ for all x; we assume this always.) △

It follows that we have for all x, a, and $b > a$,

$$P(X > x) = 1 - F_X(x).$$

and

(3) $$P(a < X \leq b) = F(b) - F(a).$$

All random variables have a distribution function. There are two principal types of random variables: the discrete, and the continuous.

(4) Discrete random variables. A *discrete random variable* takes values only in some countable subset D of $\mathbb{R}$. Very commonly this subset D is a subset of the integers. Then the probability that X takes some given value x in D is denoted by

$$f(x) = P(X = x) = P(V_x),$$

where $V_x = \{\omega : X(\omega) = x\}$ is the event that $X = x$. The function $f(x)$ may be called the *probability function* of X, or the *probability mass function* of X, or the *probability distribution* of X. It may also be denoted by $f_X(x)$ to avoid ambiguity. △

Here are some familiar and important examples of discrete random variables and their distributions.

(5) Binomial distribution. If you perform n independent trials with probability p of success at each trial, and probability $1 - p$ of failure, then the number X of successes has the distribution

(6) $$f_X(x) = \binom{n}{x} p^x (1-p)^{n-x}, \quad 0 \leq x \leq n.$$

We often refer to this as the *Binomial distribution*, and denote it by $B(n, p)$ or Binomial (n, p).

It is important and useful to consider the case when n is large, and p small; to do this we set $np = \lambda$, where λ is fixed, and allow $n \to \infty$. Then it is an easy exercise for you to show that in the limit we obtain the:

(7) **Poisson distribution.** The distribution of X is

(8) $$f_X(x) = \frac{e^{-\lambda}\lambda^x}{x!}, \quad x = 0, 1, 2, 3, \ldots.$$

We may refer to this as Poisson (λ).

Considering an unlimited sequence of trials gives rise to another key random variable.

(9) **Geometric distribution.** In a sequence of independent trials with probability p of success at each trial, and probability $1 - p$ of failure, let X be the number of trials up to and including the first success. Then X has distribution

(10) $$f_X(x) = (1 - p)^{x-1} p, \quad x = 1, 2, 3, \ldots.$$

We may refer to this as Geometric (p).

It is easy to see that any countable collection of positive numbers $(f(n), n \in \mathbb{Z})$, such that $\sum_n f(n) = 1$, is the distribution of some discrete random variable. Further examples appear in the exercises and problems.

We like to work with the probability function $f_X(x)$ of X because of the simple but crucial result that for any subset C of the possible values D of X,

(11) $$P(X \in C) = \sum_{x \in C} f_X(x).$$

In particular,

$$F_X(x) = P(X \leq x) = \sum_{y \leq x} f_X(y).$$

Thus we can obtain the distribution function $F(\cdot)$ from the probability function $f(\cdot)$, and vice versa.

Note finally that a remarkably useful discrete random variable is also one of the simplest. Let $P(X = 1) = p$, and $P(X = 0) = 1 - p$. Then X is called either an *indicator*, or a *Bernoulli random variable*, which we may refer to as Bernoulli (p).

We now turn to the second principal type of random variable.

(12) **Continuous random variables.** A random variable that is not discrete is said to be *continuous* if its distribution function $F(x)$ may be written in

the form

(13) $$F_X(x) = \int_{-\infty}^{x} f_X(u) \, du,$$

for some non-negative function $f_X(x)$ defined for all x. Then $f_X(x)$ is called the *density function* (or simply, *density*) of X. We may denote it by $f(x)$ if there is no risk of ambiguity. It is analogous to the probability mass function $f(x)$ of a discrete random variable. △

Here are some important examples.

(14) **Uniform density.** If you pick a point at random in some interval (a, b), with the usual naive interpretation of 'at random', then it is easy to see that the position X of the point satisfies

$$F(x) = P(X \le x) = \frac{x-a}{b-a} = \int_a^x \frac{1}{b-a} \, dx, \quad a < x < b.$$

Thus, X has density

$$f(x) = \begin{cases} \dfrac{1}{b-a}, & \text{when } a < x < b, \\ 0, & \text{otherwise.} \end{cases}$$

We digress for a moment here to point out that if $b - a < 1$, then $f(x) > 1$ for $a < x < b$. By contrast, for the probability mass function $f(x)$ of a discrete random variable, $f(x) \le 1$ always.

In general it is clear that any non-negative function $f(x)$ having the property that

$$\int_{-\infty}^{\infty} f(x) \, dx = 1$$

is the density of a random variable. We conclude our list of examples with the three most important densities.

(15) **Exponential density.** For any constant $\lambda > 0$,

(16) $$f(x) = \begin{cases} \lambda e^{-\lambda x}, & x \ge 0, \\ 0, & x < 0. \end{cases}$$

We may refer to this as Exponential (λ).

(17) **Normal density.** For any constants μ and $\sigma^2 > 0$,

(18) $$f(x) = \frac{1}{(2\pi\sigma^2)^{1/2}} \exp\left(-\frac{1}{2\sigma^2}(x-\mu)^2\right), \quad -\infty < x < \infty.$$

We may refer to this as $N(\mu, \sigma^2)$. One special case must be mentioned. If $\mu = 0$ and $\sigma^2 = 1$, this is the *standard Normal Density*. For this we use the special notation

$$\phi(x) = \frac{1}{\sqrt{2\pi}} \exp\left(-\frac{1}{2}x^2\right), \tag{19}$$

$$\Phi(x) = P(X \leq x) = \int_{-\infty}^{x} \phi(u)\, du. \tag{20}$$

We use these special symbols partly because the standard normal density is so important, and partly because the density $f(x)$ of $N(\mu, \sigma^2)$ cannot be integrated simply in terms of familiar elementary functions. One result of this is that it is not trivial to show $\phi(x)$ actually is a density; see Problem 1 for some not-too-hard proofs.

We like to work with the density $f(x)$ because (analogously to the probability mass function $f(x)$ in the discrete case) it has the key property that for any interval (a, b),

$$P(a < X \leq b) = \int_a^b f(u)\, du. \tag{21}$$

Furthermore it follows from the defining property (13) of $f(x)$ that

$$f(x) = \frac{dF}{dx}, \tag{22}$$

except perhaps at those places where dF/dx does not exist. At these points we merely set $f(x) = 0$. (There are extreme cases where this cavalier attitude is insufficient, but none are in this book.) Thus the density can be obtained from the distribution and vice versa. Note that it is sometimes convenient to write (22) more informally as

$$P(x < X < x + h) \simeq f(x)h, \quad \text{for small } h.$$

Furthermore, if A is any subset of $\mathbb{R}$ comprising a countable union of disjoint intervals, then

$$P(X \in A) = \int_{x \in A} f(u)\, du. \tag{23}$$

For technical reasons, we cannot expect (23) to be true for every set A (see the Appendix to Chapter 2 for some background to this). So we make the convention that whenever we write down an expression like $P(X \in A)$, the set A is of the type for which (23) is indeed true.

These facts and functions are particularly useful when we come to consider simple transformations and functions of random variables. A few important examples will suffice to give you the key ideas.

(24) Example. Scaled and shifted normal random variables. Let X be $N(\mu, \sigma^2)$, as defined in (17), and define $Y = aX + b$, for constants a and b. Show that Y is $N(a\mu + b, a^2\sigma^2)$.

Solution. If $a = 0$ then the result (like the distribution of Y), is trivial. Suppose that $a > 0$. The following type of argument is standard, and we use it often:

$$F_Y(y) = P(Y \leq y) = P(aX + b \leq y)$$
$$= F_X\left(\frac{y-b}{a}\right).$$

Differentiating for y yields the density of Y

$$f_Y(y) = \frac{1}{a} f_X\left(\frac{y-b}{a}\right)$$
$$= \frac{1}{(2\pi\sigma^2 a^2)^{1/2}} \exp\left\{-\frac{1}{2a^2\sigma^2}(y - b - a\mu)^2\right\}.$$

Comparing this with definition (18) of a normal density proves what we want, and repeating the working with $a < 0$ completes the solution.

An important corollary of this is that when X is $N(\mu, \sigma^2)$, $Y = (X - \mu)/\sigma$ is $N(0, 1)$. That is to say, Y has the standard normal density $\phi(x)$. ○

(25) Example. Let U be a random variable uniformly distributed on $(0, 1)$, and define $Y = -(1/\lambda) \log U$. Show that when $\lambda > 0$, Y is Exponential (λ).

Solution. Proceeding as in the previous example, for $y \geq 0$,

$$P(Y \leq y) = P\left(-\frac{1}{\lambda} \log U \leq y\right) = P(\log U > -\lambda y)$$
$$= P(U \geq e^{-\lambda y}) = 1 - e^{-\lambda y}.$$

Now differentiating for y yields the required density given in (15). ○

Exercises
(a) **Median.** The number m is called a median of X if

$$P(X < m) \leq \tfrac{1}{2} \leq P(X \leq m).$$

Show that every random variable has at least one median, and give an example of a random variable with more than one median.

Find the medians of the Binomial, Geometric, Poisson, Exponential, and Normal random variables.

(b) **Negative Binomial distribution.** You perform a sequence of independent trials with probability p of success at each trial, and probability $1 - p$ of failure. Let X be the number of trials up to and including the nth success. Show that

$$P(X = x) = \binom{x-1}{n-1} p^n (1-p)^{x-n}, \quad x = n, n+1, n+2, \ldots .$$

(c)* **Cauchy density.** For what value of c is $f(x) = c/(1+x^2)$, $-\infty < x < \infty$, a density?

(d) **Gamma density.** Show that

$$f(x) = \begin{cases} \lambda^n x^{n-1} e^{-\lambda x}/(n-1)!, & x > 0, \\ 0, & x \leq 0 \end{cases}$$

is a density. We may refer to this as Gamma (λ, n); some people call it Gamma (n, λ), so be careful.

(e) **Beta (m, n) density.** Show that

$$f(x) = \frac{(m+n-1)!}{(m-1)!(n-1)!} x^{n-1} (1-x)^{m-1}, \quad 0 \leq x \leq 1$$

is a density for positive integers m and n.

(f) Let X be Exponential (λ). Show that λX is Exponential (1).

1.4 Random vectors

It is often necessary and desirable, or simply interesting, to consider the joint effects of several random variables together. Provided always that they are defined on the same sample space Ω (together with a suitable probability function P defined on an appropriate event space $\mathcal{F}$), we shall find it convenient to use the following natural functions. Once again, we distinguish the discrete and continuous cases:

(1) **Joint probability mass function.** If the random variables $X_1, \ldots, X_n$ are discrete then their *joint mass function* is the probability function

(2) $$f(x_1, \ldots, x_n) = P(X_1 = x_1, \ldots, X_n = x_n),$$

where each x_i runs over the possible values of the respective X_i.

To save time, trees, and sanity, we often denote $(X_1, \ldots, X_n)$ by $\mathbf{X}$, using the standard bold notation for vectors; likewise we write for the possible values of $\mathbf{X}$,

$$\mathbf{x} = (x_1, \ldots, x_n).$$

Furthermore, $f(\mathbf{x})$ is often referred to simply as the *distribution* of $\mathbf{X}$. It has the same key property as the distribution $f(x)$ of one random variable; if C is some subset of the possible values D of $\mathbf{X}$, then

(3) $$P(\mathbf{X} \in C) = \sum_{\mathbf{x} \in C} f(\mathbf{x}).$$

Having defined joint distributions in the general case, henceforth we shall (for obvious reasons) restrict our attention in this section to the special case when $n = 2$. Extending definitions and results to larger collections as required is generally trivial, but also very tedious. You can do it.

Marginals. One very important consequence of the key property (3) is that we can recover the separate distributions of X and Y from a knowledge of their joint distribution $f(x, y)$. Since the event $\{X = x\}$ can be written as

$$\{X = x\} = \bigcup_{y \in D} \{X = x, Y = y\} = \bigcup_{y \in D} \left(\{X = x\} \cap \{Y = y\}\right),$$

we have

(4) $$f_X(x) = P(X = x) = \sum_y P(X = x, Y = y) = \sum_y f(x, y).$$

Likewise

(5) $$f_Y(y) = \sum_x f(x, y).$$

If you visualize $f(x, y)$ as a matrix and put its row sums and column sums at the appropriate points of adjacent sides, you will appreciate why $f_X(x)$ and $f_Y(y)$ are sometimes called *marginal distributions*.

(6) Example. Trinomial distribution. Let X and Y have the joint distribution

$$f(x, y) = \frac{n!}{x! y! (n - x - y)!} p^x q^y (1 - p - q)^{n - x - y},$$

where $x \geq 0$, $y \geq 0$, $0 \leq p, q \leq 1$, and $x + y \leq n$. Then this is called the *trinomial distribution*, and the marginal is

$$f_X(x) = \sum_{y=0}^{n-x} f(x, y) = \binom{n}{x} p^x (1-p)^{n-x},$$

which is Binomial (n, p). ○

Now we turn from the discrete to the continuous case:

(7) **Joint density function.** The pair of random variables (X, Y) is said to be (jointly) continuous with *density* $f(x, y)$ if, for all a, b, c, d,

(8) $$P(\{a < X < b\} \cap \{c < Y \leq d\}) = \int_a^b \int_c^d f(x, y) \, dy \, dx.$$

In particular this supplies the *joint distribution function* of X and Y, denoted by $F(x, y)$, as

$$F(x, y) = \int_{-\infty}^x \int_{-\infty}^y f(u, v) \, dv \, du.$$

It follows that, analogously to the case of a single random variable, we may obtain the density $f(x, y)$ from $F(x, y)$ by differentiation.

(9) $$f(x, y) = \begin{cases} \dfrac{\partial^2 f}{\partial x \partial y}, & \text{where the derivative exists,} \\ 0, & \text{elsewhere.} \end{cases}$$

The density has the same key property as in the discrete case, if C is some subset of the domain D of (X, Y)

(10) $$P((X, Y) \in C) = \iint_{(x,y) \in C} f(x, y) \, dx \, dy.$$

Just as in the discrete case, the key property (8) enables us to recover the separate distributions of X and Y from a knowledge of their joint density. We write

$$P(a < X \leq b) = P(\{a \leq X \leq b\} \cap \{-\infty < Y < \infty\})$$
$$= \int_a^b \int_{-\infty}^{\infty} f(x, y) \, dy \, dx, \quad \text{by (8)}$$
$$= \int_a^b f_X(x) \, dx,$$

where

(11) $$f_X(x) = \int_{-\infty}^{\infty} f(x,y)\,dy$$

is seen to be the marginal density of X, following a glance at (1.3.21). Likewise

(12) $$f_Y(y) = \int_{-\infty}^{\infty} f(x,y)\,dx.$$

Since distributions of random variables supply probabilities, it is not surprising that the concept of independence is important here also.

(13) **Independence of random variables.** The pair X and Y are *independent* if, for all x and y,

(14) $\quad P(X \leq x, Y \leq y) = F(x,y) = F_X(x)F_Y(y) = P(X \leq x)P(Y \leq y).$

△

Independence of larger collections of random variables, and pairwise independence, may be defined in the same way as for events; we omit the details.

The above definition holds for random variables of any type, and it is straightforward to deduce that X and Y are independent if for any two sets A and B

(15) $$P(X \in A, Y \in B) = P(X \in A)P(Y \in B).$$

More particularly, if X and Y are jointly continuous, then differentiating (14) reveals that they are independent if, for all x and y,

(16) $$f_{X,Y}(x,y) = f_X(x)f_Y(y).$$

Finally, we note that in many cases of interest the region C in the key identity (10) is determined by the values of the components X and Y. This vague general remark is made clear by example.

(17) **Example.** Let X and Y be independent exponential random variables with parameters λ and μ, respectively. Find $P(X < Y)$.

Solution. The joint density is $f = \lambda\mu \exp(-\lambda x - \mu y)$, for $x > 0$ and $y > 0$. Hence

$$P(X < Y) = \iint_{0<x<y} \lambda\mu \exp(-\lambda x - \mu y) \, dx \, dy$$

$$= \int_0^\infty \int_x^\infty \lambda\mu \exp(-\lambda x - \mu y) \, dy \, dx$$

$$= \int_0^\infty \lambda \exp(-\lambda x - \mu x) \, dx$$

$$= \frac{\lambda}{\lambda + \mu}.$$

We shall see a neater method for this in (1.7.19). ○

Exercises

(a) Let X and Y have joint distribution $F(x, y)$, and suppose that $a < b$, $c < d$. Show that

$$P(a < X \leq b, c < Y \leq d) = F(b,d) + F(a,c) - F(a,d) - F(b,c).$$

(b)* Show that any joint distribution function $F(x, y)$ has the properties
  (i) $F(x, y)$ is non-decreasing in x and y, as x and/or y increases.
  (ii) $F(-\infty, -\infty) = 0$, $F(\infty, \infty) = 1$.

Verify that the function

$$F(x, y) = \begin{cases} 1 - \exp(-xy), & x \geq 0, y \geq 0, \\ 0, & \text{otherwise,} \end{cases}$$

satisfies (i) and (ii). Is it a distribution function?

(c) Let X and Y be independent Exponential random variables with parameters λ and μ, respectively. Show that

$$\mu P(X \leq t < X + Y) = \lambda P(Y \leq t < X + Y).$$

1.5 Transformations of random variables

Just as for individual random variables, we are often interested in functions of jointly distributed random variables. As usual, techniques and notation differ in the discrete and continuous cases; we consider the continuous case first. In particular we may seek the distribution of some function $Z = g(X, Y)$ of the random vector (X, Y). (Note in passing that this function $g(\cdot, \cdot)$ must be such that $g(X, Y)$ is indeed a random variable.) At a formal level this is straightforward; for example, let $Z = g(X, Y)$ and write, using (1.4.10),

$$F_Z(z) = P(Z \leq z) = P(g(X,Y) \leq z)$$

(1)
$$= \iint_{x,y:\, g(x,y) \leq z} f(x,y)\, dx\, dy.$$

This slightly formidable-looking integral takes on simpler forms in many important special cases.

(2) **Example. Sum of continuous random variables.** If X and Y have joint density $f(x, y)$ and their sum is $Z = X + Y$, then by (1)

$$P(Z \leq z) = \int_{x,y:\, x+y \leq z} f(x,y)\, dx\, dy$$

$$= \int_{x=-\infty}^{\infty} \int_{y=-\infty}^{z-x} f(x,y)\, dx\, dy$$

$$= \int_{u=-\infty}^{z} \int_{v=-\infty}^{\infty} f(v, u-v)\, dv\, du,$$

where we substituted $u = x+y$ and $v = x$. Now differentiating with respect to z gives the density of Z.

(3)
$$f_Z(z) = \int_{-\infty}^{\infty} f(v, z-v)\, dv.$$

If X and Y are independent then $f = f_X(x) f_Y(y)$, and

(4)
$$f_Z(z) = \int_{-\infty}^{\infty} f_X(v) f_Y(z-v)\, dv.$$
○

(5) **Example. Normal sum.** If X and Y are independent $N(0, 1)$, then using (4) shows easily that the sum $Z = X + Y$ has density

$$f_Z(z) = \int_{-\infty}^{\infty} \frac{1}{2\pi} \exp\left\{-\frac{1}{2}v^2 - \frac{1}{2}(z-v)^2\right\} dv = \frac{1}{2\sqrt{\pi}} e^{-(1/4)z^2}.$$

Hence Z is $N(0, 2)$. (We did the integral by first completing the square in the exponent.) ○

Sometimes we do not need to do the integral:

(6) **Example.** Let X and Y be independent exponential random variables with parameters λ and μ, respectively. Find the density of the smaller of X and Y

$$Z = X \wedge Y = \min\{X, Y\}.$$

Solution.

$$\begin{aligned} P(Z > z) &= P(\{X > z\} \cap \{Y > z\}) \\ &= P(X > z)P(Y > z), \quad \text{by the independence} \\ &= e^{-(\lambda+\mu)z}. \end{aligned}$$

Hence Z is exponential with parameter $\lambda + \mu$.

There is an interesting corollary to this: Z is independent of the event $\{X < Y\}$. The proof is an exercise for you. ○

More generally, we often wish to know the joint distribution of several functions of the components of the random vector X. The following result in two dimensions is central, and readily generalized.

(7) **Change of variables.** Suppose that the random variables U and V are given by $U = u(X, Y)$, and $V = v(X, Y)$, where X and Y have joint density $f(x, y)$. Assume that the transformation

$$u = u(x, y), \quad v = v(x, y)$$

is one–one, and hence invertible as

$$x = x(u, v), \quad y = y(u, v).$$

Assume also that $x(u, v)$ and $y(u, v)$ have continuous derivatives in both arguments and define

$$J(u, v) = \frac{\partial x}{\partial u}\frac{\partial y}{\partial v} - \frac{\partial y}{\partial u}\frac{\partial x}{\partial v} \neq 0.$$

Then it is a basic result, using the calculus of several variables, that U and V have joint density.

(8) $$f_{U,V}(u, v) = f(x(u, v), y(u, v))|J(u, v)|.$$

(9) **Example.** Let $U = X + Y$, $V = X/(X + Y)$, where X and Y have joint density $f(x, y)$. Then $x = uv$, and $y = u(1 - v)$. Hence

$$J(u, v) = -uv - (1 - v)u = -u,$$

and from (8), U and V have joint density

$$f_{U,V}(u,v) = f(uv, u(1-v))u.$$

Let us note at this point that if we integrate with respect to V, we obtain the marginal density of the sum $U = X + Y$

$$f_U(u) = \int_{\mathbb{R}} f(uv, u(1-v))u\, dv = \int_{\mathbb{R}} f(z, u-z)\, dz$$

as we noted earlier in (3).

Turning to an interesting special case, suppose that X and Y are independent, Gamma (λ, n) and Gamma (λ, m), respectively. Then the joint density of U and V is

(10)
$$\begin{aligned}
f(u,v) &= \frac{\lambda^n e^{-\lambda uv}(uv)^{n-1}}{(n-1)!} \frac{\lambda^m e^{-\lambda u(1-v)}\{u(1-v)\}^{m-1}}{(m-1)!} u \\
&= \frac{\lambda^{m+n} e^{-\lambda u} u^{m+n-1}}{(m+n-1)!} \frac{(m+n-1)!}{(m-1)!(n-1)!} v^{n-1}(1-v)^{m-1},
\end{aligned}$$
$$0 \leq v \leq 1,\ 0 \leq u < \infty.$$

Hence U and V are independent, U being Gamma $(\lambda, m+n)$ and V being Beta (n, m). (See Exercises 1.3(d) and 1.3(e).) ○

Second in this section, we turn to discrete random variables. In this case the distribution of a function $Y = g(X, Y)$ of the random vector (X, Y) is given by a sum. For example, if $Z = X + Y$,

(11)
$$P(Z = z) = \sum_{x,y:x+y=z} f(x,y) = \sum_y f(z-y, y).$$

(12) **Example.** If X and Y are independent, being Poisson (λ) and Poisson (μ), respectively, then

$$P(X + Y = z) = \sum_{y=0}^{z} \frac{\lambda^{z-y} e^{-\lambda}}{(z-y)!} \frac{\mu^y e^{-\mu}}{y!}, \quad \text{by (11)}$$

$$= \frac{(\lambda + \mu)^z e^{-(\lambda+\mu)}}{z!}, \quad \text{by the Binomial theorem.}$$

Thus $X + Y$ is Poisson $(\lambda + \mu)$. ○

(13) **Example. Sum of Bernoullis.** Let X_r be Bernoulli (p), $1 \leq r \leq n$, where the X_r are independent. Then it is clear from our definition (1.3.5) that $Y = \sum_1^n X_r$ is Binomial (n, p). ○

Finally in this section, we introduce another important concept. Note that a joint distribution determines its marginals uniquely, but given marginals may arise from a wide choice of joint distributions. This freedom to choose is often useful, and when we exploit this possibility the technique is often called *coupling*; the joint distribution that we fix is called the *coupling distribution*. Here is an example.

(14) **Poisson approximation.** Recall that we derived the Poisson distribution in (1.3.7) as an approximation to the Binomial distribution; recall also that the Binomial distribution arises as the sum of independent Bernoulli (p) random variables; see (13). This result is a little limited in scope; in fact it can often be extended to cover sums of dependent Bernoulli random variables having different parameters $(p_r; r \geq 1)$. This is one reason for the great importance of the Poisson distribution in practice.

We prove a simple version of the theorem here, retaining the assumption of independence. The key idea is to connect each Bernoulli random variable to an appropriate Poisson random variable, using the following ingenious coupling.

Let X_r be Bernoulli (p_r), where the X_r are independent. Let Y_r be Poisson (p_r), where the Y_r are independent. We choose a joint distribution $f(x, y)$ for X_r and Y_r as follows:

$$f(0, 0) = 1 - p_r,$$

$$f(1, 0) = \exp(-p_r) - 1 + p_r,$$

$$f(1, y) = \frac{p_r^y \exp(-p_r)}{y!}, \qquad y \geq 1,$$

$$f(0, y) = 0, \qquad y \geq 1.$$

It is easy for you to check that the marginal distributions are correct; note that X_r and Y_r are not independent with this joint distribution. First we calculate

$$P(X_r \neq Y_r) = f(1, 0) + P(Y_r \geq 2)$$
$$= p_r - p_r \exp(-p_r) \leq p_r^2.$$

Now define $S = \sum_{r=1}^{n} X_r$ and $T = \sum_{r=1}^{n} Y_r$. Then for any subset C of the integers $\{1, \ldots, n\}$

(15)
$$\begin{aligned}
|P(S \in C) - P(T \in C)| &\leq P(S \neq T) \\
&\leq P(X_r \neq Y_r, \text{ for some } r) \\
&\leq \sum_{r=1}^{n} P(X_r \neq Y_r), \quad \text{by Boole's inequality} \\
&\leq \sum_{r=1}^{n} p_r^2.
\end{aligned}$$

This result (and its more difficult generalization to dependent Bernoulli random variables) is sometimes called the law of rare events, or the law of small numbers. For example, if each X_r is Bernoulli (λ/n), and each Y_r is Poisson (λ/n), then S is Binomial $(n, (\lambda/n))$ and T is Poisson (λ). Equation (15) then implies that

$$|P(S = k) - P(T = k)| \leq \sum_{r=1}^{n} \left(\frac{\lambda}{n}\right)^2 = \frac{\lambda^2}{n}$$
$$\to 0, \quad \text{as } n \to \infty.$$

This gives us a uniform bound on the rate of convergence of the Binomial distribution to the Poisson, which is attractive and useful.

Exercises
(a) Show that if X is $N(\mu, \sigma^2)$, and Y is $N(\nu, \tau^2)$, where X and Y are independent, then $X + Y$ is $N(\mu + \nu, \sigma^2 + \tau^2)$.
(b)* Show that if X and Y are independent $N(0, 1)$ then $Z = X/Y$ has the Cauchy density

$$f(z) = \frac{1}{\pi} \frac{1}{1 + z^2}, \quad -\infty < z < \infty.$$

[Hint: Consider the map $W = Y$, $Z = X/Y$.]

Deduce that Z^{-1} also has the Cauchy density.

(c) If X and Y are independent continuous random variables, show that $U = XY$ and $V = X/Y$ have respective densities

$$f_U(u) = \int_{\mathbb{R}} f_X(v) f_Y\left(\frac{u}{v}\right) \frac{1}{|v|} dv$$

and
$$f_V(v) = \int_{\mathbb{R}} f_X(uv) f_Y(u) |u|\, du.$$

(d) If X and Y are independent Binomial $B(n, p)$ and $B(m, p)$, respectively, show that $X + Y$ is Binomial $(m + n, p)$.

(e) **Coupling**
   (i) Let X and Y have continuous distributions $F_X(x)$ and $F_Y(y)$. Show that if $Z = F_Y^{-1}(F_X(X))$, Z has the distribution $F_Y(z)$.
   (ii) Suppose that the continuous distribution functions $F_X(x)$ and $F_Y(y)$ are such that $F_X(x) \leq F_Y(x)$ for all x. Show that there exist random variables X and Y, having the distributions F_X and F_Y, respectively, such that $P(Y \leq X) = 1$. [Hint: Use part (i).]

1.6 Expectation and moments

If a certain warehouse contains n sacks of potatoes, and, for $k \geq 0$, $n(k)$ sacks weigh k pounds, then the average weight of a sack is

$$w = \frac{1}{n} \sum_k k n(k) = \sum_k \frac{n(k)}{n} k = \sum_k f(k) k,$$

where $f(k) = n(k)/n$ is the proportion of sacks weighing k pounds. The analogy between proportion and probability, mentioned earlier, now makes the following definition natural.

(1) **Expectation.** Let X be a discrete random variable with distribution $f(k)$. The *expected value* (or *mean*) of X is denoted by EX and defined by

(2) $$EX = \sum_k k f(k),$$

provided that $\sum_k |k| f(k)$ is finite. (If this condition fails to hold then X has no finite mean.) △

You may, if you wish, think of this as a weighted average of the possible values of X, where the weights are $f(k)$.

For continuous random variables the same weighted average naturally takes the form of an integral, and we make this definition.

(3) **Expectation.** Let X be a continuous random variable with density $f(x)$. The *expected value* (or *mean*) of X is denoted by EX and defined by

(4) $$EX = \int_{\mathbb{R}} x f(x) \, dx$$

provided that $\int_{\mathbb{R}} |x| f(x) \, dx$ is finite. △

Here are two examples

(5) **Geometric mean.** When $f_X(k) = p(1-p)^{k-1}$, $k \geq 1$,

$$EX = \sum_{k=1}^{\infty} k p (1-p)^{k-1} = \frac{1}{p}.$$ ○

(6) **Exponential mean.** When $f_X(x) = \lambda e^{-\lambda x}$, $x \geq 0$,

$$EX = \int_0^\infty x \lambda e^{-\lambda x} \, dx = \frac{1}{\lambda}.$$ ○

It is sometimes convenient to have an expression for the mean in terms of the distribution function $F_X(x)$ of X. The following is therefore useful.

(7) **Tail theorem.** Let X be non-negative with distribution function $F_X(x)$. Then

(8) $$EX = \int_0^\infty P(X > x) \, dx = \int_0^\infty \{1 - F_X(x)\} dx.$$

Proof. Define the indicator random variable

$$I(x) = \begin{cases} 1, & \text{if } X > x, \\ 0, & \text{if } X \leq x \end{cases}$$

with expected value $EI = P(X > x)$. Then

$$\int_0^\infty (1 - F_X(x)) \, dx = \int_0^\infty EI(x) \, dx = E \int_0^\infty I(x) \, dx = E \int_0^X dx = EX.$$

The interchange in the order of integration and expectation is justified by the non-negativity of the integrand. □

Now compare the following example with (5) and (6).

(9) Example

(a) If X is Geometric (p), then $P(X > x) = (1-p)^{[x]}$, so
$$EX = \int_0^\infty (1-p)^{[x]}\, dx = \sum_{k=0}^\infty (1-p)^k = \frac{1}{p}.$$

(b) If X is Exponential (λ), then $P(X > x) = e^{-\lambda x}$, so
$$EX = \int_0^\infty e^{-\lambda x}\, dx = \frac{1}{\lambda}.$$
□

In previous sections, we have considered the distributions of functions of one or more random variables; it is thus natural to investigate the expected value of such functions. At a formal level this is fairly straightforward: if $Y = g(X)$ then the expected value of Y may be defined as above in terms of the distribution of Y, provided that Y is either continuous or discrete. However, we can avoid this difficulty, and the chore of calculating $f_Y(y)$, by use of the following result.

(10) Theorem. Let the random variables X and Y satisfy $Y = g(X)$, where $g(\cdot)$ is a real-valued function on $\mathbb{R}$.

(a) If X is discrete, then
$$EY = \sum_x g(x) f_X(x),$$
provided that $\sum_x |g(x)| f_X(x) < \infty$.

(b) If X is continuous, then
$$EY = \int_{-\infty}^\infty g(x) f_X(x)\, dx,$$
provided that $\int_\mathbb{R} |g(x)| f_X(x)\, dx < \infty$.

Proof of (a).

$$\begin{aligned}
EY &= \sum_y y P(Y=y), & &\text{by definition} \\
&= \sum_y y \sum_{x:g(x)=y} P(X=x), & &\text{since } Y = g(X) \\
&= \sum_x g(x) P(X=x), & &\text{since } \sum_x |g| P(X=x) < \infty.
\end{aligned}$$

The proof of part (b) is similar but relatively lengthy, and we omit it. □

An entirely natural and analogous result is true for functions of random vectors. Thus if X and Y are discrete, with the obvious notation,

(11) $$Eg(X,Y) = \sum g(x,y) f(x,y)$$

and if X and Y are jointly continuous

(12) $$Eg(X,Y) = \iint g(x,y) f(x,y) \, dx \, dy,$$

provided that the sum and integral converge absolutely. (In future we assume this absolute convergence without further remark.)

These results have numerous important consequences; we highlight these:

Linearity of E. If EX and EY exist, then for constants a and b

(13) $$E(aX + bY) = aEX + bEY.$$

Independent case. If X and Y are independent, then for functions g and h

(14) $$E\{g(X)h(Y)\} = Eg(X)Eh(Y),$$

whenever both sides exist. The proof is an exercise for you.

Among the many functions of random variables X and Y in which we shall be interested, the powers and product of X and Y are particularly important, and their expectations get a special name and notation:

(15) **Moments**

(a) The kth *moment* of X is $\mu_k = EX^k$.
(b) The kth *central moment* of X is $\sigma_k = E(X - EX)^k$.

In particular μ_1 is the mean $\mu = EX$, and σ_2 is called the *variance* and denoted by σ^2 or var X. Thus

$$\sigma^2 = E(X - \mu)^2 = \text{var } X$$

and for the second moment

(16) $$EX^2 = \text{var } X + (EX)^2 = \sigma^2 + \mu^2.$$

△

For jointly distributed random variables X and Y we are often concerned with

(17) **Covariance and correlation.** The *covariance* of X and Y is

(18) $$\text{cov}(X, Y) = E[(X - EX)(Y - EY)].$$

Their *correlation coefficient* (or *correlation*), is

(19) $$\rho(X, Y) = \frac{\text{cov}(X, Y)}{(\text{var } X \text{ var } Y)^{1/2}}.$$

$\triangle$

It is an exercise for you to show that if X and Y are independent then $\rho(X, Y) = 0$. The converse is not true in general.

(20) **Example. Normal distributions.** Let X be $N(0, 1)$, with density

$$f(x) = \frac{1}{\sqrt{2\pi}} \exp\left(-\frac{1}{2}x^2\right), \quad -\infty < x < \infty.$$

Clearly $EX = 0$, by symmetry, and hence $\text{var } X = EX^2 = 1$, by an elementary integration by parts.

Now let Y be $N(\mu, \sigma^2)$. We know from Example (1.3.24) that we can write $Y = \mu + \sigma X$, so immediately we find $EY = \mu$ and $\text{var } Y = \sigma^2$. This explains our choice of notation for this random variable. $\bigcirc$

We return to jointly distributed normal random variables in (1.9).

Exercises
(a) Let X, Y, and Z be random variables. Show that

$$\text{cov}(X, Y + Z) = \text{cov}(X, Y) + \text{cov}(X, Z).$$

(b)* Let $(X_1, \ldots, X_n)$ be independent and identically distributed with finite variance; set $M = (1/n) \sum_{i=1}^{n} X_i$. Show that $\text{cov}(M, X_r - M) = 0$, $1 \leq r \leq n$.

(c) Show that the variance of the sum of independent random variables is equal to the sum of their variances.

(d) **Inequality.** Let $g(x)$ and $h(x)$ be increasing functions. Show that, if the expectations exist,

$$E(g(X)h(X)) \geq Eg(X)Eh(X).$$

[Hint: Let Y be independent of X, having the same distribution as X, and consider $E\{(g(X) - g(Y))(h(X) - h(Y))\}$.]

1.7 Conditioning

In considering some random vector, we may know the value of one or more of its components, or there may be some constraints on their values; or we may wish to consider the effect of such knowledge or constraints. This leads us to define conditional distributions; as usual we confine ourselves to the simplest cases, leaving the obvious but time-consuming extensions as exercises.

In the discrete case matters are elementary and straightforward.

(1) Conditional mass function. If X and Y are jointly discrete, then the *conditional mass function* of X given Y is defined, for all y such that $f_Y(y) > 0$, by

$$f_{X|Y}(x \mid y) = P(X = x \mid Y = y) = \frac{P(X = x, Y = y)}{P(Y = y)}$$

(2)
$$= \frac{f_{X,Y}(x, y)}{f_Y(y)}.$$

Likewise, the conditional distribution of Y given X is

$$f_{Y|X}(y \mid x) = \frac{f(x, y)}{f_X(x)}, \quad \text{when } f_X(x) > 0.$$
△

Note that the conditional distribution $f_{X|Y}(x \mid y)$ is indeed a probability distribution in the sense that it is non-negative and sums to unity:

$$\sum_x f_{X|Y}(x \mid y) = \sum_x \frac{f(x, y)}{f_Y(y)}$$

$$= \frac{f_Y(y)}{f_Y(y)}, \quad \text{by (1.4.5)}$$

$$= 1.$$

It may therefore have an expected value as defined in the previous section; we display this important concept.

(3) Conditional expectation. The *conditional expectation* of X given that $Y = y$ is denoted by $E(X \mid Y = y)$ and defined as

(4)
$$E(X \mid Y = y) = \sum_x x f_{X|Y}(x \mid y),$$

whenever the sum is absolutely convergent. Any other conditional moment is defined similarly in the obvious way. △

Just as the conditional distribution is indeed a probability distribution, conditional expectation has all the properties of ordinary expectation, such as:

$$E(X + Y \mid Z = z) = E(X \mid Z = z) + E(Y \mid Z = z),$$

together with conditional moments and conditional correlation. We refrain from listing them all again; you do it.

The case when X and Y are jointly continuous is not so elementary or straightforward. We may proceed by analogy:

(5) **Conditional density and expectation.** If X and Y are jointly continuous then the *conditional density* of X given Y is defined by

(6) $$f_{X|Y}(x \mid y) = \frac{f(x, y)}{f_Y(y)}, \quad \text{when } f_Y(y) > 0.$$

The *conditional expectation* of X given $Y = y$ is

(7) $$E(X \mid Y = y) = \int_{\mathbb{R}} x f_{X|Y}(x \mid y) \, dx,$$

provided the integral converges absolutely. Conditional moments are defined similarly in the obvious way.

The conditional distribution function of X given $Y = y$ is

$$F_{X|Y}(x \mid y) = \int_{-\infty}^{x} f_{X|Y}(u \mid y) \, du. \qquad \triangle$$

The alert student will notice, however, that we have not used the elementary concept of conditional probability here, as we did in (2), and furthermore equations (6) and (7) make it appear that we are conditioning on an event of probability zero, which was excluded from Definition (1.2.1). Fortunately these objections do not invalidate our heuristic definition above, which can be justified by placing it in a deeper and more complicated theory of random variables and their distributions. We omit this.

However, note that $f_{X|Y}(x \mid y)$ is indeed a probability density, in that it is non-negative and $\int_{\mathbb{R}} f_{X|Y}(x \mid y) \, dx = 1$.

Furthermore, it shares all the other useful properties of the conditional mass function such as:

(8) **The partition lemma.** Let X and Y be jointly distributed.

(a) If X and Y are discrete, then

(9) $$f_X(x) = \sum_y f_{X|Y}(x \mid y) f_Y(y).$$

(b) If X and Y are continuous, then

(10) $$f_X(x) = \int_{\mathbb{R}} f_{X|Y}(x \mid y) f_Y(y) \, dy.$$

Proving (9) and (10) is an easy exercise for you.

These results and their obvious extensions are often useful; here is an illustration of the method:

(11) **Example.** Let X and Y be independent continuous random variables. Then

$$P(X < Y) = \int_{\mathbb{R}} P(X < Y \mid Y = y) f_Y(y) \, dy$$

$$= \int_{\mathbb{R}} F_X(y) f_Y(y) \, dy, \quad \text{by the independence.}$$

○

Likewise, the conditional expectation derived from a conditional density shares the properties of the expected value of a conditional mass function, such as linearity. In particular, in the discrete case we can write

$$EX = \sum_x x f_X(x) \, dx = \sum_x \sum_y x f_{X|Y}(x \mid y) f_Y(y) \, dy$$

(12) $$= \sum_y E(X \mid Y = y) f_Y(y) \, dy.$$

Similarly, in the continuous case

(13) $$EX = \int_{\mathbb{R}} E(X \mid Y = y) f_Y(y) \, dy.$$

Observe the key fact that $E(X \mid Y = y)$ is a function of the value y of Y, so that if we leave the value of Y unspecified we obtain a function of Y that we denote by $E(X \mid Y)$. This is naturally a random variable, and using it with a glance at Theorem (1.6.10) has the gratifying advantage that both the rather unwieldy equations (12) and (13) may be written in the same concise form as

(14) $$EX = E(E(X \mid Y)).$$

This simple, even trivial-seeming, identity is one of the most important in this book. Let us give some demonstrations of its use.

(15) **Example (11) revisited.** We define the indicator random variable

$$I = \begin{cases} 1, & \text{if } X < Y, \\ 0, & \text{otherwise.} \end{cases}$$

Then $E(I \mid Y = y) = P(X < y) = F_X(y)$, and therefore

$$P(X < Y) = EI = E\{E(I \mid Y)\}, \quad \text{by (14)}$$
$$= EF_X(Y)$$
$$= \int_{-\infty}^{\infty} F_X(y) f_Y(y) \, dy.$$

If X and Y are not independent then the same line of argument yields the inevitably less simple answer

$$P(X < Y) = \int_{-\infty}^{\infty} \int_{-\infty}^{y} f(u, y) \, du \, dy.$$

Of course, both these results also follow directly from the key property (1.4.10), but it is as well to try new techniques on simple problems first. ○

Now try this.

(16) **Example: Random sum.** Let $(X_r; r \geq 1)$ be a collection of independent identically distributed random variables with finite variance, and let N be independent of the Xs, non-negative, and integer-valued. Define

$$S = \sum_{r=1}^{N} X_r,$$

where, here and elsewhere, the empty sum is zero. Find ES and var S.

Solution. It would be most inadvisable to attempt to find these expectations using the distribution of S. We use conditional expectation, conditioning on N, thus:

$$ES = E(E(S \mid N)) = E\left\{E\left(\sum_{r=1}^{N} X_r \mid N\right)\right\}$$

(17)
$$= E\left\{\sum_{r=1}^{N} E(X_r \mid N)\right\} = E(NEX_1), \quad \text{by the independence}$$
$$= ENEX_1.$$

Now recall that by definition the conditional variance of X given Y is

$$\text{var}(X \mid Y) = E(X^2 \mid Y) - \{E(X \mid Y)\}^2.$$

Hence, proceeding as we did in the first part, and using (1.6.16),

$$ES^2 = E\{E(S^2 \mid N)\}$$

$$= E\left\{\text{var}\left(\sum_{r=1}^{N} X_r \mid N\right) + \left[E\left(\sum_{r=1}^{N} X_r \mid N\right)\right]^2\right\}$$

$$= E(N \text{ var } X_1) + E(N^2(EX_1)^2), \quad \text{by independence}$$

$$= EN \text{ var } X_1 + \text{var } N(EX_1)^2 + (EN)^2(EX_1)^2.$$

Therefore, using (17),

(18) $\qquad \text{var } S = EN \text{ var } X_1 + \text{var } N(EX_1)^2.$

For another example, let us revisit (1.4.17).

(19) **Example.** The random variables X and Y are independent and exponential with parameters λ and μ, respectively. Then by conditional expectation

$$P(X < Y) = E e^{-\mu X} = \frac{\lambda}{\lambda + \mu}, \quad \text{by (1.6.10)}.$$

We continue this section with a list of three key properties of conditional expectation. These are readily proved for random variables that are either discrete (use (4)), or continuous (use (7)). The really important fact is that they hold far more generally for suitable random variables of any type, for which conditional expectation can be defined. Establishing this would take us too far afield, but see Section 2.9 for a brief description of some of the issues. Here is the first:

(20) **Pull-through property.** For any random variable $g(Y)$,

(21) $\qquad E(Xg(Y) \mid Y) = g(Y)E(X \mid Y),$

from which we find, using (1.7.14), that

(22) $\qquad E[\{E(X \mid Y) - X\}g(Y)] = 0.$

This remarkable property can in fact be made the basis of a far more general definition of $E(X \mid Y)$, but we do not pursue the details.

Another important point is that conditional expectation is naturally and usefully extended to include conditioning on vectors, in the obvious way. Thus, if X, Y, and Z are jointly discrete,

$$(23) \qquad f(x \mid Y = y, Z = z) = \frac{f(x, y, z)}{f_{Y,Z}(y, z)}$$

and then $E(X \mid Y = y, Z = z) = \sum_x x f(x \mid Y = y, Z = z)$. Use of densities gives a similar result for jointly continuous random variables.

This idea leads to the second:

(24) **Tower property**. If X, Y, and Z are jointly distributed, then

$$E\big(E(X \mid Y, Z) \mid Y\big) = E(X \mid Y).$$

The proof in the easy special cases (when X, Y, and Z are either all discrete or all continuous) is an exercise for you.

Third, this natural and superficially obvious result is often useful.

(25) **One–one maps**. If $g(Y)$ is a one–one invertible function then

$$E(X \mid Y) = E(X \mid g(Y)).$$

Finally, we extend the conditional form of independence introduced for events in 1.2.8(d):

(26) **Conditional independence**. Random variables X and Y are said to be *conditionally independent* given Z if their conditional joint distribution factorizes thus:

$$(27) \qquad F(x, y \mid z) = F(x \mid z) F(y \mid z). \qquad \triangle$$

Exercises

(a) Let X be Exponential (λ). Show that

$$P(X > x + y \mid X > y) = P(X > x), \quad \text{for } x, y > 0.$$

This is called the *lack-of-memory property*, and turns out to be rather important. It can be shown that if X is non-negative, and has this property, then it must be an exponential random variable.

(b)* Let X and Y be independent Poisson (λ) and Poisson (μ), respectively. Show that

$$E(X \mid X + Y) = \frac{\lambda}{\lambda + \mu}(X + Y).$$

[Hint: Find the distribution of X conditional on $X + Y = n$.]

(c) Let X, Y, and Z be jointly distributed, and of the same type (discrete or continuous). Show that

$$E(E(X \mid Y, Z) \mid Z) = E(X \mid Z).$$

1.8 Generating functions

The Theorems (1.6.11) and (1.6.12) given above enable us to calculate the expected value of any suitable functions of random variables and vectors. One group of such functions is of particular importance. The first is this:

(1) **Moment generating function.** The *moment generating function* (mgf) of the random variable X is given by

$$M_X(t) = E\, e^{tX},$$

for all real t where the expected value exists. △

It is particularly convenient if $M_X(t)$ exists in an interval of the form $(-a, a)$, $a > 0$, for then we can write, inside the circle of convergence,

$$E\, e^{tX} = E \sum_{r=0}^{\infty} \frac{t^r X^r}{r!} = \sum_{r=0}^{\infty} \frac{t^r}{r!} E X^r.$$

Thus the rth derivative of $M_X(t)$ at $t = 0$ is the rth moment of X,

(2) $$EX^r = M_X^{(r)}(0),$$

explaining the name of the function. We often refer to it as the mgf, for obvious reasons. △

Where we have jointly distributed random variables, we may occasionally find it useful to deal with joint generating functions, whose definition is natural and easily extended to higher dimensions.

(3) **Joint mgf.** For jointly distributed X and Y the *joint mgf* is

$$M(s, t) = E \exp(sX + tY)$$

for all s and t such that the expectation exists. △

We like mgfs for three main reasons: (4), (5), and (7) below.

(4) **Uniqueness.** If $M_X(t) < \infty$ for $t \in (-a, a)$ where $a > 0$, then the moments of X determine the distribution of $F_X(x)$ of X uniquely.

That is to say, there is a unique $F_X(x)$ having this $M_X(t)$ as its mgf. Furthermore, all the moments of X are finite and given by (2). We omit the proof of this.

(5) **Factorization.** If X and Y are independent then by (1.6.14) we have

(6) $$M(s, t) = E(e^{sX+tY}) = M_X(s)M_Y(t).$$

Furthermore, it can be shown that if the joint mgf factorizes in this way, then X and Y are independent. Also, it follows immediately from (6) that we obtain the mgf of the sum $X + Y$ as

$$M_{X+Y}(t) = M_X(t)M_Y(t).$$

We use this key result frequently in later work.

(7) **Continuity theorem.** Let $M_n(t)$ be a sequence of mgfs, such that, as $n \to \infty$, $M_n(t) \to M(t)$, where $M(t)$ is the mgf of a distribution $F(x)$. Then if $M_n(t)$ is the mgf of $F_n(x)$, we have

$$F_n(x) \to F(x), \quad \text{as } n \to \infty$$

at all x where $F(x)$ is continuous. We omit the proof of this, but use it below in the central limit theorem.

Here are some examples of important mgfs.

(8) **Normal mgf.** Let X be $N(0, 1)$. Then for $t \in \mathbb{R}$,

$$\begin{aligned} M_X(t) &= \int_{-\infty}^{\infty} \frac{1}{\sqrt{2\pi}} e^{-(1/2)x^2} e^{tx} \, dx \\ &= e^{(1/2)t^2} \int_{-\infty}^{\infty} \frac{1}{\sqrt{2\pi}} \exp\left\{-\frac{1}{2}(x-t)^2\right\} dx \\ &= e^{(1/2)t^2}, \end{aligned}$$

since the final integrand is the $N(t, 1)$ density. If Y is $N(\mu, \sigma^2)$, we obtain $M_Y(t)$ by using the fact that Y has the same distribution as $\mu + \sigma X$, thus

(9) $$\begin{aligned} M_Y(t) = E\, e^{(\mu+\sigma X)t} &= e^{\mu t} M_X(\sigma t) \\ &= \exp\left\{\mu t + \tfrac{1}{2}\sigma^2 t^2\right\}. \end{aligned}$$

(10) **Exponential mgf.** If X is Exponential (λ), then

$$M_X(t) = \int_0^\infty e^{tx} \lambda e^{-\lambda x}\, dx$$

$$= \frac{\lambda}{\lambda - t}, \quad \text{if } t < \lambda.$$

This mgf does not exist for $t \geq \lambda$. ○

(11) **Poisson mgf.** If X is Poisson (λ), then for $t \in \mathbb{R}$,

$$M_X(t) = \sum_{k=0}^\infty e^{kt} \lambda^k e^{-\lambda}/k! = \exp\{-\lambda(1 - e^t)\}.$$

○

Note the appearance of e^t in this expression. It is easy to see that all integer-valued random variables will have e^t as an argument in their mgf, so for notational simplicity one often uses this generating function instead:

(12) **Probability generating function.** The *probability generating function* (pgf) of the integer-valued random variable X is given by

$$G_X(s) = \mathrm{E} s^X = \sum_n s^n f_X(n).$$

Naturally, $G_X(e^t) = M_X(t)$. △

We like the pgf for much the same reasons as we like the mgf; its only drawback is that it is not of much use for continuous random variables. A classic example of when it is useful is this:

(13) **Example. Random sums revisited.** Using the notation of Example (1.7.16), let $(X_r; r \geq 1)$ be independent with common mgf $M_X(t)$, and let N be non-negative, integer-valued, and independent of the Xs, with pgf $G_N(s)$. Define $S = \sum_{r=1}^N X_r$, and calculate its mgf as follows:

(14)
$$M_S(t) = \mathrm{E}\, e^{tS} = \mathrm{E}(\mathrm{E}(e^{tS} \mid N))$$
$$= \mathrm{E}\{(\mathrm{E}\, e^{tX_1})^N\}$$
$$= G_N(M_X(t)).$$

By differentiating this, and after some algebra, you may derive (1.7.18) as an exercise. ○

As we have seen in considering the exponential density, $M_X(t)$ may not exist for some values of t, but worse is possible. Consider the Cauchy density

(15)
$$f(x) = \frac{1}{\pi} \frac{1}{1 + x^2}, \quad -\infty < x < \infty.$$

For this density $M_X(t)$ exists only at $t = 0$, where $M_X(0) = 1$; for this and similar densities, the mgf is clearly useless. We avoid this difficulty by defining one final generating function.

(16) **Characteristic function.** The function

$$\phi_X(t) = \mathrm{E}e^{itX} = \mathrm{E}(\cos tX + \mathrm{i}\sin tX), \quad \text{for } t \in \mathbb{R}, \text{ where } \mathrm{i}^2 = -1$$

is called the *characteristic function*, or cf, of the random variable X. △

This shares many of the attractive properties of the mgf listed above, and the cf has the further advantage that it always exists, because

$$|\mathrm{E}e^{itX}| \leq \mathrm{E}|e^{itX}| = 1.$$

For example, the Cauchy density in (15), which has no mgf, can be shown to have the characteristic function

(17) $$\phi(t) = \mathrm{E}e^{itX} = \int_{-\infty}^{\infty} e^{itx} \frac{1}{\pi} \frac{1}{1+x^2}\, dx = e^{-|t|}.$$

In finding other characteristic functions the following result is very useful.

(18) **Continuation.** If X has an mgf $M_X(t)$ that exists in an interval $(-a, a)$, $a > 0$, then its characteristic function is given by

$$\phi_X(t) = M_X(\mathrm{i}t), \quad t \in \mathbb{R}.$$

We omit the proof, but note that this immediately gives from (8) that the standard normal density has characteristic function

(19) $$\phi_X(t) = \exp\left(-\tfrac{1}{2}t^2\right).$$

Conversely, if $\phi_X(z)$ is a differentiable function of the complex variable z for $|z| < a > 0$, then $M_X(t)$ exists for $|t| < a$, $\phi_X(t) = M_X(\mathrm{i}t)$, and the distribution of X is uniquely determined by its moments; see (4).

Exercises
(a) Show that $\mathrm{E}X^{-1} = \int_0^\infty M_X(-t)\, dt$, when both sides exist.
(b) Let X be Binomial $B(n, p)$. Show that the pgf of X is $(1 - p + ps)^n$.
(c)* The random variables X and Y are defined in terms of the random variable N which is Poisson (λ): X is Binomial (N, p) and

$Y = N - X$. Show that X and Y are independent, and find their distributions.

(d)* Let $(X_r; r \geq 1)$ be independent Exponential (λ), and let N be Geometric (p), and independent of the Xs. Find the distribution of $S = \sum_{r=1}^{N} X_r$.

1.9 Multivariate normal

One way of interpreting our results about normal random variables is this: we may choose to define $N(\mu, \sigma^2)$ either as the random variable having density $(2\pi)^{-1/2} \exp\{-(1/2)((x-\mu)/\sigma)^2\}$, or as the random variable $Z = \mu + \sigma X$, where X has the standard normal density $\phi(x)$. (And we exclude trivial or singular cases by requiring $0 < \sigma < \infty$.)

It turns out that we have an even greater choice of definitions when we turn to consider generalizations in the multivariate case. Perhaps the most important property of independent normal random variables is that their sum is also normally distributed. Let us consider the joint density of two such sums,

(1) $$U = aX + bY,$$
(2) $$V = cX + dY,$$

where X and Y are independent $N(0, 1)$. Theorem (1.5.7) easily shows that U and V have a joint density of the form

(3) $$f(u,v) = \frac{1}{2\pi\sigma\tau(1-\rho^2)^{1/2}} \exp\left\{-\frac{1}{2(1-\rho^2)}\left(\frac{x^2}{\sigma^2} - \frac{2\rho xy}{\sigma\tau} + \frac{y^2}{\tau^2}\right)\right\}$$

for appropriate constants ρ, σ, and τ. These constants are significant because, as you can show with some algebra

$$\sigma^2 = a^2 + b^2 = \text{var } U,$$
$$\tau^2 = c^2 + d^2 = \text{var } V,$$
$$\rho = \frac{ac + bd}{\sigma\tau} = \rho(U, V).$$

We may refer to this density as Binormal (ρ, σ, τ). If we consider U/σ and V/τ, these have variance unity, and the corresponding density is called the *standard bivariate normal density*, or Binormal (ρ), for brevity; set $\sigma = \tau = 1$ in (3) to see it.

Thus, as in the univariate case, in the bivariate case we may define a Binormal (ρ, σ, τ) random vector either as having the joint density (3), or as the random vector

$$Z_1 = (aX + bY), \quad Z_2 = (cX + dY),$$

where X and Y are independent $N(0, 1)$. As in the univariate case, we may exclude trivial and singular examples by requiring that $\sigma > 0$, $\tau > 0$, and $ad - bc \neq 0$; this last constraint excludes the possibility that Z_1 and Z_2 are linearly dependent.

Of course the choice of a, b, c, and d for a given (ρ, σ, τ) is not unique, there are infinitely many. In particular, if $ac + bd = 0$, then Z_1 and Z_2 are also independent random variables since their joint density factorizes. This turns out to be another characterization of (3); thus:

(4) **Lancaster's theorem.** If X and Y are independent and there is a non-trivial linear transformation $U = aX + bY$, $V = cX + dY$ such that U and V are independent, then all four variables are normal. We omit the proof.

A particularly useful choice of a, b, c, d is this: if we have a pair (U, V) with the Binormal (ρ) density then we can represent them in the form

(5) $$U = X,$$
(6) $$V = \rho X + \sqrt{1 - \rho^2} Y,$$

where X and Y are independent $N(0, 1)$. This makes it easy to perform many otherwise tricky tasks; for example, let us find the joint mgf of U and V:

$$M(s, t) = \mathrm{E} e^{sU + tV}$$

(7) $$= \mathrm{E} \exp\left\{(s + \rho t)X + \sqrt{1 - \rho^2} tY\right\}, \quad \text{by (5) and (6)}$$

$$= \exp\left\{\tfrac{1}{2}(s + \rho t)^2\right\} \exp\left\{\tfrac{1}{2}(1 - \rho^2)t^2\right\}, \quad \text{by the independence}$$

$$= \exp\left\{\tfrac{1}{2}\{s^2 + 2\rho st + t^2\}\right\}.$$

More generally, if (W, Z) have the Binormal (ρ, σ, τ) density centred at (μ, ν), we can represent (W, Z) by

(8) $$W = \mu + \sigma X,$$

(9) $$Z = \nu + \tau \left(\rho X + \sqrt{1 - \rho^2} Y\right).$$

Note that var $W = \sigma^2$, var $Z = \tau^2$, and $\rho(W, Z) = \rho$. It is now trivial to find the joint mgf of W and Z using (7), (8), and (9):

$$M_{W,Z}(s,t) = \exp\left\{\mu s + \nu t + \tfrac{1}{2}(s^2\sigma^2 + 2\rho\sigma\tau st + \tau^2 t^2)\right\}$$

(10) $\quad = \exp\left\{sEW + tEZ + \tfrac{1}{2}(s^2 \text{var } W + 2st\text{cov}(W,Z) + t^2 \text{var } Z)\right\}.$

From this we deduce these three important corollaries:

(i) Binormal random variables are independent if and only if they are uncorrelated, for in this case (10) factorizes, but not otherwise.

(ii) The Binormal (ρ, σ, τ) density of W and Z is determined by the means, variances, and covariance of W and Z, by inspection of (10).

(iii) The marginal densities of a Binormal density are normal; set $s = 0$ or $t = 0$ in (10) to see this.

Finally, we observe that our first remark above, about sums of independent normal random variables being normal, also serves to characterize the multivariate normal distribution. That is to say, if a pair of random variables (X, Y) has the property that $aX + bY$ is normal for all (a, b), then they have a bivariate normal density.

Of all these definitions, this one is the easiest, and perhaps the most useful, so this is our definition of choice in n dimensions.

(11) **Multinormal density.** Let $(X_r; 1 \leq r \leq n)$ be independent $N(0, 1)$. Then (for suitable constants μ_r and a_{rj}) the set

(12) $$Z_r = \mu_r + \sum_{j=1}^{n} a_{rj} X_j, \quad 1 \leq r \leq n$$

is said to have a multivariate normal distribution. We may call this Multinormal $(\boldsymbol{\mu}, \mathbf{A})$, with an obvious notation. If $\mu_r = 0$ for all r, then it is said to be centred at the origin, and if var $Z_r = 1$ for all r, then it is *standard*. The joint mgf is now readily obtained using (12), just as in the Binormal case; this is an exercise for you. We draw the same conclusion, viz: multinormal random variables are independent if and only if they are uncorrelated.

After some algebra you will also discover that their joint density takes the form

$$f(\mathbf{x}) \propto \exp\{-\tfrac{1}{2}Q(\mathbf{x} - \boldsymbol{\mu})\},$$

where $Q(\mathbf{x})$ is a quadratic form given by

$$Q(\mathbf{x}) = \mathbf{x}V^{-1}\mathbf{x}'$$

and $\mathbf{V}$ is the variance–covariance matrix of $\mathbf{X}$. We omit the details.

Exercises
(a) Prove that the function defined on $\mathbb{R}$ by $f(x) = c \exp(-x^2/2)$ can be a probability density function for an appropriate value of c.

It is interesting that in fact $c^{-1} = \sqrt{2\pi}$. No proof of this is entirely trivial, but different people prefer different proofs according to their personal knowledge of elementary techniques in calculus. Here are three for you to try.

(i) Note that
$$c^{-2} = \int_{-\infty}^{\infty} e^{-x^2/2} dx \int_{-\infty}^{\infty} e^{-y^2/2} dy$$
$$= \int_{-\infty}^{\infty}\int_{-\infty}^{\infty} e^{-(x^2+y^2)/2} dx\, dy = I, \text{ say}.$$

Evaluate the integral I by changing the variables to plane polars.

(ii) Note that I is the volume enclosed by the (x, y) plane and the surface of revolution obtained by rotating $x^2 = -2\log z$ around the z-axis for $0 < z \leq 1$. Hence
$$I = \int_0^1 \pi(x(z))^2 dz = -\int_0^1 \pi 2 \log z\, dz,$$
and evaluating I yields c.

(iii) Let
$$f(x) = \left\{\int_0^x e^{-v^2/2} dv\right\}^2$$
and
$$g(x) = 2\int_0^1 \frac{e^{-x^2(1+t^2)/2}}{1+t^2} dt.$$

Show that $df/dx = -(dg/dx)$, and deduce that $f + g = \pi/2$ by evaluating $f(0)+g(0)$. Note that $\lim_{x\to\infty} f(x)+g(x) = 1/4$ to yield the required result.

(b)* Let X and Y be Binormal (ρ). Show that
$$E(X \mid Y) = \rho Y$$
and
$$\text{var}(X \mid Y) = 1 - \rho^2.$$

Problems

1.* Let X be a non-negative continuous random variable. Show that, for integers $n \geq 1$,

$$EX^n = n \int_0^\infty x^{n-1}(1 - F_X(x))\, dx.$$

If $g(0) = 0$, and $g'(x) > 0$, show that $Eg(X) = \int_0^\infty g'(x) P(X > x)\, dx$. When X is integer-valued, show that

$$EX^n = \sum_{r=0}^\infty \{(r+1)^n - r^n\}(1 - F_X(r)).$$

2.* Let X and Y be independent, being Exponential (λ) and Exponential (μ), respectively. Define

$$U = \min\{X, Y\} = X \wedge Y,$$
$$V = \max\{X, Y\} - \min\{X, Y\} = X \vee Y - X \wedge Y,$$
$$I = \begin{cases} 1, & \text{if } X < Y \\ 0, & \text{if } X \geq Y. \end{cases}$$

(Thus, I is the indicator of the event $\{X < Y\}$.)
(i) Show that U is independent of the pair (I, V).
(ii) Find the distribution of V, and of $W = (X-Y)^+ = \max\{0, X-Y\}$.

3. Let $(X_r; r \geq 1)$ be independent random variables, such that X_r is Exponential (λ_r), where $0 < \sum_{r=1}^\infty \lambda_r < \infty$. Let $Y = \inf_{r \geq 1} X_r$. Show that with probability 1 there is a unique N such that $X_N = Y$, and

$$P(N = r) = \lambda_r \bigg/ \sum_{r=1}^\infty \lambda_r.$$

Prove that Y and N are independent.

4.* Let $(X_r; r \geq 1)$ be independent Exponential (λ) random variables, and define $T_n = \sum_{r=1}^n X_r$. Show that T_n has the so-called Gamma (λ, n) density given by

$$f(x) = \frac{\lambda^n x^{n-1} e^{-\lambda x}}{(n-1)!}, \quad x > 0.$$

5.* Show that if $\operatorname{var} X = 0$, then, with probability 1, X is constant.
6. Show that $E(X - a)^2$ is smallest when $a = EX$.

7.* **Cauchy–Schwarz inequality.** Show that for random variables X and Y

$$\{E(XY)\}^2 \leq E(X^2)E(Y^2),$$

with equality if and only if $P(sX - tY = 0) = 1$, for some constants s and t. [Hint: Consider $E\{(sX - tY)^2\}$.]

Deduce Schlömilch's inequality

$$E|X - EX| \leq \operatorname{var} X.$$

8.* **Jensen's inequality.** Let $g(x)$ be convex, which is to say that for all a, there is m such that

$$g(x) \geq g(a) + (x - a)m, \quad \text{for all } x.$$

Show that for such a function

$$Eg(X) \geq g(EX), \quad \text{when } EX < \infty.$$

9.* Let X be a bounded positive random variable and define $M(r) = \{E(X^r)\}^{1/r}$, $r > 0$. Show that $M(r)$ is an increasing function of r. [Hint: $x \log x$ is convex.]

10.* (a) You have two cubic dice; one is fair and the other may be biased. If their respective scores when rolled are X and Y, $(1 \leq X, Y \leq 6)$, show that $Z = X + Y$, (modulo 6), is uniformly distributed.

(b) The game of craps calls for two fair dice. Suppose you have a fair pair and a biased pair, but you do not know which is which. Show how you can still play, by using the dice appropriately.

11.* Let $(X_r; 1 \leq r \leq n)$ be independently distributed on $(0, 1)$, and set

$$Z = \sum_{r=1}^{n} X_r, \quad \text{(modulo 1)}.$$

Show that Z is uniform on $(0,1)$ if at least one X_r is uniform on $(0,1)$.

12.* **Negative Binomial distribution.** (a) Let $(X_r; 1 \leq r \leq n)$ be independent Geometric (p) random variables, and define

$$Y = \sum_{r=1}^{n} X_r.$$

Show that

$$P(Y = k) = \binom{k-1}{n-1} p^n (1-p)^{k-n}, \quad k \geq n \geq 1.$$

(b) Suppose that you are flipping a biased coin that shows a head with probability p. Find the distribution of the number of flips required until you first have n heads.

13.* (a) Let X be Normal, $N(\mu, \sigma^2)$, and let $g(x)$ be differentiable and such that $Eg'(X) < \infty$. Show that

$$E\{(X - \mu)g(X)\} = \sigma^2 Eg'(X).$$

(b) Let X and Y have a Binormal distribution. Show that

$$\text{cov}(g(X), Y) = Eg'(X)\text{cov}(X, Y).$$

14.* Let $X_1, \ldots, X_n$ be independent and uniformly distributed on $[0, t]$. Let $X_{(1)}, \ldots, X_{(n)}$ denote the values of $X_1, \ldots, X_n$ taken in increasing order $0 \leq X_{(1)} \leq X_{(2)} \leq \cdots \leq X_{(n)} \leq t$. Show that the joint density of $X_{(1)}, \ldots, X_{(n)}$ is

$$f = n! t^{-n}, 0 \leq x_{(1)} \leq \cdots \leq x_{(n)} \leq t.$$

15. You pick a personal pattern of heads and tails of length $L < \infty$. Now you flip a coin repeatedly until your pattern appears. If N is the number of flips required, show that $EN < \infty$. [Hint: Do not try to find the distribution of N.]

16.* (a) Flash, Ming, and Zarkov land their three spaceships (named A, B, and C) independently and uniformly at random on the surface of the planet Mongo, which is a sphere centre O.
 (i) Find the density of the angle AOB. Denote the angle AOB by θ. The ships A and B can communicate by radio if and only if $\theta < \pi/2$, with a similar constraint for any other pair of ships.
 (ii) If $\theta = x$, show that the probability that ship C can communicate with neither A nor B is $(\pi - x)/2\pi$.
 (b) Show that the probability that no ship can communicate with any other is $(4\pi)^{-1}$.

17. **Matsumoto–Yor property.** Let X and Y be independent with respective densities

$$f_X(x) = f_1(x; a, b) = k_1 x^{-c-1} \exp(-ax - b/x), \quad x > 0,$$
$$f_Y(x) = f_2(x; a) = k_2 x^{c-1} \exp(-ax), \quad x > 0,$$

where a, b, c, k_1, k_2 are non-negative constants. Define $U = (X+Y)^{-1}$ and $V = X^{-1} - (X+Y)^{-1}$. Show that U and V are independent, with respective densities $f_1(u; b, a)$ and $f_2(v; b)$. [Hint: Recall (1.5.7) and calculate $J = u^{-2}(u+v)^{-2}$.]

2 Introduction to stochastic processes

I am Master of the Stochastic Art...
  Jonathan Swift, Right of Precedence between Physicians and Civilians *(1720)*

He got around to talking stochastic music and digital computers with one technician.
  Thomas Pynchon, V. (1963)

2.1 Preamble

We have looked at single random variables, and finite collections of random variables $(X_1, \ldots, X_n)$, which we termed random vectors. However, many practical applications of probability are concerned with random processes evolving in time, or space, or both, without any limit on the time (or space) for which this may continue. We therefore make the following definitions, confining ourselves in the first instance to processes evolving in one dimension that we call time, for definiteness.

(1) **Stochastic processes.** A *stochastic process* is a collection of random variables $(X(t): t \in T)$ where t is a parameter that runs over an index set T.

In general we call t the *time-parameter* (or simply the time), and $T \subseteq \mathbb{R}$. Each $X(t)$ takes values in some set $S \subseteq \mathbb{R}$ called the *state space*; then $X(t)$ is the state of the process at time t. △

For example $X(t)$ may be the number of emails in your in-tray at time t, or your bank balance on day t, or the number of heads shown by t flips of some coin. Think of some more examples yourself. There is also, of course, some underlying probability space Ω and probability function P; since we are not concerned here with a general theory of random processes, we shall not need to stress this part of the structure.

Since any stochastic process is simply a collection of random variables, the name 'Random Process' is also used for them, and means exactly the same thing. (The word 'stochastic' (meaning 'random') comes from a Greek root $\sigma\tau`o\chi os$ meaning 'to aim at' or 'to guess'. The adjective $\sigma\tau o\kappa\alpha\sigma\tau\iota\kappa\grave{o}s$ was used, for example by Plato, to mean 'proceeding by guesswork'.)

Stochastic processes are characterized by three principal properties:

(a) the state space;
(b) the parameter set;

(c) the dependence relations between (that is to say, the joint distribution of) the various random variables $X(t)$.

Of course, since $X(t)$ is a random variable, the state space may be discrete or continuous. Note that $X(t)$ *itself* may be a random vector! For example, consider a particle performing a simple symmetric random walk on the plane integer lattice. (That is to say, at each step it changes its x-coordinate by ± 1, or its y-coordinate by ± 1, each with probability $\frac{1}{4}$.) Then the process is the sequence of vectors

$$\{(X(n), Y(n)): n \in \mathbb{Z}^+\}.$$

The parameter may be discrete, for example, $T = \mathbb{Z}^+$, or $T = \mathbb{Z}$; in which case we customarily use n to denote the time and call $X(n)$ a discrete-time process. When T is an interval in the real line, typically $T = [0, \infty)$, then $X(t)$ is called a continuous-time process.

The dependence relations between the random variables of a particular stochastic process are usually suggested by some practical problem; they are a formal statement about the modelling assumptions that we make in order to analyse the process and make predictions in the real world. Typically the process is specified by determining its local behaviour, and our analysis is directed to discovering its global behaviour. Here is an example to illustrate what we mean by that.

(2) **Example. Queue.** In real queues, customers arrive at a service point and hope to be served before it closes. If we are to model this system to predict, say, the chances of the customer being successful in this, we must make a number of modelling assumptions. For example,

(i) How does the server deal with customers?
(ii) How fast do customers enter the queue?

We must also set up a mathematical framework, in which these assumptions can be formulated.

One possible model of interest could be specified like this:

(a) The number of customers in the queue at time t is $X(t), t \geq 0$; $X(0) = 0$.
(b) The times between the arrivals of successive customers are independent identically distributed Exponential (λ) random variables.
(c) Customers are served in order of arrival.
(d) Their service times are independent random variables, that are Exponential (μ) and independent of the arrivals process.

From this local description one hopes to discover, for example, the probability of being served before nightfall if you arrive at noon to see a dozen folk in front of you, and so on. We return to this problem in Chapter 4. ○

This is a convenient moment to remark that there are two ways of looking at a random process $X(t)$. To see this informally, consider, for example, the queue above.

On the one hand, if you are a manager, your interest will perhaps lie in the overall properties of $X(t)$ as specified by its probability distributions, such as

(i) The probability that more than c customers are left unserved at the end of the day.
(ii) The probability that the server spent more than n hours idle for lack of customers.

On the other hand, if you are a customer you have little or no interest in such matters; your only concern is with your single experience of the queue.

Formally, the distinction between these two viewpoints can be expressed like this: first,

(i) We can study $X(t)$ by considering its joint distributions.

$$(3) \qquad F(x_1, t_1; \ldots; x_n, t_n) = P(X(t_1) \leq x_1; \ldots; X(t_n) \leq x_n)$$

for any collection of times $(t_1, \ldots, t_n)$. Alternatively,

(ii) We can regard each $X(t)$ as a function on the sample space Ω; for any given $\omega \in \Omega$ we obtain $X(t, \omega)$. For this ω, as time passes, we get a particular sequence $(X(t, \omega); t \in T)$. This trajectory is called a *realization* or *sample path* of the process $X(t)$. We may study these.

Naturally, we shall often wish to consider the properties of sample paths hand-in-hand with the joint distributions of the process. However, in more advanced work this may be problematical because the joint distributions may not tell us everything about sample paths. One may attempt to sidestep this difficulty by defining $X(t)$ in terms of its sample paths, but this is not easy in general and we do not attempt it here. There is a brief discussion of some of these questions, and methods for answering them, in Section (2.9).

Here is a simple but important example of how one particular sample path property can be studied.

(4) **Example. Infinitely often.** Suppose $(X_n; n \geq 1)$ is some stochastic process; let A_n be some event of interest, for example the event that X_n exceeds

some value a. We are often interested in knowing whether only finitely many of $(A_n; n \geq 1)$ occur, or not. The event that infinitely many of the A_n occur is expressed as

$$\{A_n \text{ i.o.}\} = \{A_n \text{ infinitely often}\}$$
$$= \bigcap_{n=1}^{\infty} \bigcup_{r=n}^{\infty} A_r.$$

It is an easy exercise for you to check that the complementary event may be written

$$\{A_n \text{ finitely often}\} = \{A_n \text{ i.o.}\}^c$$
$$= \bigcup_{n=1}^{\infty} \bigcap_{r=n}^{\infty} A_r^c.$$

Finally, we note that while it is often convenient to define a stochastic process by giving its joint distributions

$$F(x_1, t_1; \ldots; x_n, t_n) = P(X(t_1) \leq x_1; \ldots; X(t_n) \leq x_n)$$

for any choice of $(t_1, \ldots, t_n)$ and $(x_1, \ldots, x_n)$, this leaves open the question of whether there is in fact a real process having these joint distributions. Usually we can construct the required process explicitly, so this is not a problem; but sometimes the construction is not so simple.

Fortunately it can be proved (Kolmogorov's theorem), that these are indeed the distributions of a real process if they satisfy these two conditions

(a) $F(x_1, t_1; \ldots; x_n, t_n; x_{n+1}, t_{n+1}) \to F(x_1, t_1; \ldots; x_n, t_n)$, as $x_{n+1} \to \infty$,
(b) $F(x_{d(1)}, t_{d(1)}; \ldots; x_{d(n)}, t_{d(n)}) = F(x_1, t_1; \ldots; x_n, t_n)$, for any permutation $(d(1), \ldots, d(n))$ of $(1, \ldots, n)$.

You will be pleased to know that this is true of all processes in this book.

Exercises
(a)* Let $(X_n; n \geq 1)$ be a collection of independent positive identically distributed random variables, with density $f(x)$. They are inspected in order from $n = 1$.
  (i) An observer conjectures that X_1 will be greater than all the subsequent $X_n, n \geq 2$. Show that this conjecture will be proved wrong with probability 1.

What is the expected number of inspections before the conjecture is proved wrong?

(ii) Let N be the index of the first X_n to exceed a

$$N = \min\{n \geq 1 : X_n > a\}.$$

Show that $EX_N = \int_a^\infty x f(x)\, dx / P(X_1 > a)$.

(iii) Let $S_n = \sum_{r=1}^N X_r$. Show that

$$ES_N = EX_1 / P(X_1 > a).$$

(b)* Let $(X_r; r \geq 1)$ be a collection of independent random variables where X_r is Exponential (λ_r). Show that
(i) If $\sum_{r=1}^\infty \lambda_r^{-1} < \infty$, then $P(\sum_r X_r < \infty) = 1$.
(ii) If $\sum_{r=1}^\infty \lambda_r^{-1} = \infty$, then $P(\sum_r X_r = \infty) = 1$.

2.2 Essential examples; random walks

Now it is a convenient moment to illustrate the special flavour of the subject by considering some simple and classic examples. This also serves to introduce important ideas and techniques in an open and transparent way. And we use these elementary processes repeatedly in later work.

(1) Example. Independent sequence. Arguably, the simplest non-trivial stochastic process is $(X_n; n \geq 0)$ such that X_n and X_m are independent and identically distributed for $m \neq n$. Repeated rolls of a die, or flips of a coin, or independent trials of any kind, supply real-world realizations of this model. ◯

This sequence is primarily used to generate other processes with more complex and interesting behaviour, such as this.

(2) Example. General random walk on the line. Let $(X_n; n \geq 1)$ be independent and identically distributed, and set

(3) $$S_n = \sum_{r=1}^n X_r + S_0.$$

Then $(S_n, n \geq 0)$ is a general random walk on the line. ◯

For another example, drawn from statistics, a classical estimator of the mean of a distribution, given n independent samples $(X_1, \ldots, X_n)$ having that distribution, is the sample mean

$$\bar{X}_n = \frac{1}{n} \sum_{r=1}^{n} X_r, \quad n \geq 1.$$

For obvious reasons we may well be interested in the long-run behaviour of S_n and $\bar{X}_n$ as n increases.

A natural way of visualizing the general random walk in Example (2) is as the path of a particle making independent and identically distributed steps, or jumps, in the real line. If it is at some point S, it next jumps to the point $S + X$, where X has the distribution of X_1. The fact that the points of $\mathbb{R}$ are ordered gives this walk some special properties, but the idea is readily extended to more general models.

(4) **Example. Random walk on a graph.** A graph is a collection of points or vertices; it is convenient to label the points using (say) the integers. We define a sequence of random variables $(X_n, n \geq 0)$ by visualizing a particle moving randomly on these vertices thus:

(a) The particle starts at X_0.
(b) If the particle is at v, its next position is selected, independently of its previous states, from a set $A(v)$ of available points according to a probability distribution over $A(v)$. ○

(5) **Example. Random flights.** Let $(X_n; n \geq 1)$ be a collection of independent unit vectors whose directions are chosen uniformly and independently. Thus, in two dimensions the direction is an angle Θ uniformly distributed on $(0, 2\pi]$; in three dimensions the direction is uniformly distributed on the unit sphere with bivariate density

$$f(\theta, \phi) = \sin \phi / 4\pi, \quad 0 \leq \phi \leq \pi, \ 0 \leq \theta \leq 2\pi.$$

The sum $S_n = \sum_{r=1}^{n} X_r$ is a random flight, first studied by Rayleigh in connexion with acoustics. ○

All these stochastic processes naturally generate others, whose importance depends on the context. For example, if C is some subset of the state space S,

(i) Let $N(t)$ be the number of occasions on which $S_n \in C$ by time t.
(ii) Let $T_r(C)$ be the instant at which the process S_n enters C for the rth time.

We discuss these processes, and others, more generally later on; for the moment let us look at them in the context of a classic special case.

(6) **Example. Simple random walk.** Here we consider a sequence of independent random variables $(X_n; n \geq 1)$ such that

$$P(X_n = 1) = p, \qquad P(X_n = -1) = q, \qquad p + q = 1, \quad n \geq 1.$$

Then $S_n = S_0 + \sum_{r=1}^n X_r$ is called *simple random walk* on the integers, starting from some integer S_0. If $p = q = \frac{1}{2}$, the walk is said to be *symmetric*. ○

This random walk arose initially as a simple model for gambling: you make a series of independent wagers, on each of which your fortune increases by 1 with probability p or decreases by 1 with probability q. Then $S_n - S_0$ is the change in your fortune after n plays. The following questions are natural in this context:

(a) What is the probability of gaining a before your losses reach b?
(b) What is the expected number of plays until one of these events occurs?

The methods used for this are exemplary.

(7) **Example. Gambler's ruin.** With S_n defined in (6), let w_k be the probability that the walk reaches $a > 0$ before it reaches $b < 0$, given that $S_0 = k$. Then, conditioning on the first step, and using the independence of steps, we have

(8) $$w_k = pw_{k+1} + qw_{k-1}, \quad b < k < a.$$

By definition, we know that $w_a = 1$ and $w_b = 0$. The general solution of (8) is given by

$$w_k = \alpha + \beta(q/p)^k, \quad p \neq q, \ \alpha, \beta \text{ constant}$$

and hence, imposing the above values for w_a and w_b, we find that

(9) $$w_k = \frac{(q/p)^b - (q/p)^k}{(q/p)^b - (q/p)^a}, \quad b < k < a, \ p \neq q.$$

The answer to the gambler's question (a), above, is therefore w_0, and

$$P(\text{gain } a \text{ before losing } b) = w_0 = \frac{(q/p)^b - 1}{(q/p)^b - (q/p)^a}.$$

When $p = q = \frac{1}{2}$, the general solution to (8) is, for any constants α and β,

(10) $$w_k = \alpha + \beta k.$$

Hence, imposing the above values for w_a and w_b, $w_k = (k-b)/(a-b)$, and

(11) $$P(\text{gain } a \text{ before losing } b) = w_0 = \frac{b}{b-a}.$$

In this case, when $p = q = \frac{1}{2}$, the answer to the gambler's second question (b), above, is easily ascertained. Let m_k be the expected number of steps until the walk first hits a or b, starting from k. By conditional expectation,

(12) $$m_k = 1 + \tfrac{1}{2}m_{k+1} + \tfrac{1}{2}m_{k-1}.$$

By definition $m_a = m_b = 0$, and it is easy to check that the solution to (12) satisfying these constraints is

(13) $$m_k = (a-k)(k-b), \quad b < k < a.$$

Hence, the answer to the gambler's query (b) is given by

$$m_0 = -ab.$$
○

The ideas in the following example are often useful.

(14) Example. Returns of a simple random walk. Let us define the indicator I_n of the event that the walk returns to the origin at time n. That is to say,

$$I_n = \begin{cases} 1, & \text{if } S_n = 0, \\ 0, & \text{otherwise} \end{cases}$$

with expectation $EI_n = P(S_n = 0) = p_{00}(n)$, say, when $S_0 = 0$. It is an easy calculation to show that this satisfies $p_{00}(2r) = \binom{2r}{r} p^r q^r$.

If N is the total number of returns that the walk ever makes to the origin, then by definition

$$N = \sum_{n=1}^{\infty} I_n.$$

Therefore,

$$EN = E\sum_{n=1}^{\infty} I_n = \sum_{n=1}^{\infty} EI_n = \sum_{n=1}^{\infty} p_{00}(n)$$

(15) $$= \sum_{r=1}^{\infty} \binom{2r}{r} p^r q^r.$$

It is a straightforward exercise to show that this final sum converges if $p \neq q$, but diverges if $p = q = \frac{1}{2}$.
○

In fact, we can use the fact that steps of a random walk are independent, to obtain a deeper result.

(16) Example. Returns of a random walk. Let $S_n = \sum_{r=1}^{n} X_r$ be a random walk on the integers started at the origin (not necessarily a simple random walk). Define the indicator

$$I_n = I(S_n = 0) = \begin{cases} 1, & \text{if } S_n = 0, \\ 0, & \text{otherwise}, \end{cases}$$

which indicates the event that the random walk revisits its starting point at the nth step. Now let η be the probability that it ever revisits its starting point, denote by $T_0(k)$ the time of the kth such return (if there are any), and let N be the total number of returns. Obviously,

(17) $$\eta = P(N \geq 1) = P(T_0(1) < \infty).$$

Next, we seek the distribution $P(N \geq n); n > 1$.

The key to discovering this is the observation that all the steps of the walk are independent. Hence, if $T_0(k) < \infty$, the progress of the walk after $T_0(k)$ is independent of its path up to $T_0(k)$, and has the same distribution as the walk originating at time zero. (This is a crucial observation, to which we return in studying Markov chains.) It follows that we may calculate

$$P(N \geq n+1) = P(N \geq n+1 \mid N \geq n)P(N \geq n)$$
(18)
$$= P(N \geq n+1 \mid T_0(n) < \infty)P(N \geq n)$$
$$= P(N \geq 1)P(N \geq n), \quad \text{by the remark above,}$$
$$= \eta P(N \geq n)$$
$$= \eta^{n+1}, \quad \text{by induction.}$$

Clearly there are two distinct cases here. If $\eta = 1$ then $P(N \geq n) = 1$ for all n, and hence the walk revisits its starting point infinitely often. If $\eta < 1$, then $EN < \infty$ and so the walk can revisit its starting point only finitely often, with probability 1.

Finally, let us return to the special case of simple random walk; we can use another important technique to find η explicitly. Let η_{jk} be the probability that the walk starting from j ever reaches k. From the structure of the walk (independent identically distributed steps), it follows that $\eta_{-1,0} = \eta_{0,1}$, and, using arguments similar to those in (18), it is clear that

(19) $$\eta_{-1,1} = \eta_{-1,0}\eta_{0,1} = \eta_{0,1}^2.$$

The key idea is to condition on the first step of the walk, obtaining

$$\eta = P(\text{ever return to } 0 \mid X_1 = 1)p + P(\text{ever return to } 0 \mid X_1 = -1)q$$
$$(20) \qquad = p\eta_{1,0} + q\eta_{-1,0} = p\eta_{1,0} + q\eta_{0,1}.$$

To find the unknown quantities on the right of this equation, we again condition on the first step to obtain

$$\eta_{0,1} = P(\text{ever reach } 1 \mid X_1 = 1)p + P(\text{ever reach } 1 \mid X_1 = -1)q$$
$$= p + q\eta_{-1,1}$$
$$= p + q\eta_{-1,0}\eta_{0,1}, \quad \text{as in (19)}$$
$$= p + q\eta_{0,1}^2.$$

This quadratic has roots

$$\eta_{0,1} = \{1 \pm (1 - 4pq)^{1/2}\}/q = \{1 \pm \{(p-q)^2\}^{1/2}\}/q,$$
$$= 1 \text{ or } p/q.$$

Either by a similar argument, or by interchanging the roles of p and q, we find that

$$\eta_{1,0} = 1 \quad \text{or} \quad q/p.$$

Without loss of generality, assume $p > q$. Then $\eta_{0,1} = 1$, because $p/q > 1$ is not a probability. For $\eta_{1,0}$, we still have two possibilities; however if $\eta_{1,0} = 1$, then substituting in (20) gives $\eta = 1$, which contradicts the fact, established in Examples (14) and (16), that $\eta < 1$ when $p \neq q$. Therefore $\eta_{1,0} = q/p$, and hence

$$\eta = p(q/p) + q = 2q.$$

Likewise, if $p < q$, $\eta = 2p$. Hence, in any case,

$$(21) \qquad \eta = 2\min\{p,q\} = 2p \wedge q = 1 - |p - q|.$$

A more detailed problem is to ask exactly when the walk returns to the origin. We need some more notation for this: write

T_{jk} is the first-passage time to k for a random walk starting at j;

formally

$$T_{jk} = \min\{n \geq 1 : S_n = k \text{ for } S_0 = j\}.$$

We write $G_{jk}(s) = \mathrm{E}s^{T_{jk}}$. Using the fact that steps of the walk are independently and identically distributed, in the same way as above we obtain, for $k > 0$,

$$G_{0k}(s) = \mathrm{E}s^{T_{0k}} = \mathrm{E}s^{T_{01}}\mathrm{E}s^{T_{12}}\cdots\mathrm{E}s^{T_{k-1,k}}$$

(22)
$$= (\mathrm{E}s^{T_{01}})^k = [G_{01}(s)]^k.$$

Conditioning on the first step, as usual,

$$G_{01}(s) = p\mathrm{E}(s^{T_{01}} \mid X_1 = 1) + q\mathrm{E}(s^{T_{01}} \mid X_1 = -1)$$

(23)
$$= ps + qs[G_{01}(s)]^2.$$

Only one root of the quadratic in (23) is the probability-generating function (pgf) of a non-negative random variable, so

$$G_{01}(s) = \{1 - (1 - 4pqs^2)^{1/2}\}/2qs.$$

Now $G_{10}(s)$ is obtained immediately by interchanging p and q.

Hence finally, once more conditioning on the first step,

$$G_{00}(s) = p\mathrm{E}(s^{T_{00}} \mid X_1 = 1) + q\mathrm{E}(s^{T_{00}} \mid X_1 = -1)$$

(24)
$$= psG_{10}(s) + qsG_{01}(s)$$
$$= 1 - (1 - 4pqs^2)^{1/2}.$$

Expanding these in powers of s yields the distribution of T_{01} and T_{00}; we may also recover our previous result in (21) by noting that

$$\eta = \mathrm{P}(T_{00} < \infty) = G_{00}(1)$$

(25)
$$= 1 - |p - q|.$$

Similar methods may be used to tackle simple random walks in two or three dimensions; see the exercises and problems. It is interesting that for the simple symmetric walk in two dimensions $\eta = 1$, but in three dimensions $\eta < 1$.

In conclusion, let us return to the symmetric random walk for which $p = q = \frac{1}{2}$. From (25), we have that T_{00} is finite with probability 1. But from (24) we have

(26)
$$G'_{00}(s) = s/(1 - s^2)^{1/2} \to \infty$$

as $s \to 1$. Hence $\mathrm{E}T_{00} = \infty$.

Exercises
(a)*
  (i) Let $(X_k; k \geq 1)$ be a sequence of trials yielding independent Bernoulli (p) random variables. Let N be the number of trials until there is a sequence of $|a|+|b|$ successive zeros. Show that $EN < \infty$.
  (ii) Let S_n be simple symmetric random walk started at the origin, and let T be the number of steps required to hit $a > 0$ or $b < 0$. Show that $T \leq N$, and deduce that $ET < \infty$.

(b) Verify the assertion in Example (14) that, when $p+q=1$,

$$\sum_{n=1}^{\infty} \binom{2n}{n} p^n q^n \begin{cases} < \infty, & \text{for } p \neq q, \\ = \infty, & \text{for } p = q = \frac{1}{2}. \end{cases}$$

(c) **Symmetric simple random walk in the plane.** This takes place on the points (j,k) with integer coordinates, $k, j \in \mathbb{Z}$. From any point (j,k) the walk steps to one of the four points $\{(j\pm 1, k), (j, k\pm 1)\}$, with equal probability. Show that the probability of returning to the origin at the $2n$th step is given by

$$p_{00}(2n) = \sum_{r=0}^{n} \frac{2n!}{\{r!(n-r)!\}^2} \left(\frac{1}{4}\right)^{2n}$$

$$= \left(\frac{1}{4}\right)^{2n} \sum_{r=0}^{n} \binom{2n}{n} \binom{n}{r} \binom{n}{n-r} = \left(\frac{1}{4}\right)^{2n} \binom{2n}{n}^2.$$

Prove that $\sum_{n=1}^{\infty} p_{00}(2n) = \infty$, and deduce that the walk returns to its starting point with probability 1. [Hint: Stirling's formula is $\lim_{n \to \infty} \{n^{n+(1/2)} e^{-n} \sqrt{2\pi}/n!\} = 1$.]

2.3 The long run

We are all familiar with the fact that many physical systems 'settle down' if allowed to run for a long time. For example, the planets seem to have settled down into nearly elliptical orbits around the sun; the temperature of any inactive conductor of heat will eventually be the ambient temperature, and so on. For a stochastic process, we cannot expect in general that it will settle down to non-random behaviour. But we are all familiar with the fact that averages of random processes may exhibit long-run regularity; consider the proportion of heads in many flips of a coin, or the proportion of boys in a long series of births.

We need to move carefully in making the idea of convergence precise for stochastic processes, since random variables are functions on Ω, having distributions and moments. There are several ways in which they might be said to converge, so we begin with some simple results that will tie all these approaches together.

(1) Markov inequality. Let X be a non-negative random variable. Then for any $a > 0$

$$P(X \geq a) \leq EX/a.$$

Proof. As we have often done before, define the indicator

$$I(a) = \begin{cases} 1, & \text{if } X \geq a, \\ 0, & \text{otherwise.} \end{cases}$$

Then by definition $aI(a) \leq X$ for all $a > 0$. Taking the expected value of each side of this inequality gives the required result

(2) $$aEI(a) = aP(X \geq a) \leq EX.$$ □

Some special cases of this are important, especially the following.

(3) Chebyshov inequality. For any random variable X

$$P(|X| \geq a) \leq EX^2/a^2, \quad a > 0.$$

Proof. Note that $|X| \geq a$ if and only if $X^2 \geq a^2$, and use the Markov inequality on X^2. □

There are further useful inequalities in the exercises and problems.

This result is also useful.

(4) Borel–Cantelli lemma. Let $(A_n; n \geq 1)$ be a collection of events, and let A be the event $\{A_n \text{ i.o.}\}$ that infinitely many of the A_n occur. If $\sum_{n=0}^{\infty} P(A_n) < \infty$, then $P(A) = 0$.

Proof. From the expression for A given in (2.1.5) we have, for all n, that

$$A \subseteq \bigcup_{r=n}^{\infty} A_r.$$

Hence,

$$P(A) \leq P\left(\bigcup_{r=n}^{\infty} A_r\right) \leq \sum_{r=n}^{\infty} P(A_r), \quad \text{by Boole's inequality, Exercise (1.1.(c))},$$
$$\to 0, \quad \text{as } n \to \infty.$$

The result follows. □

There is a partial converse to this; we omit the proof.

(5) Second Borel–Cantelli lemma. If $(A_n; n \geq 1)$ is a collection of independent events, and $\sum_{n=1}^{\infty} P(A_n) = \infty$, then $P(A_n \text{ i.o.}) = 1$.

Now let us turn to considering the long-run behaviour of a stochastic process $(X_n; n \geq 0)$, in particular the question of its possible convergence as $n \to \infty$.

Recall that a sequence x_n of real numbers is said to converge to a limit x, as $n \to \infty$, if $|x_n - x| \to 0$, as $n \to \infty$. Clearly, when we consider X_n with distribution $F_n(x)$, the existence of a limit X with distribution $F(x)$ must depend on the properties of the sequences $|X_n - X|$, and $|F_n(x) - F(x)|$. We therefore define the events

$$A_n(\epsilon) = \{|X_n - X| > \epsilon\}, \quad \text{where } \epsilon > 0.$$

(6) Summation lemma. This gives a criterion for a type of convergence called *almost sure convergence*. It is straightforward to show that, as $n \to \infty$,

(7) $$P(X_n \to X) = 1,$$

if and only if finitely many $A_n(\epsilon)$ occur, for any $\epsilon > 0$. From the Borel–Cantelli lemma (4) above, a sufficient condition for this is $\sum_{n=0}^{\infty} P(A_n(\epsilon)) < \infty$. (We shall soon see an application of this when we consider the Strong Law of Large Numbers.) □

If (7) holds, then we say X_n converges with probability 1, or *almost surely*, and write

$$X_n \xrightarrow{\text{a.s.}} X, \quad \text{as } n \to \infty.$$

This is a strong type of convergence, and we shall often be satisfied with weaker kinds:

(8) Convergence in probability. If, for all $\epsilon > 0$,

$$P(A_n(\epsilon)) = P(|X_n - X| > \epsilon) \to 0, \quad \text{as } n \to \infty,$$

then X_n is said to *converge in probability* to X. We may write $X_n \xrightarrow{P} X$. △

It is trivial to see that almost sure convergence implies convergence in probability; formally

$$X_n \xrightarrow{a.s.} X \implies X_n \xrightarrow{P} X.$$

However, the inequalities (1) and (3) above (and others to be proved later) supply a different method of approach to establishing convergence in probability.

Convergence in mean square. From Chebyshov's inequality we have

(9) $$P(|X_n - X| > \epsilon) \leq E|X_n - X|^2/\epsilon^2.$$

Therefore, if we can show that $E|X_n - X|^2 \to 0$ as $n \to \infty$, it follows that $X_n \xrightarrow{P} X$. This is often a very convenient way of showing convergence in probability, and we give it a name: if $E|X_n - X|^2 \to 0$ as $n \to \infty$, then X_n is said to *converge in mean square* to X. We may write $X_n \xrightarrow{m.s.} X$. △

However, even this weaker form of convergence sometimes fails to hold; in the last resort we may have to be satisfied with showing convergence of the distributions $F_n(x)$. This is a very weak form of convergence, as it does not even require the random variables X_n to be defined on a common probability space.

(10) **Convergence in distribution.** If $F_n(x) \to F(x)$ at all the points x such that $F(x)$ is continuous, then X_n is said to *converge in distribution*. We may write $X_n \xrightarrow{D} X$. △

It can be proved that convergence in probability implies convergence in distribution, but not conversely.

Finally, since any distribution is determined by its characteristic function, the following result is no surprise. We omit the proof.

(11) **Continuity theorem.** Let $(X_n; n \geq 1)$ be a sequence of random variables with characteristic functions $(\phi_n(t); n \geq 1)$. If $\phi_n(t) \to \phi(t)$ as $n \to \infty$, where $\phi(t)$ is the characteristic function of a distribution $F(x)$, then

$$F_n(x) = P(X_n \leq x) \to F(x), \quad \text{as } n \to \infty$$

at points where $F(x)$ is continuous. This key result is useful because it is often easier to manipulate generating functions than the distributions themselves.

The following applications of these ideas turn out to be particularly useful. Define $S_n = \sum_{r=1}^{n} X_r$, where the X_r are independent and identically distributed. Then we have these three classic results:

(12) **Central limit theorem.** If $\mathrm{E}X_r = \mu$ and $0 < \mathrm{var}\, X_r = \sigma^2 < \infty$, then, as $n \to \infty$,

$$\mathrm{P}\left(\frac{S_n - n\mu}{(n\sigma^2)^{1/2}} \leq x\right) \to \Phi(x),$$

where $\Phi(x)$ is the standard normal distribution.

(13) **Weak law of large numbers.** If $\mathrm{E}X_r = \mu < \infty$, then for $\epsilon > 0$, as $n \to \infty$,

$$\mathrm{P}\left(\left|\frac{S_n}{n} - \mu\right| > \epsilon\right) \to 0.$$

(14) **Strong law of large numbers.** As $n \to \infty$,

$$\frac{S_n}{n} \xrightarrow{\text{a.s.}} \mu$$

for some finite constant μ, if and only if $\mathrm{E}|X_r| < \infty$, and then $\mu = \mathrm{E}X_1$.

The central limit theorem is the principal reason for the appearance of the normal (or 'bell-shaped') distribution in so many statistical and scientific contexts. The first version of this theorem was proved by Abraham de Moivre before 1733.

The laws of large numbers supply a solid foundation for our faith in the usefulness and good behaviour of averages. In particular, as we have remarked above, they support one of our most appealing interpretations of probability as long-term relative frequency. The first version of the weak law was proved by James Bernoulli around 1700; and the first form of the strong law by Emile Borel in 1909.

We do not prove any of these results in their strongest or most general forms, which require lengthy and technical arguments. But we do need one useful new bit of notation: if $g(x)$ and $h(x)$ are functions of x, we write $g(x) = \mathrm{o}(h(x))$ as $x \to 0$, if $\lim_{x \to 0}\{g(x)/h(x)\} = 0$.

Proof of (12). Assume that X_r has finite moments of all orders. Then for all finite t

$$\mathrm{E}\exp(it(X_r - \mu)/\sigma) = 1 + it\mathrm{E}\left(\frac{X_r - \mu}{\sigma}\right) - \frac{1}{2}t^2\mathrm{E}\left(\frac{X_r - \mu}{\sigma}\right)^2 + \mathrm{o}(t^2)$$

$$= 1 - \frac{1}{2}t^2 + \mathrm{o}(t^2).$$

Hence, by the independence of the Xs,

$$\operatorname{E} \exp\left\{it\frac{(S_n - n\mu)}{(n\sigma^2)^{1/2}}\right\} = \operatorname{E}\left\{\exp\left(i\frac{t}{\sqrt{n}}(X_r - \mu)/\sigma\right)\right\}^n$$

(15)
$$= \left\{1 - \frac{1}{2}\frac{t^2}{n} + o\left(\frac{t^2}{n}\right)\right\}^n$$

$$\to \exp\left(-\frac{1}{2}t^2\right), \quad \text{as } n \to \infty.$$

The central limit theorem follows using (11) and (1.8.19).

If you prefer, you can follow the same method using moment-generating functions (mgfs) and the result (1.8.7). □

Proof of (13). Assume that $\operatorname{var} X_r < \infty$. By Chebyshov's inequality (3)

$$\operatorname{P}\left(\left|\frac{S_n}{n} - \mu\right| > \epsilon\right) \le \operatorname{E}\left(\frac{S_n}{n} - \mu\right)^2 \Big/ \epsilon^2$$

(16)
$$= \operatorname{var} X_1/(n\epsilon^2), \quad \text{by the independence}$$

$$\to 0, \quad \text{as } n \to \infty.$$
□

Proof of (14). Assume that X_r has finite mgf $M(t)$ in $[-a, a]$, for some $a > 0$. Now let $Y_r = X_r - \mu - \epsilon$, $\epsilon > 0$, where $\mu = \operatorname{E} X_1$. Since $\operatorname{E} Y_r < 0$ it follows that the mgf $M_Y(t) = \operatorname{E} \exp[(X_r - \mu - \epsilon)t]$ has negative slope at $t = 0$. Since it exists in $[-a, a]$ there is therefore some value v, $0 < v < a$, such that $M_Y(v) < b < 1$. Therefore we can write

$$\operatorname{P}\left(\frac{S_n - n\mu}{n} > \epsilon\right) = \operatorname{P}(\exp\{v(S_n - n\mu - n\epsilon)\} > 1)$$

(17)
$$\le \operatorname{E}\exp\{v(S_n - n\mu - n\epsilon)\}, \quad \text{by (1)}$$

$$= \{M_Y(v)\}^n < b^n.$$

Similarly there is some constant c, $0 < c < 1$, such that

$$\operatorname{P}((S_n - n\mu)/n < -\epsilon) < c^n.$$

Hence
$$\sum_n \operatorname{P}(|S_n - n\mu|/n > \epsilon) < \sum_n (b^n + c^n) < \infty$$

and by Lemma (6) we have $S_n/n \xrightarrow{\text{a.s.}} \mu$, as $n \to \infty$. □

Finally, in this section we return to reconsider the relationships between the various types of convergence of a sequence $(X_n; n \geq 0)$ of random variables. We have proved that $X_n \xrightarrow{\text{a.s.}} X$ implies that $X_n \xrightarrow{P} X$; and also that $X_n \xrightarrow{\text{m.s.}} X$ implies that $X_n \xrightarrow{P} X$. It can be shown that $X_n \xrightarrow{P} X$ implies that $X_n \xrightarrow{D} X$.

In general the converse statements are false, but there are some important special cases in which they do hold. This is the most important.

(18) **Dominated convergence.** Suppose that $X_n \xrightarrow{P} X$, and $|X_n| \leq Z$ for all n, where $EZ < \infty$. Then $E|X_n - X| \to 0$.

Proof. Let $Z_n = |X_n - X|$. Because $|X_n| \leq Z$, we have $|Z_n| \leq 2Z$. For any event A we denote the indicator that A occurs by $I(A)$:

$$I(A) = \begin{cases} 1, & \text{if } A \text{ occurs,} \\ 0, & \text{if } A^c \text{ occurs.} \end{cases}$$

Then for $\epsilon > 0$,

$$E|Z_n| = E(Z_n I(Z_n \leq \epsilon)) + E(Z_n I(Z_n > \epsilon))$$
(19) $$\leq \epsilon + 2E(ZI(Z_n > \epsilon)).$$

Next, for $y > 0$, we can write

$$E(ZI(Z_n > \epsilon)) = E(ZI(Z_n > \epsilon, Z > y)) + E(ZI(Z_n > \epsilon, Z \leq y))$$
$$\leq E(ZI(Z > y)) + yP(Z_n > \epsilon).$$

Now choose y and then n as large as we please to make the right-hand side arbitrarily small. But we can then choose ϵ as small as we please in (19) to find that $E|Z_n| \to 0$ as $n \to \infty$, as required. □

Here are two other useful results of the same type that we state without proof.

(20) **Monotone convergence.** If $X_n \xrightarrow{\text{a.s.}} X$ as $n \to \infty$, where the sequence is monotone, then $EX_n \to EX$.

(21) **Pratt's lemma.** If $U_n < X_n < W_n$, where $EU_n \to EU$, $EW_n \to EW$, $E|U|E|W| < \infty$, and $X_n \xrightarrow{P} X$, then

$$E|X_n - X| \to 0, \quad \text{as } n \to \infty.$$

Exercises

(a)* **One-sided Chebyshov inequality.**
  (i) Let X have mean 0; show that for $\epsilon > 0$
  $$P(X \geq \epsilon) \leq \frac{\operatorname{var} X}{\operatorname{var} X + \epsilon^2}.$$

  (ii) Deduce that if m is a median of X, where X has mean μ and variance σ^2, then
  $$(m - \mu)^2 \leq \sigma^2.$$

(b) **Chernov inequality.** Let X have mgf $M(t)$. Show that for $\epsilon > 0$,
$$P(X \geq \epsilon) \leq e^{-\epsilon t} M(t), \quad \text{for } t > 0.$$

Deduce that
$$P(X \geq \epsilon) \leq \inf_{t>0} e^{-\epsilon t} M(t)$$
and evaluate the right-hand side when X is $N(0, 1)$.
Now consider the function $g(x) = \frac{1}{2} \exp(-\frac{1}{2}x^2) - 1 + \Phi(x)$, where $\Phi(x)$ is the distribution function of $N(0, 1)$. Find the maximum of $g(x)$ for $x > 0$, and by considering $g(x)$ as $x \to 0$ and as $x \to \infty$ show that
$$P(X \geq \epsilon) \leq \tfrac{1}{2} \exp\left(-\tfrac{1}{2}\epsilon^2\right), \quad \text{for } \epsilon > 0.$$

(c) If $X_n \xrightarrow{D} c$, as $n \to \infty$, where c is constant, show that $X_n \xrightarrow{P} c$.

(d) **Convergence in rth mean.** If $E|X_n - X|^r \to 0$ as $n \to \infty$, for $r \geq 1$, then we write $X_n \xrightarrow{r} X$. Show that for $r < s$, $X_n \xrightarrow{s} X \Rightarrow X_n \xrightarrow{r} X$. [Hint: use the result of Problem 1.9.]

2.4 Martingales

We remarked in Section 2.1 that the various types of stochastic process are each characterized by dependence relations between their random variables. In this section we examine a class of processes that may be thought of as an abstract model for fair games. The key point about fair games, of whatever type, is that your expected fortune after any bet should equal your fortune before. This is our definition of fairness, actually.

We set this notion out formally thus as follows; make sure you have read the remarks about conditional expectation in (1.7.20)–(1.7.25).

(1) **Martingale.** The sequence $(M_n; n \geq 0)$ is a *martingale* if, for all n,

(a) $E|M_n| < \infty$.
(b) $E(M_{n+1} \mid M_0, \ldots, M_n) = M_n$.

More generally, we may define a martingale in terms of an auxiliary process. That is to say, given $(Y_n; n \geq 0)$, we replace (b) by

(c) $E(X_{n+1} \mid Y_0, \ldots, Y_n) = X_n$.
Then X_n is said to be a martingale with respect to Y_n.

The second condition (b) asserts (in the gaming context) that your expected fortune is the same after your nth bet as before. We can rearrange (b) to read

$$E(M_{n+1} - M_n \mid M_0, \ldots, M_n) = 0,$$

which says that your expected winnings (net of your stake) on your next bet have mean zero, given the results of your past wagers. If you are playing in a real casino, of course this assumption is not true; we should replace (b) with

(d) $\qquad\qquad E(M_{n+1} \mid M_0, \ldots, M_n) \leq M_n.$

Then $(M_n; n \geq 0)$ is called a *supermartingale*. For the casino's fortune X_n, naturally, (b) is replaced by

(e) $\qquad\qquad E(X_{n+1} \mid X_0, \ldots, X_n) \geq X_n.$

In this case, $(X_n; n \geq 0)$ is called a *submartingale*. △

These gambling models may seem to be rather flimsy and frivolous reasons for studying martingales, but it turns out that they crop up all over the place. Not only are they discovered in models for almost every form of random process, but they also form a cornerstone of modern financial mathematics. This is, of course, because they encapsulate one idea of 'fair value'. But see Example (27), which shows that 'fair' is not always the same as 'sensible'.

Here are some preliminary examples of how martingales turn up; we shall see many more.

(2) **Example.** Let $(X_n, n \geq 0)$ be a collection of independent random variables with finite respective means $(\mu_n; n \geq 0)$.

(a) Let $S_n = \sum_{r=0}^{n}(X_r - \mu_r)$. Then S_n is a martingale. We see this because $E|S_n| \leq E \sum_r |X_r - \mu_r| < \infty$, and $E(X_r - \mu_r) = 0$.

(b) Let $M_n = \prod_{r=0}^{n}(X_r/\mu_r)$. Then M_n is similarly a martingale, provided that $\mu_n \neq 0$ for all n. ○

(3) **Example.** Let $(S_n; n \geq 0)$ be the simple symmetric random walk defined in Section (2.3). Then $M_n = S_n^2 - n$ is a martingale.

To see this, we check that

$$E|M_n| \leq ES_n^2 + n = \frac{n}{4} + n < \infty.$$

Also, using the independence of the steps X_n,

$$E(M_{n+1} \mid M_0, \ldots, M_n) = E(S_{n+1}^2 - n - 1 \mid S_0^2, \ldots, S_n^2)$$
$$= E(S_n^2 + 2X_{n+1}S_n + X_{n+1}^2 - n - 1 \mid S_0^2, \ldots, S_n^2)$$
$$= S_n^2 - n = M_n.$$ ○

To appreciate the beauty of martingales, we need one more concept that also arises naturally in the context of gambling, namely, stopping times. Of course, you cannot gamble for ever, but when do you stop? Obvious stopping times are when:

- you run out of money;
- the casino runs out of money (this seems improbable);
- you have to leave to catch your flight home;

and so on; think of some yourself. A moment's thought reveals that all these stopping times, and other reasonable rules for stopping, cannot be decided by your fortune *after* you stop. Some rules may depend on your fortunes up to the time of stopping. But you cannot reasonably obey a rule such as:

- stop just before you have a loss.

We make this formal as follows.

(4) **Stopping time.** If $(X_n; n \geq 0)$ is a stochastic process, then the non-negative integer-valued random variable T is a *stopping time* for X if the event $\{T = n\}$ depends only on $(X_0, \ldots, X_n)$, and does not depend on $(X_{n+k}; k \geq 1)$. We may express this otherwise by saying that the indicator of the event that $T = n$, $I(T = n)$, is a function only of $X_0, \ldots, X_n$. △

Now we can see one of the reasons why martingales turn out to be so useful.

(5) **Simple systems theorem.** Let $(X_n; n \geq 0)$ be a martingale, and let T be a stopping time for X. Define

(6) $$Z_n = \begin{cases} X_n, & \text{if } n \leq T, \\ X_T, & \text{otherwise,} \end{cases}$$

so Z_n is essentially X_n stopped at T. Then Z_n is a martingale with respect to X_n.

Proof. Define the indicator, which depends only on $X_0, \ldots, X_n$,

$$I(T > n) = \begin{cases} 1, & \text{if } T > n, \\ 0, & \text{otherwise.} \end{cases}$$

Then, by inspection, we can write

(7) $$Z_{n+1} = Z_n + (X_{n+1} - X_n)I(T > n).$$

To see this, consider the two cases: $T \leq n$, in which case $Z_{n+1} = X_T = X_T + 0$; or $T > n$, in which case $Z_{n+1} = X_{n+1} = X_n + X_{n+1} - X_n$.

Next we notice that by their definitions, both Z_n and I are functions of $X_0, \ldots, X_n$ only. Hence, by the pull-through property (1.7.20),

$$E(Z_{n+1} \mid X_0, \ldots, X_n) = Z_n + I(T > n)E(X_{n+1} - X_n \mid X_0, \ldots, X_n)$$
$$= Z_n.$$

Finally, we write Z_n in the form

(8) $$Z_n = \sum_{r=0}^{n-1} X_r I(T = r) + X_n I(T \geq n),$$

where $I(T = r)$ is the indicator of the event that $T = r$. Since $0 \leq I \leq 1$, it follows immediately that

$$E|Z_n| \leq \sum_{r=0}^{n} E|X_r| < \infty.$$

This verifies both conditions for Z_n to be a martingale, as required. Thus

$$EZ_n = EZ_0 = EX_0. \qquad \square$$

The reason we called this the simple systems theorem is that many simple gambling systems take the form: 'Play until you have reached a target gain, and then stop'. The theorem asserts that if you play a fair game with this system, your fortune will still be a martingale.

However, and this is important, unless there are restrictions on T, or X, or both, it does *not* follow that

$$EX_T = EX_0.$$

(9) **Example. Quit when you are ahead.** Suppose that we consider the simple symmetric random walk, shown to be a martingale in Exercise (a) below. Since we know that the walk reaches 1 with probability 1, if this is your game, and your rule is to quit when you are ahead, you guarantee to win every time you play. Indeed, since the walk visits any point $k > 0$ with probability 1, you can be sure of winning any amount you choose. Unfortunately, the expected time to do so is infinite, and your possible losses in the meantime are unbounded. You need infinite credit and time to guarantee your gain; but not many casinos offer this. See (27) for an even more foolish system. ○

Formally, we see, from the definition of Z_n in (6), that if $P(T < \infty) = 1$, then as $n \to \infty$,

$$(10) \qquad Z_n \xrightarrow{\text{a.s.}} X_T.$$

It is natural to ask for conditions which ensure that $EX_T = EX_0$. One sufficient condition is supplied immediately by the dominated convergence theorem (DCT), (2.3.18), thus:

(11) **Elementary optional stopping.** Let X_n, T, and Z_n, be defined in (4), (5), and (6). If there is a random variable Y such that $EY < \infty$ and $|Z_n| < Y$ for all n, then as $n \to \infty$ we have from (10)

$$(12) \qquad |X_T - Z_n| \to 0$$

and by the DCT it follows that $E|X_T - Z_n| \to 0$. Now

$$|EX_0 - EX_T| = |EZ_n - EX_T| \leq E|Z_n - X_T|.$$

Therefore,
$$|EX_0 - EX_T| \leq \lim_{n \to \infty} E|Z_n - X_T| = 0.$$

Hence, $EX_0 = EX_T$. □

We state two further popular and useful conditions without proof.

(13) **Optional stopping.** Let $(X_n; n \geq 0)$ be a martingale and T a stopping time for X. Then
$$EX_T = EX_0,$$
if either

(i) T is bounded; or

(ii) $ET < \infty$ and for some constant $c < \infty$,
$$E(|X_{n+1} - X_n| \mid X_0, \ldots, X_n) < c.$$

Here is a classic application of optional stopping.

(14) **Wald's equation.** Let $S_n = \sum_{r=1}^{n} X_r$ be a random walk started at the origin, with $EX_r = \mu < \infty$. Let T be a stopping time for S_n, with $ET < \infty$. Then $ES_T = \mu ET$.

Proof. $M_n = S_n - n\mu$ is easily shown to be a martingale; see (2). Furthermore

$$E(|M_{n+1} - M_n| \mid S_0, \ldots, S_n) = E|X_{n+1} - \mu|$$
$$\leq E|X_{n+1}| + |\mu| = c < \infty.$$

Hence, by the second part of the optional stopping theorem (13), we have

$$EM_T = E(S_T - \mu T) = EM_0 = 0.$$

The result follows. □

In addition to stopping gracefully, martingales are also subject to some rather powerful and useful inequalities. Here is a typical result:

(15) **Maximal inequalities.** Let X_n be a non-negative martingale, $n \geq 0$, and let $x > 0$. Define

$$U = \max_{n \geq 0} X_n$$

and

$$V_n = \max_{k \leq n} X_k.$$

Then we have the inequalities

(16) $$P(U \geq x) \leq EX_0/x$$

and

(17) $$EV_n^2 \leq 4EX_n^2.$$

Proof of (16). Define $T = \min\{m : X_m \geq x\}$ to be the first time the martingale surpasses x. Then $T \wedge n = \min\{n, T\}$ is a bounded stopping time, and therefore

$$EX_{T \wedge n} = EX_0 = EX_n.$$

Now using the indicators $I(T \leq n)$ and $I(T > n)$, with an obvious notation

$$EX_n = EX_{T \wedge n} = E\{X_{T \wedge n} I(T \leq n)\} + E\{X_{T \wedge n} I(T > n)\}$$
$$\geq xEI(T \leq n) + E(X_n I(T > n)), \quad \text{because } X_T \geq x.$$

Hence, because X_n is non-negative

$$xP(T \leq n) \leq E\{X_n I(T \leq n)\}$$
(18)
$$\leq EX_n, \quad \text{since } X_n > 0$$
$$= EX_0, \quad \text{since } X_n \text{ is a martingale.}$$

Therefore, because $V_n \geq x$ if and only if $T \leq n$,

$$P(V_n \geq x) \leq E(X_n I(T \leq n))/x$$
(19)
$$\leq EX_n/x$$
$$= EX_0/x$$

and (16) follows on letting $n \to \infty$.

Proof of (17). Using the tail integral for expectation

$$EV_n^2 = 2\int_0^\infty xP(V_n > x)\,dx$$

(20) $\quad \leq 2\int_0^\infty E(X_n I(V_n \geq x))\,dx, \quad \text{by (19)}$

$$= 2E\left(\int_0^{V_n} X_n\,dx\right) = 2E(X_n V_n), \quad \text{since the integrand is positive}$$

$$\leq 2(EX_n^2)^{1/2}(EV_n^2)^{1/2}, \quad \text{by the Cauchy–Schwarz inequality,}$$
Problem 1.7.

Now dividing by $(EV_n^2)^{1/2}$ gives the result. There is an application of (17) in (2.7.24). □

(21) **Example. Gambling again.** Suppose the non-negative martingale X_n is indeed the sequence of values of a gambler's wealth playing only fair games, with no credit, where the stake is always less than the existing fortune X_n. Without loss of generality, let $X_0 = 1$. Then the maximal inequality (16) shows that the probability that this wealth *ever* reaches the level $x \geq 1$ is less than x^{-1}. For example, we have that no matter what fair games you play, for whatever stakes, (with no credit), the chance of ever doubling your money is less than $\frac{1}{2}$. ○

Finally, we note that martingales also have the useful property of converging under rather weak conditions. Here is one example.

(22) **Martingale convergence.** Let $(Y_n; n \geq 0)$ be a martingale such that $EY_n^2 \leq K < \infty$ for all n. Then Y_n converges in mean square as $n \to \infty$.

Proof. For $r > i$

$$\begin{align}
E(Y_r \mid Y_0, \ldots, Y_i) &= E(E(Y_r \mid Y_{r-1}, \ldots, Y_0) \mid Y_0, \ldots, Y_i) \\
&= E(Y_{r-1} \mid Y_0, \ldots, Y_i) \\
&= Y_i
\end{align} \tag{23}$$

on iterating. Hence

$$E(Y_r Y_i) = E(E(Y_r Y_i \mid Y_0, \ldots, Y_i)) = EY_i^2. \tag{24}$$

Therefore, for $i \leq j \leq k$,

$$E\{(Y_k - Y_j)Y_i\} = E(Y_k Y_i) - E(Y_j Y_i) = 0. \tag{25}$$

It follows from (24) and (25) that

$$0 \leq E(Y_k - Y_j)^2 = EY_k^2 - EY_j^2.$$

Thus $(EY_n^2; n \geq 1)$ is a bounded, non-decreasing, and therefore convergent sequence. Hence, we have

$$E(Y_k - Y_j)^2 \to 0,$$

as $j, k \to \infty$, immediately from the above. It can be shown (after some work that we omit) that this entails the existence of a random variable Y such that $Y_k \xrightarrow{\text{m.s.}} Y$, as $k \to \infty$. Indeed, it can also be shown that $Y_n \xrightarrow{\text{a.s.}} Y$, but it is beyond our scope to venture further into this field.

It is noteworthy that the above proof included the following small but useful result

(26) **Orthogonal increments.** If $(X_n; n \geq 0)$ is a martingale, then it has orthogonal increments in the sense that for $i \leq j \leq k \leq m$

$$E\{(X_m - X_k)(X_j - X_i)\} = 0. \qquad \square$$

Finally, we mention a classic cautionary example.

(27) **The martingale.** This is a well-known (or notorious) gambling system, that runs as follows: you bet \$1 at evens; if it wins you quit. If it loses, you double your bet to \$2; if it wins you quit. If it loses, you double the stake to \$4, and so on.

This system has the advantage that it guarantees you will win \$1, after a finite number of bets (having a Geometric ($\frac{1}{2}$) distribution). Unfortunately it is easy to see that the expected size of your winning bet is infinite!

Exercises

(a)* Let $(S_n; n \geq 0)$ be simple random walk. Show that the following are martingales:

   (i) S_n, if the walk is symmetric.

   (ii) $S_n - n(p-q)$, in any case.

   (iii) $(q/p)^{S_n}$.

   (iv) $S_n^2 - 2\mu n S_n + (n\mu)^2 - n\sigma^2$, where $\mu = p - q$, $\sigma^2 = 1 - \mu^2$.

(b)* Let $(S_n; n \geq 0)$ be simple random walk started at the origin: $S_0 = 0$. Let T be the number of steps until the walk hits a or b, where $b < 0 < a$ are integers. Let A and B be the events that $S_T = a$ or $S_T = b$, respectively. Use the martingales in Exercise (a), together with the optional stopping theorem, to find $P(A)$, $P(B)$, and ET. Find var T when $a = \infty$ and $p < q$. How would you choose to find var T when $a < \infty$?

(c) **Polya's urn.** A box initially contains a red ball and a blue ball. A ball is removed at random, its colour noted, and it is put back in the box together with a fresh ball of the same colour. Let R_n be the number of red balls in the box after n repetitions of this procedure.

   (i) Show that $R_n/(n+2)$ is a martingale.

   (ii) Let N be the number of balls drawn until the first blue ball is seen. Show that $E\{(N+2)^{-1}\} = \frac{1}{4}$.

   (iii) Show that $P(R_n/(n+2) \geq \frac{3}{4}$ for some $n) \leq \frac{2}{3}$.

2.5 Poisson processes

In real life, as in many scientific problems and industrial applications, we are often faced with an unending random sequence of events or happenings of some type. Here are some examples:

(1) **Email.** Emails arrive at your address unremittingly, but at random and unpredictable times. A random and unpredictable proportion are junk.

(2) **Shop.** Customers enter a shop at random times, and occasionally make random purchases.

(3) **Cable.** As a cable, or any other continuous product—yarn, steel strip, filament—is produced, flaws occur at random along its length.

(4) **Pests.** Wasps, midges, mosquitoes, and assorted Arachnida puncture you unpredictably at random.

Think of some examples yourself. We model such sequences using a counting process $(N(t); 0 \leq t < \infty)$, where $N(t)$ must obviously have these properties in general:

(5) **Counting process.** $N(t); t \geq 0$ is a *counting process* if it is non-negative, integer-valued, and non-decreasing in t. Then $N(t) - N(s)$ is the number counted in $(s, t]$. △

By far the most important counting process is this one.

(6) **Poisson process.** The counting process $N(t)$ is a *Poisson process with intensity* λ if $N(0) = 0$, and

(a) $N(t+s) - N(s)$ is Poisson (λt), $0 < s < s + t$;
(b) For any $0 \leq t_0 < t_1 < \cdots < t_n < \infty$, $N(t_0), N(t_1) - N(t_0), \ldots, N(t_n) - N(t_{n-1})$ are independent. This is called the property of *independent increments*. △

Of course our first task is to devote some time to justifying our claim about the supremacy of the Poisson process. Remember that we are seeking to characterize 'completely random' sequences of events. For definiteness, consider the sequence of meteorite impacts within 100 miles of where you are, which is about as random as anything could be.

Suppose you start a clock at time 0, and there are no hits up to time t. What is the probability that there will be no hits in $(t, t+s)$? Since meteorites do not (as far as we can tell) in any way coordinate their arrivals, this must be the same as the probability of no hits in $(0, s)$. But we remarked in Exercise (1.7(a)) that the only distribution with this lack-of-memory property is the exponential density. We therefore make this definition.

(7) **Meteorite counting.** Let $(X_r; r \geq 1)$ be independent and Exponential (λ); define $T_n = \sum_{r=1}^{n} X_r$, with $T_0 = 0$, and set

$$N(t) = \max\{n : T_n \leq t\}.$$

Then T_n is the time of arrival of the nth meteorite, and $N(t)$ is the number that have arrived by time t. △

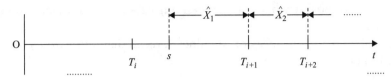

Figure 2.1 The diagram defines the intervals $\hat{X}_1, \hat{X}_2, \hat{X}_3, \ldots$ between events of the counting process $\hat{N}(t)$.

Obviously $N(t)$ is a counting process; less obviously, we have this:

(8) **Theorem.** $N(t)$ is a Poisson process.

Proof. By definition,

$$P(N(t) = n) = P(T_n \leq t) - P(T_{n+1} \leq t).$$

Now you showed in Problem (1.4) that T_n has a Gamma (λ, n) density, so the right-hand side is

$$\int_0^t \frac{\lambda^n}{(n-1)!} x^{n-1} e^{-\lambda x} - \frac{\lambda^{n+1}}{n!} x^n e^{-\lambda x} \, dx = e^{-\lambda t} \frac{(\lambda t)^n}{n!},$$

which is the Poisson (λt) distribution. Next, we show that $\hat{N}(t) = N(s+t) - N(s)$ is independent of $(N(u); 0 \leq u \leq s)$, and $\hat{N}(t)$ has the same distribution as $N(t)$. Let us consider $\hat{N}(t)$ conditional on the event $\{N(s) = i\}$; there is a diagram above (Figure 2.1).

Conditional on $N(s) = i$, and the values of $T_1, \ldots, T_i$, the lack-of-memory property of X_{i+1} shows that $\hat{X}_1$ is Exponential (λ). Furthermore, it is independent of $\hat{X}_2 = X_{i+2}$, and indeed all of $(X_{i+k}; k \geq 3)$; and therefore

$$\hat{N}(t) = \max\left\{ n : \sum_{r=1}^n \hat{X}_r \leq t \right\}$$

has the same distribution as $N(t)$, independent of $N(s)$ and $T_1, \ldots, T_i$. Removing the conditioning on the values of $T_1, \ldots, T_i$ and then the value of $N(s)$, leaves this statement intact. This is what we set out to prove, and it follows immediately that $N(t)$ has independent increments. □

Note that you can prove conversely that if $N(t)$ is a Poisson process of intensity λ, then the times between the jumps of $N(t)$ are independent and Exponential (λ). (An easy exercise for you.)

Remarkably, we can also generate the Poisson process by looking at (say) meteorite hits in a different way; that is locally, rather than globally. (This alternative approach turns out to be very useful in later work.) The basis of the idea is to look at what may happen during an arbitrary small interval $(t, t+h)$.

To make precise our ideas of 'smallness', recall that a function $f(h)$ of h is said to be $o(h)$ if

(9) $$f(h)/h \to 0, \quad \text{as } h \to 0.$$

We write $f = o(h)$ in such a case (with a small abuse of notation).

Now suppose, as usual, that $N(t)$ counts the arrivals during $[0, t]$. The essential properties of meteorites mentioned above lead naturally to the following assumptions about $N(t+h) - N(t)$.

(10) **Independence.** The arrivals in $(t, t+h)$ are independent of arrivals in $(0, t)$.

(11) **Constant intensity.** The chance of arrivals is the same for any interval of length h
$$P(N(t+h) - N(t) \geq 1) = \lambda h + o(h).$$

(12) **Rarity.** Multiple arrivals are unlikely in $(t, t+h)$, that is,
$$P(N(t+h) - N(t) > 1) = o(h).$$

These modelling assumptions result in this.

(13) **Theorem.** The counting process satisfying (10), (11), and (12) is a Poisson process.

Proof. The independence of increments is one of our assumptions; we have only to prove that $N(t)$ has a Poisson distribution.

Let $f_n(t) = P(N(t) = n)$. Then by conditional probability

$$f_n(t+h) = (1 - \lambda h + o(h)) f_n(t) + (\lambda h + o(h)) f_{n-1}(t) + o(h), \quad n \geq 1$$

and
$$f_0(t+h) = (1 - \lambda h + o(h)) f_0(t).$$

Rearranging, dividing by h, and allowing $h \to 0$, yields

$$\frac{d f_n(t)}{dt} = -\lambda f_n(t) + \lambda f_{n-1}(t), \quad n \geq 1,$$

$$\frac{d f_0(t)}{dt} = -\lambda f_0(t).$$

We may solve these equations either by induction, or by introducing the generating function $G(s, t) = E s^{N(t)} = \sum_{n=0}^{\infty} s^n f_n(t)$.

Summing the set, where each is multiplied by the corresponding term s^n, gives

(14) $$\frac{\partial G}{\partial t} = \lambda(s-1)G$$

with solution $G = \exp\{\lambda t(s-1)\}$, so that $N(t)$ is Poisson (λt), as required. □

Note that you can easily show conversely that a Poisson process satisfies (10), (11), and (12).

Finally, it is interesting and informative to see yet another constructive reason why $N(t)$ must have the Poisson distribution, rather than some other discrete distribution.

Suppose that λ meteorites arrive during an epoch E of duration L. What is the distribution of the number X that arrive during the relatively brief interval $I = (s, s+t) \in E$, where $nt = L$?

All possible times in E are the same to a random meteorite, so the chance of any such meteorite arriving in I is L/n. They arrive independently, so X has the Binomial $(\lambda, L/n)$ distribution. For large n we have shown above that this is approximately Poisson $(\lambda L/n)$ = Poisson (λt).

In fact this final motivation for the Poisson process is also canonical.

(15) **Theorem.** Let $N(t)$ be a Poisson process. Conditional on $N(t) = n$, the positions of these n events in $[0, t]$ are independently and uniformly distributed on $[0, t]$.

Proof. We use the fact that a joint density f of continuous random variables is characterized by the property that

$$P(x_1 \leq X_1 \leq x_1 + h_1; \ldots; x_n \leq X_n \leq x_n + h_n) = (f + r)h_1, \ldots, h_n,$$

where $r \to 0$ as $\max\{h_1, \ldots, h_n\} \to 0$.

Now, for the events of a Poisson process of intensity λ, at times $S_1, S_2, S_3, \ldots$, we have

$$P(t_1 < S_1 \leq t_1 + h_1; \ldots; t_n < S_n \leq t_n + h_n \leq t \mid N(t) = n)$$

$$= \prod_{j=1}^{n} h_j \{\lambda^n e^{-\lambda t_1} e^{-\lambda(t_2 - t_1)} \ldots e^{-\lambda(t_n - t_{n-1})} e^{-\lambda(t - t_n)} + r\} / \{e^{-\lambda t}(\lambda t)^n / n!\}$$

$$= \{n! t^{-n} + r\} h_1 \ldots h_n.$$

The result follows when we recall that the joint density of the order statistics of n independent Uniform $(0, t)$ random variables is $f = n! t^{-n}$. (This was Problem 1.14.) □

Exercises

(a)* Emails arrive according to a Poisson process with intensity λ. Any such arriving email is independently junk with probability σ.
  (i) Given that n emails arrive in $(0, t)$, find the distribution of the amount of junk.
  (ii) Show that arrivals of junk and nonjunk form independent Poisson processes.
  (iii) Following the arrival of a junk email, find the distribution of the number of emails until the next item of junk.

(b) At time $t > 0$, let $A(t)$ be the time since the most recent event of a Poisson process of intensity λ. Find the distribution of $A(t)$, and calculate $\mathrm{E}A(t)$.

(c) Let $N_1(t)$ and $N_2(t)$ be independent Poisson processes with respective intensities λ and μ. Show that $N_1(t) + N_2(t)$ is a Poisson process of intensity $\lambda + \mu$.

(d) Show that the Poisson process as defined in (6) has independent Exponential (λ) interarrival times as defined in (7).

2.6 Renewals

In (2.5.5) we defined the general notion of a counting process $N(t)$; and in (2.5.7) we introduced an important special type with Exponential (λ) intervals between jumps: the Poisson process. Of course, many practical counting processes do not have exponentially distributed intervals, so we make this more general definition.

(1) **Renewal process.** Let $(X_r; r \geq 1)$ be independent with common distribution $F(x)$, such that $\mathrm{P}(X_r > 0) > 0$. We set $S_n = \sum_{r=1}^{n} X_r$, with $S_0 = 0$, and define the *renewal process* $N(t)$ by

(2) $$N(t) = \max\{n : S_n \leq t\}, \quad t \geq 0.$$

The mean $m(t) = \mathrm{E}N(t)$ is called the *renewal function*. △

The name renewal process is used because of an enormous range of practical applications of the following type. You insert a new component into some operating system; for example: a tyre on your car; a metal ball and socket in your hip; a battery in your remote; a lamp in a socket; a sole on your shoe, and so on. These all have the property that they fail after a random time X_1 (which is most unlikely to be Exponential (λ)), and are replaced immediately by a similar new component which may be assumed to fail after an independent but identically distributed interval X_2, and so on. Then $N(t)$ is the number of occasions that the item has been renewed by time t.

In fact renewal theory arises in many other more important contexts than simply replacing components. For example,

(3) **Random walks.** Because of the independence of steps, the times at which the walk revisits its starting point form a renewal process. ○

(4) **Queues: (2.1.2) revisited.** Let $N(t)$ be the number of occasions by time t on which the queue empties; that is to say the server despatches a customer and finds nobody in the queue waiting for service. The lack-of-memory property of the exponential density ensures that the intervals between such events are independent, and $N(t)$ is a renewal process. ○

Formulate some more examples yourself, such as counting the number of cars passing a checkpoint, or the number of times a machine is repaired, and so on.

(5) **Distribution of $N(t)$.** By definition of $N(t)$, we have the important fact that $N(t) \geq n$ if and only if $S_n \leq t$. (We noted this already for the Poisson process above.) Hence $P(N(t) \geq n) = P(S_n \leq t)$, and so

(6) $$P(N(t) = n) = P(S_n \leq t) - P(S_{n+1} \leq t).$$

This relation is often useful, though the right-hand side rarely reduces to a simple closed form.

(7) **Moments of $N(t)$.** First of all, we have for the renewal function $m(t)$, by the tail sum theorem,

(8) $$EN(t) = \sum_{r=1}^{\infty} P(N(t) \geq r) = \sum_{r=1}^{\infty} P(S_r \leq t).$$

Unfortunately it is not usually the case that $P(S_r \leq t)$ has a nice closed-form expression, so (8) is less useful than it may seem. We must develop other lines of approach, for example,

(9) **Theorem.** When $EX_r > 0$, $N(t)$ has finite moments for all $t < \infty$.

Proof. We prove that $EN(t) < \infty$; the more general result is an easy exercise for you.

First note that because $EX_r > 0$, there is some $\epsilon > 0$ such that $P(X_r \geq \epsilon) \geq \epsilon$. Define a new renewal process $M(t)$ with interarrival times

$(Z_r; r \geq 1)$ where

(10) $$Z_r = \epsilon I(X_r \geq \epsilon) = \begin{cases} \epsilon, & \text{if } X_r \geq \epsilon, \\ 0, & \text{otherwise.} \end{cases}$$

Note that by construction this renewal process must count more renewals in any interval than $N(t)$; that is, $M(t) \geq N(t)$. But it is also easy to see that $M(t)$ comprises a Geometric (p) number of renewals at each of the points $(r\epsilon : r \geq 0, r\epsilon \leq t)$, where $p = P(X_r \geq \epsilon)$. Thus

$$EN(t) \leq EM(t) = \frac{1}{P(X_r \geq \epsilon)} \left[\frac{t}{\epsilon}\right] < \infty.$$

□

Now that we know $EN(t)$ is finite for $EX_r > 0$, we can recall Wald's equation from (2.4.14) to give this.

(11) **Theorem.** $ES_{N(t)+1} = EX_1 E(N(t) + 1)$.

Proof. By inspection, $N(t) + 1$ is a stopping time with respect to the martingale $S_n - nEX_1$. Now use (2.4.14). (Note that $N(t)$ is *not* a stopping time.)

Alternative proof. It is instructive to see the direct argument, without appealing to optional stopping theorems. Simply note that the indicator

$$I(N(t) + 1 \geq m) = I(N(t) \geq m - 1)$$

is not a function of X_m. Hence

$$ES_{N(t)+1} = \sum_{m=1}^{\infty} E(X_m I(N(t) + 1 \geq m))$$

(12) $$= \sum_{m=1}^{\infty} EX_m E I(N(t) + 1 \geq m),$$

by the pull-through property (1.7.20),

$$= EX_1 \sum_{m=1}^{\infty} P(N(t) + 1 \geq m)$$

$$= EX_1 E(N(t) + 1), \quad \text{by the tailsum lemma for expectation.}$$

□

We can illustrate these results using our unusually tractable friend, the Poisson process.

(13) **Poisson process.** Recall the results of Section 2.5. We know that $N(t)$ is Poisson (λt), so it follows immediately that $E(N(t))^r < \infty$; in particular $m(t) = EN(t) = \lambda t$.

By the lack-of-memory property of the exponential distribution

$$E(S_{N(t)+1}) = t + E(X_1) = t + \frac{1}{\lambda}$$
$$= \frac{1}{\lambda}(\lambda t + 1) = EX_1 E(N(t) + 1).$$

□

Another line of approach that can be useful is given by the following.

(14) **Theorem.** The renewal function $m(t)$, defined in (1), satisfies the renewal equation

(15) $$m(t) = F(t) + \int_0^t m(t-x)\,dF(x).$$

Proof. By conditional expectation

(16) $$EN(t) = E(EN(t) \mid X_1).$$

There are two cases:

(a) If $X_1 > t$, then $E(N(t) \mid X_1) = 0$.
(b) If $X_1 = x < t$, then

$$E(N(t) \mid X_1 = x) = 1 + EN(t-x),$$

since the counting process starts afresh from x. Substituting these into (16) gives (15). □

We can illustrate this for the Poisson process where $F(t) = 1 - e^{-\lambda t}$: now (15) becomes

(17) $$m(t) = 1 - e^{-\lambda t} + \int_0^t m(t-x)\lambda e^{-\lambda x}\,dx.$$

It is easy to check that $m(t) = \lambda t$ is indeed the solution of (17).

As we have remarked, such explicit closed-form solutions are rare, so we lower our sights to study simply the asymptotic behaviour of $N(t)$ and $m(t)$ as $t \to \infty$. First we need an elementary result:

(18) **Lemma.** Whenever $P(X_r < \infty) = 1$, we have

$$N(t) \to \infty, \quad \text{as } t \to \infty.$$

Proof. Since $N(t)$ is monotone non-decreasing, as $t \to \infty$, either $N(t) \to \infty$, or $N(t) \to Z < \infty$.

The latter case occurs if and only if at least one of $X_1, \ldots, X_{Z+1}$ is infinite. But

$$P\left(\bigcup_{r=1}^{Z+1} \{X_r = \infty\}\right) \le \sum_{r=1}^{\infty} P(X_r = \infty) = 0.$$

$\square$

We obtain two trivial but useful consequences of this: as $t \to \infty$,

(19)
$$\frac{N(t)+1}{N(t)} \xrightarrow{\text{a.s.}} 1$$

and, by the strong law of large numbers, (2.3.14),

(20)
$$\frac{S_{N(t)}}{N(t)} \xrightarrow{\text{a.s.}} EX_1.$$

After all these preliminaries we finally arrive at two key results.

(21) **First renewal theorem.** With probability 1, as $t \to \infty$,

$$N(t)/t \to 1/EX_1.$$

Proof. By definition, $S_{N(t)} \le t < S_{N(t)+1}$, therefore

(22)
$$\frac{S_{N(t)}}{N(t)} \le \frac{t}{N(t)} \le \frac{N(t)+1}{N(t)} \frac{S_{N(t)+1}}{N(t)+1}.$$

Letting $t \to \infty$, and using (18), (19), and (20), gives the required result. $\square$

The next theorem is not so easy, despite its name.

(23) **Elementary renewal theorem.** As $t \to \infty$,

(24)
$$m(t)/t \to 1/EX_1.$$

Proof. This proof has two halves. First, by definition, $t < S_{N(t)+1}$. Taking expectations, and using (11), we find $t < EX_1(EN(t) + 1)$. Hence

(25)
$$\lim_{t \to \infty} \frac{m(t)}{t} \ge \lim_{t \to \infty} \left(\frac{1}{EX_1} - \frac{1}{t}\right) = \frac{1}{EX_1}.$$

For the second half of the proof, define a new sequence of interarrival times

(26) $$Y_r = a \wedge X_r = \begin{cases} X_r, & \text{if } X_r < a, \\ a, & \text{if } X_r \geq a \end{cases}$$

with associated renewal process $N_a(t)$, having mean $m_a(t)$. Because $Y_r \leq X_r$, it is clear that for all t we have $N_a(t) \geq N(t)$. By definition, $S_{N_a(t)} \leq t$, and therefore

$$t \geq \mathrm{E} S_{N_a(t)} = \mathrm{E}(S_{N_a(t)+1} - Y_{N_a(t)+1})$$

$$= (m_a(t) + 1)\mathrm{E} Y_1 - \mathrm{E}(Y_{N_a(t)+1}), \quad \text{using (11)}$$

$$\geq (m(t) + 1)\mathrm{E} Y_1 - a.$$

Hence

$$\lim_{t \to \infty} \frac{m(t)}{t} \leq \lim_{t \to \infty} \left\{ \frac{1}{\mathrm{E} Y_1} + \frac{a - \mathrm{E} Y_1}{t \mathrm{E} Y_1} \right\} = \frac{1}{\mathrm{E} Y_1}$$

(27) $$\to \frac{1}{\mathrm{E} X_1}, \quad \text{as } a \to \infty.$$

Combining (25) and (27) proves (24). □

Renewal theory as sketched above turns up in a surprising variety of contexts, but we can make it even more useful and applicable by extending it a little. In real life (and therefore desirably in our models) renewals often involve rewards or costs. For convenience we regard a cost as a negative reward, and denote the sequence of rewards by $(R_j; j \geq 1)$, where R_j is the reward associated with the jth renewal. We assume that the R_j are independent and identically distributed, with finite mean; for each j, R_j may depend on X_j, but is independent of $(X_i; i \neq j)$. Here are some simple examples:

Taxi. Let X_j be the duration of the jth hire of a taxi, and R_j the fare paid.

Queue. Let X_j be the interarrival time between successive buses, and R_j the size of the queue that boards.

Machine. Let X_j be the interval between successive breakdowns, and R_j the cost of repairs.

Lamps. Let R_j be the cost of installing the jth lamp, which lasts for X_j.

Think of some more yourself. We call these *renewal–reward* processes, and we are naturally interested in the cumulative reward $C(t)$ up to time t.

The above examples illustrate that this may accumulate in several ways:

Terminally. The 'reward' is collected at the end of any interval. Then

$$(28) \qquad C(t) = \sum_{n=1}^{N(t)} R_n.$$

Initially. The 'reward' is collected at the start of any interval. (For example, you have to pay for the new lamp before you fit it, not when it fails.) Then

$$(29) \qquad C_i(t) = \sum_{n=1}^{N(t)+1} R_n.$$

Partially. Rewards may accrue during any interval, so that at time t one has accumulated

$$(30) \qquad C_p(t) = \sum_{n=1}^{N(t)} R_n + P_{N(t)+1}$$

using an obvious notation, in which P_n denotes partial reward.

Intuitively one expects that there will not be too much difference (or any) in the long-run behaviour of cumulative rewards in these three cases, and this turns out to be correct, under fairly mild assumptions about the properties of rewards. Note that in each case, the expected value of the cumulative reward is called the *reward function*, and denoted, respectively, by $c(t) = EC(t)$, $c_i(t) = EC_i(t)$, $c_p(t) = EC_p(t)$.

Of course we are interested in the rate of reward, especially in the long term. We have this:

(31) **Renewal–reward theorem.** Assume that $0 < EX_r < \infty$, and $E|R_n| < \infty$. Then as $t \to \infty$

$$(32) \qquad C(t)/t \xrightarrow{\text{a.s.}} ER_1/EX_1.$$

In advance of proving this result, we take a moment to see what it means intuitively. In this context it is traditional to call the interval between successive renewals a *cycle*.

The right-hand side of (32) is the ratio

$$(33) \qquad \frac{\text{expected reward accruing in a cycle}}{\text{expected duration of a cycle}}.$$

The theorem asserts that this is the same as the long-run average reward from the renewal–reward process.

If your intuition is sufficiently well-developed, a few moments' thought will convince you that this is a natural and attractive result.

Proof of (31). By the strong law of large numbers and the first renewal theorem,

(34) $$C(t)/t = \frac{\sum_{n=1}^{N(t)} R_n}{N(t)} \frac{N(t)}{t} \xrightarrow{\text{a.s.}} \frac{ER_1}{EX_1}, \quad \text{as } t \to \infty.$$

□

Exactly the same argument with $N(t) + 1$ replacing $N(t)$ shows that

(35) $$C_i(t)/t \xrightarrow{\text{a.s.}} ER_1/EX_1.$$

In the case of partial rewards, we need to assume some form of good behaviour; for example, if rewards accrue in a monotone manner over any interval, then (34) and (35) imply that also $C_p(t)/t \xrightarrow{\text{a.s.}} ER/EX$.

Next we turn to corresponding results for the respective reward functions. For $c_i(t)$ things are relatively easy:

(36) **Theorem.** If $E|R_n| < \infty$ and $0 < EX_n < \infty$, then

(37) $$c_i(t)/t \to ER_n/EX_n, \quad \text{as } t \to \infty.$$

Proof. $N(t) + 1$ is a stopping time for the sequence $(R_j; j \geq 1)$; therefore, by Wald's equation, (2.4.14),

$$c_i(t) = EC_i(t) = E \sum_{n=1}^{N(t)+1} R_n = E(N(t) + 1)ER_n.$$

Dividing by t, allowing $t \to \infty$, and recalling the elementary renewal theorem (23), gives the result. □

In this case, it is the corresponding property for $c(t)$ that is trickier to prove in general.

(38) **Theorem.** If $0 < EX_n < \infty$, and $E|R_n| < \infty$, then, as $t \to \infty$,

(39) $$c(t)/t \to ER_n/EX_n.$$

Proof. First we prove a simpler version that is often useful. Assume that rewards are uniformly bounded: $|R_j| \leq K < \infty$ for some constant K. Then

$$|C(t)/t| = \left| \sum_{n=1}^{N(t)} R_n/t \right| \leq KN(t)/t.$$

But, as $t \to \infty$, $EN(t)/t \to 1/EX_1$, by the elementary renewal theorem, and $C(t)/t \xrightarrow{a.s.} ER_n/EX_n$. Hence, the required result follows by dominated convergence; see (2.3.21).

Turning to the general case, we have from the argument in (36) that

$$C(t)/t = \left\{ E \sum_{n=1}^{N(t)+1} R_n - ER_{N(t)+1} \right\} / t$$

$$= (m(t) + 1)ER_1/t - E(R_{N(t)+1})/t, \quad \text{using (2.4.14)}.$$

The result follows immediately when we show that $E(R_{N(t)+1})/t \to 0$ as $t \to \infty$. This seems blindingly obvious, but proving it is not totally trivial. First note that for $\epsilon > 0$

$$\sum_{n=1}^{\infty} P(|R_n|/n > \epsilon) = \sum_{n=1}^{\infty} P(|R_1|/\epsilon > n)$$

$$\leq \int_0^{\infty} P(|R_1|/\epsilon > x) \, dx \leq E|R_1|/\epsilon < \infty.$$

Therefore, by Lemma (2.3.4), $|R_n|/n > \epsilon$ only finitely often as $n \to \infty$. It follows that $R_n/n \xrightarrow{a.s.} 0$, as $n \to \infty$, and, further, that

$$\frac{1}{n} \max_{1 \leq k \leq n} R_k \xrightarrow{a.s.} 0.$$

Now, writing

$$R_{N(t)+1}/t = \frac{R_{N(t)+1}}{N(t)+1} \frac{N(t)+1}{t},$$

we can use the above, and the first renewal theorem, to find that $R_{N(t)+1}/t \xrightarrow{a.s.} 0$ as $t \to \infty$.

Finally, we note that

$$|R_{N(t)+1}/t| \leq \frac{1}{t} \sum_{n=1}^{N(t)+1} |R_n|$$

and the right-hand side has expected value

$$E|R_n|(m(t)+1)/t \to ER_n/EX_n, \quad \text{as } t \to \infty.$$

The result follows by dominated convergence (2.3.21). The analogous result for monotone partial rewards is now a straightforward corollary. □

Here is a typical example of a renewal–reward theorem in action.

(40) Inspection paradox. In many real-life renewal processes we may well be interested in the time since the last renewal, called the *age* or *current life*, and the time until the next, called the *excess life*. These are denoted by $A(t)$ and $E(t)$, respectively; by definition, at any time $t > 0$,

$$A(t) = t - S_{N(t)},$$
$$E(t) = S_{N(t)+1} - t.$$

Here $S_n = \sum_{r=1}^{n} X_r$ are the renewal times, as usual.

Define a reward process by $C(t) = \sum_{n=1}^{N(t)} R_n$, where

$$R_n = \min\{y, X_n\}.$$

Now define the indicator

$$I(t) = \begin{cases} 1, & \text{if } A(t) \leq y, \\ 0, & \text{otherwise.} \end{cases}$$

We see that $EI(t) = P(A(t) \leq y)$ and

$$\int_{S_{n-1}}^{S_n} I(u)\, du = \min\{y, X_n\} = R_n.$$

Hence

$$\int_0^t I(u)\, du = \sum_{n=1}^{N(t)} R_n + P_{N(t)+1},$$

where $P_{N(t)+1}$ is a partial reward given by $\int_{S_{N(t)}}^{t} I(u)\, du$.

Here the rewards are uniformly bounded (by y), so both forms of the renewal–reward theorem are true, and

$$ER_n = \int_0^y P(X_n > x)\, dx.$$

In particular,

$$\lim_{t\to\infty} EC(t)/t = \lim_{t\to\infty} \int_0^t EI(u)\,du/t$$

(41)
$$= \lim_{t\to\infty} \int_0^t P(A(u) < y)\,du/t$$

$$= ER_n/EX_n, \quad \text{by the renewal–reward theorem}$$

$$= \int_0^y P(X_n > x)\,dx/EX_n.$$

It can be shown (we defer the proof) that the distribution of $A(t)$ does converge to a limiting distribution $F_A(y)$ as $t \to \infty$. Then, from (41),

$$F_A(y) = \int_0^y P(X_n > x)\,dx/EX_n$$

and differentiating gives the limiting density of $A(t)$

$$f_A(y) = P(X_n > y)/EX_n$$

with mean value

$$EA = \int_0^\infty yP(X_n > y)\,dy/EX_n = \frac{(1/2)EX_n^2}{EX_n}.$$

A similar argument applied to the indicator

$$J(t) = \begin{cases} 1, & \text{if } E(t) \le y, \\ 0, & \text{otherwise}, \end{cases}$$

shows that the limiting distribution of the excess life is the same as that of the age, as $t \to \infty$. It follows that

$$\lim_{t\to\infty} \frac{1}{t}\int_0^t EA(u) + EE(u)\,du = \frac{EX_n^2}{EX_n} \ge EX_n$$

with strict inequality if var $X_n > 0$.

In words, this says that in the long run, as $t \to \infty$, the expected duration of the renewal cycle in progress at time t is strictly greater than the expected duration of a renewal cycle. This sounds paradoxical, and is called the *inspection paradox*.

In fact it is not paradoxical when one thinks clearly about what is going on. The point is that, if we take the average over all time of the age (or total life—or excess life) of the component in service at time t, then this average attaches greater weight to longer intervals. In statistical terms the sampling is biased by the length of the intervals, and this is called *length biasing*. Very roughly speaking, any given time t is more likely to be found in a longer interval than a short one. ○

Finally, we note that in many practical applications the first interval of a renewal process may have a different distribution from all the rest.

(42) **Delayed renewal.** If $S_n = \sum_{r=1}^{n} X_n$, where the $(X_n; n \geq 2)$ are identically distributed, and independent of X_1 as well as each other, then $N_d(t) = \max\{n : S_n \leq t\}$ is a delayed renewal process. It may give rise to a *delayed renewal–reward process*. Clearly if X_1 has finite mean, the asymptotic results above remain unchanged; we omit the details. △

Exercises
For brevity, we refer to the events of a renewal process as arrivals, so that any X_r is an interarrival time.

(a) Let $N(t)$ be a renewal process such that interarrival times are Uniform $(0,1)$. For $0 \leq t \leq 1$, use (14) to show that the renewal function is $m(t) = e^t - 1$. Deduce that if

$$M = \min\{n : S_n > 1\}, \quad \text{then } ES_M = e/2.$$

(b)* Consider a renewal process with interarrival density $f(x), 0 < x < \infty$. Let $b(a)$ be the expected time required until an interval of length $a > 0$ is first observed to be free of arrivals. Show that

$$b(a) = a + \int_0^a t f(t)\, dt \bigg/ \int_a^\infty f(x)\, dx.$$

When the renewal process is a Poisson process, show that $b(a) = (e^{\lambda a} - 1)/\lambda$.

(c) Let M be the number of interarrival times that occur before the first such interarrival time which is larger than the time until the first arrival. Find $P(M > n), n \geq 2$, and hence calculate EM.

2.7 Branching processes

Many real physical processes have (at least approximately) the following structure: they comprise a single type of entity, each of which may produce a random number of further entities of the same kind. How does the size of the population behave? The entities may be neutrons, individuals in a population, cells, and so on. It is traditional to call them *particles*, and specify their behaviour thus:

(1) **Branching process.** Initially there is a single particle. Each particle gives rise to a family of particles that replace it in the next generation; family sizes are independent and identically distributed random variables with pgf $\mathcal{G}(s)$. The size of the nth generation is denoted by Z_n, with pgf $\mathcal{G}_n(s) = \mathrm{E}s^{Z_n}$.

That is to say, $Z_0 = 1$, and $\mathrm{E}s^{Z_1} = \mathcal{G}(s) = \mathcal{G}_1(s)$. We denote the mean and variance of a family by μ and $\sigma^2 > 0$, respectively, so that $\mu = \mathcal{G}'(1)$, and

$$\sigma^2 = \mathcal{G}''(1) + \mathcal{G}'(1) - [\mathcal{G}'(1)]^2.$$

We require $\sigma^2 > 0$ to exclude trivial cases. By construction, if $Z_n = 0$ for some n, then $Z_m = 0$ for all $m \geq n$; the particles are said to be *extinct*.

The keys to answering questions about the behaviour of Z_n are the following crucial identities: for the first, let $X(i)$ be the size of the family left by the ith member of the nth generation. Then

$$(2) \qquad Z_{n+1} = \sum_{i=1}^{Z_n} X(i),$$

where the empty sum is zero, as always. For the second, let $Y(i)$ be the number of descendents, after n generations, of the ith particle in the first family. Then

$$(3) \qquad Z_{n+1} = \sum_{i=1}^{Z_1} Y(i).$$

We use these results repeatedly; let us first find the moments of Z_n. By conditional expectation, using the independence of family sizes, equation (2) gives

$$(4) \qquad \mathrm{E}Z_{n+1} = \mathrm{E}\left\{\mathrm{E}\left\{\sum_{i=1}^{Z_n} X(i)|Z_n\right\}\right\} = \mathrm{E}(Z_n \mu).$$

Hence, iterating, $\mathrm{E}Z_n = \mu^n$. Likewise, or by recalling the expression (1.7.18) for the variance of a random sum,

$$\mathrm{var}\, Z_{n+1} = \mu \,\mathrm{var}\, Z_n + \sigma^2 (\mathrm{E}Z_n)^2$$
$$= \mu\, \mathrm{var}\, Z_n + \sigma^2 \mu^{2n}.$$

Solving this recurrence relation yields

(5) $$\text{var } Z_n = \begin{cases} n\sigma^2, & \text{if } \mu = 1, \\ \dfrac{\mu^{n-1}(1-\mu^n)\sigma^2}{1-\mu}, & \text{if } \mu \neq 1. \end{cases}$$

Let us turn to the question of extinction; we denote the probability that the process is extinct at or before the nth generation by η_n. By definition

(6) $$\eta_n = \mathcal{G}_n(0)$$

and the probability of ultimate extinction is

(7) $$\eta = \lim_{n \to \infty} \eta_n,$$

which certainly exists since η_n is a non-decreasing sequence of probabilities. Initially $\eta_0 = 0$, since $Z_0 = 1$. When $\mu < 1$, the question has an easy answer. By Markov's inequality

$$P(Z_n \geq 1) \leq EZ_n = \mu^n \to 0, \quad \text{as } n \to \infty.$$

Hence ultimate extinction is certain in this case:

$$\eta = \lim_{n \to \infty} P(Z_n < 1) = 1.$$

When $\mu \geq 1$, we must work a little harder. Let $\hat{s}$ be the smallest non-negative solution of $x = \mathcal{G}(x)$. By conditional probability, using the independence of families,

(8) $$\eta_{n+1} = E\{P(Z_{n+1} = 0 \mid Z_1)\} = \mathcal{G}(\eta_n), \quad n \geq 0.$$

Since $\mathcal{G}(x)$ is non-decreasing in $[0, 1]$, we have

$$\eta_1 = \mathcal{G}(0) \leq \mathcal{G}(\hat{s}) = \hat{s}$$

and hence

(9) $$\eta_2 = \mathcal{G}(\eta_1) \leq \mathcal{G}(\mathcal{G}(\hat{s})) = \mathcal{G}(\hat{s}) = \hat{s}.$$

Repeating this, and using induction, gives $\eta_n \leq \hat{s}$ for all n. But allowing $n \to \infty$ in (8) shows that $\eta = \mathcal{G}(\eta)$, so that η is some root of $\mathcal{G}(x) = x$. It follows that $\eta = \hat{s}$.

Further information about the evolution of Z_n can be obtained in principle from its pgf. Once again we use (2) and conditional expectation to give

$$\mathcal{G}_{n+1}(s) = \mathrm{E}s^{Z_{n+1}} = \mathrm{E}\{\mathrm{E}s^{\sum_{r=1}^{Z_n} X(r)} \mid Z_n\}$$
(10)
$$= \mathrm{E}\{\mathcal{G}(s)\}^{Z_n} = \mathcal{G}_n(\mathcal{G}(s)).$$

Thus $\mathcal{G}_n(s)$ is the nth iterate of the function $\mathcal{G}(\cdot)$. Unfortunately, this iterate can rarely be evaluated explicitly; see the exercises for one tractable case.

We may also be interested in T, the total number of particles that ever exist, defined in terms of $T_n = \sum_{r=0}^n Z_r$ by

$$T = \lim_{n\to\infty} \sum_{r=0}^n Z_r = \lim_{n\to\infty} T_n.$$

The limit exists, since the sequence is non-decreasing, but it may be infinite if extinction does not occur. For brevity we may call it the *tree* of the initial particle.

(11) Total population. For the tree T, we have:

(a) The generating function $G_T(s) = \mathrm{E}s^T$ satisfies $G_T(s) = s\mathcal{G}(G_T(s))$.

(b) For $\mu < 1$ the mean and variance of T are

(12) $\qquad \mathrm{E}T = (1-\mu)^{-1}$ and $\operatorname{var} T = \sigma^2(1-\mu)^{-3}$.

Proof.

(a) Let $T(i)$ be the tree of the ith member of the first family Z_1. Then

$$T = 1 + \sum_{i=1}^{Z_1} T(i),$$

where the $T(i)$ are independent having the same distribution as T. The first result follows immediately, using the same method as that for (10).

(b) Recalling the expressions in (1.7.16) for the mean and variance of a random sum gives

$$\mathrm{E}T = 1 + \mu \mathrm{E}T,$$
(13) $\qquad \operatorname{var} T = \mu \operatorname{var} T + \sigma^2 (\mathrm{E}T)^2.$

Assuming that $\operatorname{var} T < \infty$, solving these equations yields the required result.

For example, if $\mathcal{G}(s) = q + ps^2$, so $\mu = 2p$, we find

(14) $$ET = \frac{1}{q - p}, \quad \text{if } q > p.$$

□

It is interesting and useful that there are many martingales associated with branching processes. Several appear in the examples and problems; here we look at one in detail, namely

(15) $$M_n = Z_n \mu^{-n}.$$

It is simple to check that this satisfies the conditions to be a martingale, and furthermore, when $\mu \neq 1$,

$$\text{var } M_n = \mu^{-2n} \text{var } Z_n = \frac{\sigma^2(1 - \mu^{-n})}{\mu(\mu - 1)}, \text{ by (5)},$$

(16) $$\leq \frac{\sigma^2}{\mu(\mu - 1)}, \quad \text{when } \mu > 1.$$

That is to say, the variance is uniformly bounded. It follows that, as $n \to \infty$, M_n converges in mean square, and hence in probability; see Theorem (2.4.24). The limit is hard to find in general; we consider one simple example in (17).

Note in passing that by the maximal inequality (2.4.17)

$$\frac{1}{4} \text{E} \max_{k \leq n} Z_k^2 \leq \text{E} Z_n^2 = \begin{cases} \mu^{2n} + \sigma^2 \mu^{n-1}(\mu^n - 1)/(\mu - 1), & \mu \neq 1, \\ n\sigma^2 + 1, & \mu = 1. \end{cases}$$

(17) **Geometric branching limit.** Since M_n defined in (15) is non-negative, we seek its mgf in the form

$$\psi_n(t) = \text{E} \, e^{-tM_n} = \mathcal{G}_n(e^{-t\mu^{-n}}).$$

Recalling that $\mathcal{G}_{n+1}(s) = \mathcal{G}(\mathcal{G}_n(s))$, we find that $\psi(t) = \lim_{n \to \infty} \psi_n(t)$ satisfies the functional equation

$$\psi(\mu t) = \lim_{n \to \infty} \psi_{n+1}(\mu t) = \lim_{n \to \infty} \mathcal{G}(\psi_n(t))$$

(18) $$= \mathcal{G}(\psi(t)).$$

Solutions to functional equations are few and far between, but in the special case when $\mathcal{G}(s) = p(1 - qs)^{-1}$, it is readily checked by substitution that for $p < q$,

$$(19) \qquad \psi(t) = \frac{q - p + pt}{q - p + qt} = \frac{p}{q} + \left(1 - \frac{p}{q}\right)^2 \frac{1}{1 - (p/q) + t}.$$

This is the mgf of a distribution being a mixture of an atom at zero (i.e. the extinction probability $\eta = p/q$), and an Exponential $(1 - p/q)$ density.

We interpret this in terms of the branching process as follows: either it becomes extinct (with probability $\eta = p/q$) or, with probability $1 - \eta$, the number of particles grows asymptotically like $W\mu^n$, where W is an exponential random variable. In the long run, Z_n goes to either zero or infinity. This information is better than nothing, but it is still not terribly informative; nor does it seem very illuminating about observed ongoing processes which, naturally, must be finite and not extinct.

Various tricks are employed to address this difficulty. First, we may seek to look at the process conditional on some event of interest. For example, consider a so-called *subcritical* process ($\mu \leq 1$) for which extinction is certain. We may look at the process Z conditional on the event that extinction has not yet occurred by time n, that is to say our interest will be directed at the conditional distribution

$$c_k(n) = P(Z_n = k \mid Z_n > 0), \quad k \geq 1.$$

In this case ($\mu \leq 1$) it can be shown that as $n \to \infty$ the distribution $(c_k(n); k \geq 1)$ converges to a distribution whose generating function $C(s)$ satisfies the functional equation

$$(20) \qquad C(\mathcal{G}(s)) = \mu C(s) + 1 - \mu.$$

We omit the proof.

On the other hand, if $\mu \geq 1$, then the total number of particles (i.e. the tree) tends to infinity with probability $1 - \eta$. It is natural to look at the tree conditional on the event that it stays finite. In this case it is convenient to consider the family size conditional on the event Q that extinction occurs:

$$P(Z_1 = k \mid Q) = P\left(\{Z_1 = k\} \bigcap Q\right)/\eta$$
$$= P(Q \mid Z_1 = k)P(Z_1 = k)/\eta$$
$$= \eta^{k-1} P(Z_1 = k), \quad k \geq 0.$$

This replaces the original family size distribution; the new pgf is $\hat{G}(s)$ where

(21) $$\hat{G}(s) = \mathcal{G}(\eta s)/\eta,$$

having mean $\hat{\mu} = \mathcal{G}'(\eta)$ and variance $\hat{\sigma}^2 = \eta \mathcal{G}''(\eta) + \mathcal{G}'(\eta) - [\mathcal{G}'(\eta)]^2$. Hence, for example, considering the tree conditional on Q, (12) gives $E(T \mid Q) = 1/(1 - \mathcal{G}'(\eta))$. In particular, if $\mathcal{G}(s) = q + ps^2$, $\eta = q/p \leq 1$, we see that $E(T \mid Q) = 1/(p - q)$.

A second approach to the problem discussed above, when $\mu \leq 1$, is to arrange for fresh particles to be added to the population at appropriate moments, thus ensuring that permanent extinction is impossible.

(22) **Refreshed branching.** Let Z_n be an ordinary branching process with $\mu < 1$. Define the refreshed process R_n thus:

(i) At each generation n a random number V_n of extra particles is added to the existing population, where the V_n are independent of each other and the process up to time n; $EV_n = a$ and var $V_n = b^2$.
(ii) Each fresh particle initiates a branching process independent of all the other particles, but having the same description and distribution as Z_n.

In this case, we assume $R_0 = 0$.

As usual, we let the family of the ith member of the nth generation have size $X(i)$. Then

(23) $$R_{n+1} = \sum_{i=1}^{R_n} X(i) + V_{n+1}.$$

Exploiting all the independence, and using conditional expectation as usual, we obtain

$$ER_{n+1} = \mu ER_n + a,$$
$$\text{var } R_{n+1} = \sigma^2 ER_n + \mu^2 \text{ var } R_n + b^2.$$

Solving these in succession, and allowing $n \to \infty$, we find that in the limit

$$ER_n \to \frac{a}{1-\mu},$$

$$\text{var } R_n \to \frac{a^2 + (1-\mu)b^2}{(1-\mu)(1-\mu^2)}.$$

It turns out that R_n itself does converge as $n \to \infty$, but further investigation is beyond our scope.

Exercises

(a)* Show that the probability of extinction of Z_n defined in (1) is 1 if and only if $\mu \leq 1$.

(b) If $\mathcal{G}(s) = q + ps^2$, show that $\eta = (1 - |p-q|)/2p$. Explain why this is the same as the probability that a simple random walk started at 1 ever reaches 0. For the total population T, calculate ET, var T, and $G_T(s)$, and explain the connexion with simple random walk started at 1.

(c) When $\mu < 1$, and $\sigma^2 < \infty$, show that var $T < \infty$.

(d) If $\mathcal{G}(s) = q/(1-ps)$, show that $\eta = (1 - |p-q|)/2p$. Explain why this is the same as in Exercise (b). Do you get the same result for $G_T(s)$? Explain the answer.

2.8 Miscellaneous models

This section contains a few examples of random processes that do not fit conveniently into our templates for other chapters, but which are nevertheless too much fun to leave out. They are not a prerequisite for anything else, so this section may be omitted if you are in haste.

(1) **Quicksort.** Computers are often required to sort a list of numbers into increasing order; in fact they do this so often that it is worth having an efficient procedure to do it (more efficient than simply looking for the smallest, then looking for the next smallest, and so on, which takes a relatively long time). The following rule works very well; we assume we have a sequence $S = (x_1, \ldots, x_n)$ of distinct numbers:

(i) If $n = 2$, put the smallest first; otherwise do this:
(ii) Pick one of the x_i at random, call it x_r.
(iii) Then compare each of the unchosen numbers with x_r, in order, placing those that are smaller in the set S_r and those that are larger in the set L_r.
(iv) Then treat S_r and L_r in the same way as S.

Let c_n be the expected number of comparisons required to sort a sequence of n numbers that may initially be presented in any order. We assume all orderings are equally likely. The first number selected, x_r, is equally likely to be any of $x_1, \ldots, x_n$; if we condition on its rank order we obtain

$$(2) \qquad c_n = n - 1 + \frac{1}{n}\sum_{r=1}^{n}(c_{r-1} + c_{n-r}) = n - 1 + \frac{2}{n}\sum_{r=1}^{n} c_{r-1}.$$

The term $n-1$ records the number of comparisons required to sort the rest into S_r and L_r; these two sets then entail c_{r-1} and c_{n-r} expected comparisons to be sorted, respectively.

Multiplying (2) by n, and subtracting the result from the next recurrence

$$(n+1)c_{n+1} = (n+1)n + \sum_{r=1}^{n+1} c_{r-1}$$

yields

$$(n+1)c_{n+1} - nc_n = 2n + 2c_n.$$

Rearranging, and iterating, gives

(3)
$$\begin{aligned}\frac{c_{n+1}}{n+2} &= \frac{2n}{(n+1)(n+2)} + \frac{c_n}{n+1}\\ &= 2\sum_{r=1}^{n+1} \frac{r-1}{r(r+1)}\\ &= 2\sum_{r=1}^{n+1} \left(\frac{2}{r+1} - \frac{1}{r}\right)\\ &= 2\sum_{r=1}^{n+1} \frac{1}{r} - \frac{4(n+1)}{n+2}.\end{aligned}$$

Hence we obtain the asymptotic expression

(4)
$$\lim_{n\to\infty} \frac{c_n}{2n\log n} = 1,$$

when we recall that

$$\lim_{n\to\infty} \frac{1}{\log n} \sum_{r=1}^n \frac{1}{r} = 1.$$

(5) Self sort. Again we have a list, of what we may suppose to be objects, labelled $l_1, \ldots, l_n$. At time t, $t \geq 1$, one of the objects is called out of the list, where l_r is called with probability p_r, $1 \leq r \leq n$. After its callout, the object is replaced at the front of the list, before the next callout. Callouts are independent. When this procedure has been running for a long period of time, what is the expected position μ of the object called for?

We calculate this by first conditioning on which object is called; suppose l_r is called out. For any $k \neq r$, l_k is before l_r in the list if l_k has been called more recently than l_r. But the probability that the last call for either was for l_k is $p_k/(p_k + p_r)$, by conditional probability. Now let I_k be the indicator of the event that l_k is before l_r in the list. The position of l_r is

$$L_r = 1 + \sum_{k \neq r} I_k.$$

Hence the expected position of l_r is

(6) $$\mathrm{E}L_r = 1 + \sum_{k \neq r} \mathrm{E}I_k = 1 + \sum_{k \neq r} \frac{p_k}{p_k + p_r}.$$

Finally, by conditional expectation,

(7) $$\mu = \sum_{r=1}^{n} p_r \mathrm{E}L_r = 1 + \sum_{r=1}^{n} p_r \sum_{k \neq r} \frac{p_k}{p_k + p_r}.$$

Note that if $p_k = 1/n$, $1 \leq k \leq n$, then $\mu = (n+1)/2$, as is otherwise obvious in this case. ○

Next, we turn from sorting things to questions about the reliability of systems.

Many complicated objects can be regarded as a system S containing a number n of individual components which work together to ensure that the assemblage functions. Examples include any electricity network, the World Wide Web, any vehicle (car, aeroplane, bicycle, horse, ...), and any body. Given that component failures are often random and unpredictable, it is natural to ask for the probability that the system is functioning. In general, this is a difficult question with complicated answers, so we make some simplifying assumptions immediately.

(1) **Binary system.** For the ith component C_i, $1 \leq i \leq n$, define the indicator of the event that C_i is working

$$X_i = \begin{cases} 1, & \text{if } C_i \text{ works,} \\ 0, & \text{otherwise.} \end{cases}$$

Let $\mathrm{E}X_i = p_i = \mathrm{P}(C_i \text{ works})$, $1 \leq i \leq n$. △

All our systems will be assumed to be binary, which is to say that questions about whether a system is working or not can be answered in terms of a function of the Xs.

(2) **Structure function.** For $\mathbf{X} = (X_1, \ldots, X_n)$ define the indicator of the event that the system S works

$$\phi(\mathbf{X}) = \begin{cases} 1, & \text{if } S \text{ works}, \\ 0, & \text{otherwise}. \end{cases}$$

△

(3) **Reliability function.** For $\mathbf{p} = (p_1, \ldots, p_n)$, where $p_i = \mathrm{E}X_i$, define the reliability function

$$R(\mathbf{p}) = \mathrm{E}\phi(\mathbf{X}),$$

which gives the probability that S works. If for all i, $p_i = p$, then we denote the reliability by $R(p)$. This is the 'equal reliability' assumption. △

(4) **Monotonicity.** It is natural to assume that any system does not perform worse when more components are working. The effect of this is to require that if $X_i \leq Y_i$ for all i, then

$$\phi(\mathbf{X}) \leq \phi(\mathbf{Y}).$$

We always assume this. △

As a useful point of notation, if $X_i \leq Y_i$, for all i, then we write $\mathbf{X} \leq \mathbf{Y}$. If for at least one i we also have $X_i < Y_i$, then we write $\mathbf{X} < \mathbf{Y}$.

(5) **Independence.** It is often natural to assume that components fail independently, which is to say that $X_1, X_2, \ldots, X_n$ are independent. In this case it is an easy exercise for you to show that when $\phi(\mathbf{X})$ is a monotone function of independent components, then $R(\mathbf{p})$ is monotone non-decreasing in each p_i, $1 \leq i \leq n$.

We shall always assume components are independent. △

Certain special systems and their structure functions are particularly important, and we shall now see that it is often useful to think of them in terms of a natural interpretation as flows in a network or graph.

(6) **The parallel, or one-out-of-n, system.** This system works if any one of the n components work. By inspection, the structure function is

$$\phi(\mathbf{X}) = \max\{X_1, \ldots, X_n\} = 1 - \prod_{r=1}^{n}(1 - X_r).$$

The name 'parallel' arises from this visualization: imagine two vertices (labelled 'source' or A, and 'sink' or B) joined by n edges, each one of which runs direct from A to B. Each edge may be open or closed, and there is 'flow' from A to B if at least one is open, and this is indicated by $\phi = 1$.

The reliability is

$$E\phi(\mathbf{X}) = 1 - \prod_{r=1}^{n}(1 - p_r), \quad \text{by independence}$$

$$= 1 - (1 - p)^n, \quad \text{if components have equal reliability.}$$

△

(7) **The series, or n-out-of-n, system.** This system works if and only if all the components work. The structure function is

$$\phi(\mathbf{X}) = \prod_{r=1}^{n} X_r = \min\{X_1, \ldots, X_n\}.$$

The name 'series' arises when we visualize a sequence of edges that may be 'open' or 'closed', joined end-to-end, and running from the source A to the sink B. There is 'flow' only if all the edges are open, which is indicated by $\phi = 1$.

The reliability is $R(\mathbf{p}) = \prod_{r=1}^{n} p_r$, which takes the value p^n when components have equal reliability. △

(8) **The k-out-of-n system.** As the name explains, this works if any subset of k out of the n components is working. △

The series and parallel structures are useful and important because any more general system can (in principle) be expressed in a representation using only these two basic blocks. To see this we need some more ideas and definitions.

(9) **Cut sets.** Let B be a set of components such that if none is working then the system fails. That is to say, if $\mathbf{x}$ is such that $x_i = 0, i \in B$, and $x_i = 1, i \in B^c$, then $\phi(\mathbf{x}) = 0$. Then B is called a *cut set*. If also $\phi(\mathbf{y}) = 1$ for any $\mathbf{y} > \mathbf{x}$, then B is called a *minimal cut set*. △

(10) **Path sets.** Let R be a set of components such that if all are working then the system works. That is to say, if $\mathbf{x}$ is such that $x_i = 1, i \in R$, and $x_i = 0, i \in R^c$, then $\phi(\mathbf{x}) = 1$. Then R is called a *path set*. If also $\phi(\mathbf{y}) = 0$ for any $\mathbf{y} < \mathbf{x}$, then R is called a *minimal path set*. △

The point of these concepts is this:

(11) **Theorem. Cuts and paths.** Let (B_i) be the set of all minimal cut sets, and (R_i) the set of all minimal path sets. Then

(12) $$\phi(\mathbf{X}) = \min_{i} \max_{r \in B_i} X_r = \max_{i} \min_{r \in R_i} X_r.$$

Proof. For the first equality, note that the system fails if and only if at least one of $\max_{r \in B_i} X_r$ is 0. For the second, observe that the system works if and only if at least one of $\min_{r \in R_i} X_r$ is 1. □

If we recall the particular structure functions in (6) and (7), it will be obvious that the first equality in (12) effectively interprets the system as an arrangement in series of the minimal cut sets, each bearing its component members in parallel. Likewise the second equality in (12) effectively represents the system as an arrangement in parallel of all the minimal path sets, each having its components in series.

One attractive application of these results is in supplying simple bounds for the reliability function.

(13) **Bounds for $R(p)$**. Let β_i be the probability that all the components in the minimal cut set B_i fail. Then, using Boole's inequality, we have

(14) $$1 - R(p) = P\left(\bigcup_i \{\text{all members of } B_i \text{ fail}\}\right) \leq \sum_i \beta_i.$$

Likewise, if ρ_i is the probability that all the components in the minimal path set R_i are working, we have

$$R(p) = P\left(\bigcup_i \{\text{all members of } R_i \text{ work}\}\right) \leq \sum_i \rho_i.$$

We illustrate this with an elementary but useful example.

(15) **Bridge system**. Consider this system represented as a 'flow' network, with equal reliability components (Figure 2.2):

The minimal cut sets are $\{1,2\}, \{4,5\}, \{2,3,4\}, \{1,3,5\}$; the minimal path sets are $\{1,4\}, \{2,5\}, \{2,3,4\}$, and $\{1,3,5\}$. Hence

$$1 - 2(1-p)^2 - 2(1-p)^3 \leq R(p) \leq 2p^2 + 2p^3.$$

The utility of the bounds depends on the value of p; it is an exercise for you to show that the actual value of $R(p)$ is $2p^2 + 2p^3 - 5p^4 + 2p^5$. ○

Better bounds of a similar type can be derived if necessary, with more work.

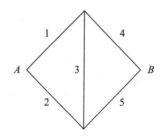

Figure 2.2 Each of the five edges is present independently with probability p. The system works if there is a continuous path from A to B.

Finally, we note that for many systems time is a crucial factor. Realistically, the probability that C_i is working at time $t \geq 0$ will be a decreasing function $p_i(t)$ of t. If the system is working at time $t = 0$, a natural question is: at what time T will it fail? In this case, for our monotone structure function ϕ,

$$P(T > t) = R(\mathbf{p}(t))$$

and so, using (1.6.7),

$$ET = \int_0^\infty R(\mathbf{p}(s))\,ds.$$

This calculation is sometimes feasible.

Exercises

(a)* Let $\phi(\mathbf{X})$ be a monotone structure function, with independent components. Show that, if $0 < p_i \leq q_i < 1$ for all i, then

$$R(\mathbf{p}) \leq R(\mathbf{q}).$$

(b) Let $R(p)$ be the reliability function of a system with independent components having the same probability p of working. Show that, for any p and q such that $0 \leq p, q \leq 1$, $R(pq) \leq R(p)R(q)$.

(c) In a k-out-of-n system each component has a lifetime that is Exponential (λ), independent of all the others. What is the probability that the system is working at time t? Show that the expected time until the system fails is $\sum_{r=k}^{n}(r\lambda)^{-1}$.

(d) **Reliability.** Consider a 2-out-of-3 system with independent components. Show that the structure function $\phi(\mathbf{X})$ is

$$X_1 X_2 + X_2 X_3 + X_3 X_1 - 2 X_1 X_2 X_3$$
$$= X_1 X_2 X_3 + X_1 X_2(1-X_3) + X_1(1-X_2)X_3 + (1-X_1)X_2 X_3$$
$$= 1 - (1 - X_1 X_2)(1 - X_2 X_3)(1 - X_3 X_1)$$
$$= \{1 - (1 - X_1)(1 - X_2)\}\{1 - (1 - X_2)(1 - X_3)\}$$
$$\quad \times \{1 - (1 - X_3)(1 - X_1)\}.$$

Interpret the three representations on the right-hand side in terms of disjoint events, minimal path sets, and minimal cut sets, respectively. In the latter two cases, sketch the corresponding diagrams with series and parallel components.

Calculate $R(\mathbf{p})$ and $R(p)$.

Compare the bounds given by (13) and (14) with the actual values.

2.9 Some technical details

> wherever we start in mathematics, whatever foundations we choose or axioms we assume,..., and however rigorously we argue, there is always some purist in the next-door college who regards our work as a mere flimsy of heuristics.[1]
>
> John M. Hammersley (*SIAM Journal*, 1974)

As remarked in the preface, this text is concerned principally with the ideas, applications, and techniques of stochastic processes and models, not with their rigorous development from first principles. Numerous mathematicians have worked during the last century to justify this carefree line of approach. However, some students may feel some uneasiness about the foundations of the subject; for them this section will point out some (not all) of the possible pitfalls, and sketch the way in which they are avoided. If you feel no anxiety in this regard, you need not read this bit.

Our first possible difficulty arises from our characterization of probability as a function $P(\cdot)$ on a family of subsets of Ω obeying the addition rules. In general, a countably additive non-negative real-valued set function is called a *measure*. When Ω is countable there are no insuperable difficulties in assigning a probability measure. However, when Ω is, for example, the unit interval $[0, 1]$, it can be shown that there are sets in $[0, 1]$ for which this cannot be done. Such sets are called *non-measurable*, and cannot be included in the event space. Fortunately this subtlety presents no practical problems.

Another difficulty arises when we consider events of probability zero, called *null events*. If A is a null event it is at the least desirable (and sometimes essential) that any subset of A should be a null event. Fortunately, in this case we can extend our event space to include such events and define a probability function on them so that they are indeed all null. This new probability space is called the *completion* of the original space.

Further problems arise when we come to consider conditional probability and conditional expectation in the context of continuous random variables and uncountable state spaces. Our naive definition of conditioning cannot be extended to condition on null events, so in more advanced work it is customary to adopt a different definition. Roughly speaking, in this new framework, we note that the principal property of conditional expectation is that $E[E(X \mid Y)] = EX$. (Recall (1.7.14).) It turns out that essentially we can define $E(X \mid Y)$ to be the function $\psi(\cdot)$ of Y such that

$$E[(\psi(Y) - X)h(Y)] = 0$$

---

[1] From the Greek meaning 'find', as is 'Eureka', whose correct transliteration is heurika.

for all random variables of the form $h(Y)$ having finite expected value. We can say 'the' function ψ, because it also turns out that if any other function $\hat{\psi}$ has this property, then $P(\psi = \hat{\psi}) = 1$.

Several further possible pitfalls arise when we start to consider random processes, especially those in continuous time. It is natural to consider such processes in the context of their finite-dimensional joint distributions $F(\mathbf{x},\mathbf{t}) = P(X(t_1) \leq x_1; \ldots; X(t_n) \leq x_n)$. But then it is equally natural to consider the process in the context of its sample paths, where one such path is denoted by

$$(X(t,\omega); t \geq 0, \omega \in \Omega).$$

One problem is that the space of sample paths may be far bigger than the product space associated with finite dimensional joint distributions. Recall, for example, the countable sequence of Bernoulli random variables; the product space for $F(\mathbf{x},\mathbf{t})$ is $\{0,1\}^n$, of cardinality 2^n. The set of all sample paths comprises the interval $[0,1]$ which is uncountable.

In continuous time, things are even worse, since $(X(t); 0 \leq t \leq 1)$ is an uncountable collection of random variables. We address part of this problem by introducing the idea of a *filtration*, that is a sequence of probability spaces $(\Omega_t, \mathcal{F}_t, P_t)$ that keep track of $(X(u); 0 \leq u \leq t)$, which is everything $X(u)$ does up to time t. But even then, it is not clear that quantities defined in terms of sample paths (such as 'the first time $X(t)$ takes the value 1', 'the time for which $X(u) = 0$ in $[0,t]$', or '$X(t) = 0$ for some $t > 0$') are random variables on this filtration. Indeed they may not be, unless we introduce some restrictions on the behaviour of sample paths.

It turns out, after a remarkable amount of hard work, that things work out the way we want them to if we confine our attention to processes that are continuous on the right, with limits from the left; such an $X(t)$ is sometimes called RCLL.

The basic plan of campaign for such processes is first to consider them at a suitable set of points that is countably infinite (the rationals, say); sort out the problems using finite-dimensional distributions, and then extend the results to the whole of $t \geq 0$ by exploiting the continuity (or right-continuity) of $X(t)$. (You can think of this informally as 'filling in the gaps'.) The details are beyond our scope here.

Problems

1. Let $A_n(\epsilon)$ be the event that $|X_n - X| > \epsilon$, and let $A(\epsilon)$ be the event that infinitely many of $(A_n(\epsilon); n \geq 1)$ occur, where $\epsilon > 0$. Prove Lemma (6), that $X_n \xrightarrow{\text{a.s.}} X$ if, for all $\epsilon > 0$, $|X_n - X| > \epsilon$ only finitely often.
[Hint: $P(X_n \not\to X) = P(\bigcup_{\epsilon > 0} A(\epsilon)) = P(\bigcup_{m=1}^{\infty} A(m^{-1}))$.]

2.* Prove the strong law of large numbers for the sequence $S_n = \sum_{r=1}^{n} X_r$, assuming that $(X_r; r \geq 1)$ are independent and identically distributed with zero mean and $EX_1^4 < \infty$. [Hint: Prove that $ES_n^4 = nEX_1^4 + 3n(n-1)E(X_1^2 X_2^2)$.]

3.* $N+1$ plates of soup are laid out around a circular dining table, and are visited by a fly in the manner of a symmetric random walk: after visiting any plate it transfers its attention to either of the neighbouring plates with equal probability $\frac{1}{2}$. Show that (with the exception of the plate first visited) each plate has probability $1/N$ of being the last visited by the fly.

What is the corresponding probability if the fly is equally likely to transfer its attention to any other plate on the table?

Let C be the number of flights until the fly first visits the last plate. Find the expected value of C in each of the above two cases. Compare their relative magnitudes as N increases. (C is known as the *exploration time* (or *covering time*) of the process.)

4.* Consider a simple symmetric random walk started at k, $(0 < k < N)$, which stops when it first hits either 0 or N.
   (a) Show that the expected number of visits to j, $(0 < j < k)$, before the walk stops is
   $$2j(N-k)/N.$$
   (b) Find the corresponding expression for $k < j < N$.
   (c) Find the distribution of the number of visits X to its starting point k, and show that X is independent of where the walk stops.

5.* Let Z_n be the size of the nth generation of a branching process with $Z_0 = 1$ and var $Z_1 > 0$, having mean family size μ and extinction probability η.
   (a) Show that $Z_n \mu^{-n}$ and η^{Z_n} are martingales.
   (b) Let $Es^{Z_n} = \mathcal{G}_n(s)$, and define H_n to be the inverse function of $\mathcal{G}_n(s)$ on $0 \leq s < 1$. Show that $\{H_n(s)\}^{Z_n}$ is a martingale with respect to Z_n.

6. Let $X_n, n \geq 0$, be a non-negative submartingale.
   (a) Show that if T is a stopping time for X such that $P(T < \infty) = 1$ then $EX_{T \wedge n} \leq EX_n$.
   (b) Show that for $x > 0$
   $$P(\max_{0 \leq k \leq n} X_k > x) \leq EX_n/x.$$

(c) Let $S_n = \sum_{r=1}^{n} X_r$ be a random walk such that $EX_r = 0$. Show that

$$P(\max_{0 \leq k \leq n} |S_k| > x) \leq n \text{var } X_1/x^2.$$

7.* **Matching martingale.** In a cloakroom there are C distinct coats belonging to C people who each pick a coat at random and inspect it. Those who have their own coat leave, the rest return the coats and try again at random. Let X_n be the number of coats remaining after n rounds. Show that $X_n + n$ is a martingale. Now let T be the number of rounds of coat-picking until everyone has the right coat. Show that $ET = C$.

8.* **Campbell–Hardy theorem.** Let $N(t)$ be a Poisson process with constant intensity λ, with events at $(S_n; n \geq 1)$. Let $S = \sum_{n=1}^{\infty} g(S_n)$, where $g(\cdot)$ is a nice function. Show that $E(S) = \lambda \int_0^{\infty} g(x)\, dx$, and var $S = \lambda \int_0^{\infty} (g(x))^2\, dx$.

9. **Insurance.** Claims are made on an insurance fund at the events $(S_n; n \geq 1)$ of a Poisson process with constant intensity λ. The rth claim is C_r, where the C_r are independent with density $f(x)$, and independent of the Poisson process. The present discounted liability of a claim C at time t is $C e^{-\beta t}$. Show that the present total discounted liability D of this claim stream

$$D = \sum_n C_n e^{-\beta S_n}$$

has mean and variance

$$ED = \lambda EC_1/\beta,$$

$$\text{var } D = \lambda E(C_1^2)/2\beta.$$

10. **Geometric branching.** Let Z_n be a branching process with family size pgf given by $Es^{Z_1} = q/(1 - ps)$, with $Z_0 = 1 = p + q$. Show (by induction) that

$$G_n(s) = Es^{Z_n} = \begin{cases} \dfrac{q(p^n - q^n - ps(p^{n-1} - q^{n-1}))}{p^{n+1} - q^{n+1} - ps(p^n - q^n)}, & p \neq q, \\[2mm] \dfrac{n - (n-1)s}{n + 1 - ns}, & p = q. \end{cases}$$

11. Let $N(t)$ be a Poisson process with intensity $\lambda(t)$, which is to say that

$$P(N(t+h) - N(t) = 1) = \lambda(t)h + o(h)$$
$$P(N(t+h) - N(t) = 0) = 1 - \lambda(t)h + o(h).$$

Show that the distribution of $N(t)$ is Poisson ($\int_0^t \lambda(s)\,ds$) and find the joint distribution of the times of the first two events of $N(t)$.

12. Let $N(t)$ be a renewal process with interarrival times $(X_n; n \geq 1)$. Let $D(t)$ be the *total life* (or *spread*) at t, given by

$$D(t) = S_{N(t)+1} - S_{N(t)} = X_{N(t)+1}.$$

By defining the reward

$$R_n = \begin{cases} X_n, & \text{if } X_n \leq x, \\ 0, & \text{otherwise,} \end{cases}$$

show that the limiting distribution of $D(t)$ as $t \to \infty$ is given by

$$F_D(x) = \left\{ xF(x) - \int_0^x F(u)\,du \right\} \Big/ EX_1,$$

where $F(x)$ is the distribution of X_1. Find the mean and variance of $F_D(x)$, and the density $f_D(x)$. Now let U be Uniform $(0, 1)$, independent of D having density $f_D(x)$. Show that the distribution of UD is the same as the limiting distribution of the age A, and excess life E, defined and derived in (40). Explain why this is an intuitively natural result.

13.* **Alternating renewal.** A machine repeatedly fails and is then repaired. The time from repair until the nth failure is Z_n with distribution $F_Z(z)$; the time from the nth failure until the completion of repair is Y_n with distribution $F_Y(y)$. All these random variables are independent.

Show that the long-run proportion of time for which the machine is working is $EZ_1/(EZ_1 + EY_1)$.

14. **Umbrella.** I have one umbrella, and my life is divided between n places; I travel to and fro on foot. If it is raining when I set out on any transfer, I take the umbrella if it is with me (ignore the chance of rain developing on the way). It rains on any trip with probability p independently of all other trips. If it is dry, I leave the umbrella at the place I set out from.

Consider two possible models for my life.
(a) I visit all n locations sequentially.

(b) I move to any other location at random, independently of the past. Show that in either case, the long-run proportion of journeys on which I get wet is
$$\frac{p(n-1)(1-p)}{1+(n-1)(1-p)}.$$

15.* Let $(X_n; n \geq 1)$ be independent Uniform $(0,1)$. Let A_n be the event that $X_n > X_r$ for all $r < n$. Show that A_n occurs infinitely often with probability 1.

16.* Let $(X_r; r \geq 1)$ be independent, non-negative, and identically distributed with infinite mean. Show that as $n \to \infty$
$$\frac{1}{n}(X_1 + X_2 + \cdots + X_n) \xrightarrow{\text{a.s.}} \infty.$$

17. **Replacement rule.** A component's lifetime is the random variable X; replacements have lifetimes that are independent and have the same distribution as X.

    You adopt the following strategy with regard to this component: you replace it immediately when it fails, or you replace it when it reaches the age a, whenever is the sooner. Any replacement costs c, but replacing a failure costs an extra d. Show that the long-run average cost of this policy is
$$\frac{c}{E(X \wedge a)} + \frac{d}{P(X \leq a)(EX \wedge a)},$$
    where $X \wedge a$ is the smaller of a and X.

18. Let S be a system with independent components having monotone structure function $\phi(\mathbf{X})$ and reliability $R(\mathbf{p})$. Suppose that each component C_i is duplicated by another component D_i, working in parallel with C_i where D_i has reliability q_i. Show that the reliability of this system is $R(\boldsymbol{\pi})$, where $\pi_i = 1 - (1 - p_i)(1 - q_i)$. Prove that $R(\boldsymbol{\pi}) \geq 1 - (1 - R(\mathbf{p}))(1 - R(\mathbf{q}))$, and interpret this result. [Hint: Because ϕ is monotone, $\phi(\mathbf{X} \wedge \mathbf{Y}) \geq \phi(\mathbf{X}) \wedge \phi(\mathbf{Y})$.]

3 Markov chains

πάντα ρεῖ, οὐδὲν μένει
[All is flux, nothing is still]

Heracleitus

3.1 The Markov property; examples

Useful models of the real world have to satisfy two conflicting requirements: they must be sufficiently complicated to describe complex systems, but they must also be sufficiently simple for us to analyse them. This chapter introduces Markov chains, which have successfully modelled a huge range of scientific and social phenomena, supported by an extensive theory telling us how Markov chains behave. Because they combine tractability with almost limitless complexity of behaviour, Markov processes, in general, are the most useful and important class of stochastic models. Their underlying rationale is the observation that many real-world processes have this property:

If you have a complete description of the state of the process at time t, then its future development is independent of its track record before t.

For examples, consider:

- Your sequence of fortunes after successive bets in a casino.
- The sequence of genes in a line of descent.
- The order of a pack of cards at successive shuffles.

Think of some more yourself. We have previously looked at several processes that have this property, for example,

(1) **Branching process (2.7.1) revisited.** The decomposition given in (2.7.2),

$$Z_{n+1} = \sum_{i=1}^{Z_n} X(i)$$

displays explicitly the fact that, if we are given Z_n, the future of the process is independent of its past. ○

(2) **Random walk on the integers.** If $(X_r; r \geq 1)$ are independent, identically distributed, integer-valued random variables then they yield a random walk

$$S_n = \sum_{r=1}^{n} X_r.$$

The representation $S_{n+1} = S_n + X_{n+1}$ shows that S_n must have the property that if we are given S_n, the future of the walk is independent of its past. ○

This idea that we have identified informally is called the *Markov property*. We proceed to a formal definition. Note that it is essentially a statement about conditional independence.

(3) **Markov chain.** Let $(X_0, X_1, \ldots)$ be a sequence of random variables taking values in some countable set S, called the *state space*. (Without loss of generality, we generally take S to be a suitable set of integers.)

Then $X = (X_0, X_1, \ldots)$ is a *Markov chain* if it has the *Markov property*:

(4) $\qquad P(X_n = k \mid X_0 = x_0, \ldots, X_{n-1} = j) = P(X_n = k \mid X_{n-1} = j)$

for all $n \geq 1$, and all $x_0, x_1, \ldots, j, k$ in S. If, for all n, we have also that

(5) $\qquad\qquad\qquad P(X_n = k \mid X_{n-1} = j) = p_{jk}$,

then the chain is called *homogeneous*. (All chains here are homogeneous unless explicitly mentioned otherwise.) The array

$$\mathbf{P} = (p_{jk}), \qquad j, k \in S,$$

is called the matrix of *transition probabilities*; the chain X is said to be Markov (**P**). We may denote X_n by $X(n)$, and we sometimes write $p_{j,k}$ or $p(j, k)$ for p_{jk} (note that some writers denote it by p_{kj}). Since p_{jk} is the conditional distribution of X_n given X_{n-1}, we must have for all j that

(6) $\qquad\qquad\qquad \sum_{k \in S} p_{jk} = 1.$

Any matrix **P** satisfying (6), with non-negative entries, is said to be *stochastic*. If in addition it is true that

(7) $\qquad\qquad\qquad \sum_{j \in S} p_{jk} = 1,$

then it is said to be *doubly stochastic*. By construction, the development of the chain is completely described by the transition matrix **P** and knowledge of the initial distribution α of X_0, where

$$\alpha_j = P(X_0 = j).$$

We explore this further in the next section; here are some examples.

(8) Random walk revisited. From (2) it is clear that the transition probabilities are

$$p_{jk} = P(X_n = k - j).$$

(9) Genetic drift. Consider a population of d individuals from a diploid species (of plants or animals). At any locus the gene may be either of two forms (alleles). A simple model assumes that the next generation also contains d individuals who get their genes by sampling with replacement from the $2d$ available from the previous generation. Let X_n be the number of genes of the first form in the nth generation. By construction X is a Markov chain with

(10) $$p_{jk} = \binom{2d}{k} \left(\frac{j}{2d}\right)^k \left(1 - \frac{j}{2d}\right)^{2d-k}, \quad 0 \leq j, k \leq 2d.$$

(11) Waiting for sixes. Let X_n be the number of rolls of a die since the last 6. By construction, and the independence of rolls, X is a Markov chain with transition probabilities

$$p_{j,j+1} = \tfrac{5}{6} \quad \text{and} \quad p_{j,0} = \tfrac{1}{6}, \quad j \geq 0.$$

(12) Markov property revisited. The Markov property that we defined in (3) is the most natural and convenient way to introduce the idea. But it is equivalent to several other characterizations which may be occasionally more convenient to take as definitive. It is left as an exercise for you to show that (4) holds if and only if any one of the following four statements is true. (Recall that $p_{jk} = p(j, k) = p_{j,k}$, and $X_n = X(n)$.)

(13)
$$P(X_0 = j_0, X_1 = j_1, \ldots, X_n = j_n)$$
$$= P(X_0 = j_0) p(j_0, j_1) p(j_1, j_2), \ldots, p(j_{n-1}, j_n),$$

for any $j_0, \ldots, j_n$ in S.

(14) $$P(X_{n+m} = k \mid X_0, \ldots, X_n = j) = P(X_{n+m} = k \mid X_n = j),$$

for any positive m and n.

(15) $$P(X(n_r) = k \mid X(n_1), \ldots, X(n_{r-1}) = j)$$
$$= P(X(n_r) = k \mid X(n_{r-1}) = j),$$

for any $n_1 < n_2 < \cdots < n_r$.

(16) $P(X_1 = k_1, \ldots, X_{r-1} = k_{r-1}, X_{r+1} = k_{r+1}, \ldots, X_n = k_n \mid X_r = j)$
$= P(X_1 = k_1, \ldots, X_{r-1} = k_{r-1} \mid X_r = j)$
$\times P(X_{r+1} = k_{r+1}, \ldots, X_n = k_n \mid X_r = j),$ for any $k_1, \ldots, k_n$ in S.

Before we introduce further examples and theory, we turn aside briefly to remark that there are several different ways of visualizing a Markov chain in the mind's eye; and, equally, several ways of presenting it formally on paper. You should choose the one that is best suited to the problem at hand and your own preferences.

For example, one may think of a Markov chain as a particle performing a random walk on the vertices of a graph. Then p_{jk} is the probability of a step to k, given that it is at j. Or one may think of it as a system being in one of a number of states; then p_{jk} is the probability of a change from state j to state k.

Because of this imagery, we may speak of a chain being in state k, or taking the value k, or visiting k. Furthermore, the interpretation as a random walk on a graph is often useful when represented as a diagram or transition graph. Some examples follow.

(17) **Flip–flop circuit.** This is a Markov chain with two states; it passes from one to the other at random. It is a model for the basic flip–flop switching circuit used (and thus named) by engineers. Such circuits originally used valves (from the Latin word valva, meaning door), and such a circuit is said to be either 'open' or 'closed'. In other contexts it may be 'on' or 'off', or one may say 'up' or 'down'.

For ease of reference the two states are simply assigned the state space $S = \{0, 1\}$. If $X_n = 0$, then

$$X_{n+1} = \begin{cases} 1, & \text{with probability } a, \\ 0, & \text{with probability } 1 - a, \end{cases}$$

independently of previous moves. In the notation above

$$p_{01} = P(X_{n+1} = 1 \mid X_n = 0) = a, \qquad p_{00} = P(X_{n+1} = 0 \mid X_n = 0) = 1 - a.$$

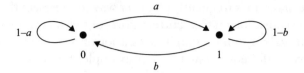

Figure 3.1 Diagram for two-state chain.

Likewise $p_{10} = b$ and $p_{11} = 1 - b$. These probabilities may be displayed as a matrix

$$(18) \qquad P = \begin{pmatrix} 1-a & a \\ b & 1-b \end{pmatrix} = \begin{pmatrix} p_{00} & p_{01} \\ p_{10} & p_{11} \end{pmatrix}$$

or as a self-explanatory diagram (Figure 3.1).

Here, the loops represent the chance of the state not changing. But since the sum of all the probabilities on arrows leaving any vertex is 1, the loops add no extra information and are often omitted. ○

(19) Discrete renewal. Let $(X_r; r \geq 1)$ be independent and integer-valued with $P(X_r > 0) = 1 = P(X_r < \infty)$. Define $S_n = \sum_{r=1}^{n} X_r$, and the renewal process

$$N(k) = \max\{n : S_n \leq k\}, \quad k \in \mathbb{Z}^+,$$

which is called *discrete renewal*, for obvious reasons.

Let $A(k) = k - S_{N(k)}$ be the age of the process; that is the time since the last renewal. It is easy to see that $A(k)$ is a Markov chain with transition probabilities defined in terms of the probability distribution F_X, and mass function f_X, of X. For $j \geq 0$:

$$(20) \qquad \begin{aligned} p_{j,j+1} &= \frac{\{1 - F_X(j)\}}{\{1 - F_X(j-1)\}} \\ p_{j,0} &= \frac{f_X(j)}{\{1 - F_X(j-1)\}}. \end{aligned}$$

In this case the diagram is unbounded (you should sketch part of it), and so of course is the transition matrix

$$P = \begin{pmatrix} 0 & 1 & 0 & \cdots \\ f_X(1) & 0 & 1 - F_X(1) & \cdots \\ \frac{f_X(2)}{1 - F_X(1)} & 0 & 0 & \ddots \\ \vdots & \vdots & \vdots & \end{pmatrix}.$$

○

Drawing the diagram is frequently an easier way to represent the chain than laying out the transition matrix. Furthermore, our visualization of the chain as a random walk around a graph gives a lot of insight into the types of behaviour we shall be interested in. For example, for some vertices v and w we may consider:

- The chance of being in v in the long run.
- The time spent at v.
- The chance of hitting v before w.
- The expected time to return to v.
- The expected time to reach w from v.
- The expected number of visits to w on an excursion from v.

Other questions of interest appear later, but we consider some simple examples to guide our intuition before we plunge into a detailed general theory.

(21) **Gambler's ruin.** At each of a sequence of independent bets your fortune either increases by \$1 with probability p, or decreases by \$1 with probability $q = 1 - p$. You stop if your fortune is ever 0 or $K > 0$. Let X_n be your fortune after the nth such bet; $0 \leq X_0 \leq K$.

The sequence $(X_n; n \geq 0)$ clearly has the Markov property; its transition probabilities are

$$p_{KK} = p_{00} = 1,$$
$$p_{j,j+1} = p, \quad p_{j,j-1} = q, \quad 1 \leq j \leq K - 1.$$

You should now display the transition matrix **P**, and sketch the transition diagram.

We have already analysed this chain extensively; it is useful to summarize the conclusions in this new context.

For brevity, recall the case when $p = q = \frac{1}{2}$: we have proved that

(i) The probability of hitting 0 before K, if you start at k, is $1 - k/K$.
(ii) The expected time to hit either of 0 or K is $k(K - k)$, starting from k.
(iii) The states $\{1, \ldots, K-1\}$ are visited only a finite number of times; with probability 1 the chain enters $\{0, K\}$ and stays there ever after. ○

When the state space is infinite, radically different behaviour may ensue.

(22) **Simple random walk revisited.** Here the position of the walk S_n is a Markov chain with

$$p_{j,j+1} = p, \quad p_{j,j-1} = q, \quad j \in \mathbb{Z}.$$

Once again you should display part of the matrix **P**, and the transition diagram. It is easy to see that if $S_0 = 0$, and $n + j$ and $n - j$ are both even, then

$$P(S_n = j) = \binom{n}{(n+j)/2} p^{(n+j)/2} q^{(n-j)/2}$$

(23)
$$\to 0, \quad \text{as } n \to \infty,$$

for any fixed j. This is true for both symmetric and asymmetric simple random walk. However, there is an interesting distinction between the symmetric and asymmetric cases with respect to other types of behaviour.

(i) If $p \neq q$ then the walk revisits 0 only finitely often. By symmetry any state of the chain is visited only finitely often; since the probability of ever returning to the starting point is $(p \wedge q)/(p \vee q)$.

(ii) If $p = q = \frac{1}{2}$, then the origin is revisited infinitely often. Since any other state is reached from the origin with probability 1, all states are visited infinitely often. However, as we showed in (2.2.26), the expected time to return to the origin is infinite.
A more subtle and interesting property of the walk is this: for any $j \neq 0$ the expected number of visits to j before returning to 0 is 1. See Exercise (3.1.d).

(iii) The walk can revisit its starting point only at an even number of steps; it is said to be *periodic*. ○

Our final example exhibits yet another type of behaviour.

(24) **Waiting for sixes.** Suppose you roll a die n times. We showed above in (11) that $(X_n, n \geq 1)$, denoting the number of rolls since the most recent 6, is a Markov chain. The transition probabilities are, for all $j \geq 0$,

$$p_{j,j+1} = \tfrac{5}{6} \quad \text{and} \quad p_{j,0} = \tfrac{1}{6}.$$

Like the symmetric random walk this chain returns to its starting point with probability 1. (To show this is an easy exercise for you.) However (unlike the simple random walk), the expected time to do so is finite. For example, the expected time for the chain to revisit 0 is just the expected number of rolls to get a 6, namely six.

Even more interestingly, we can write down the distribution of X_n immediately; suppose the chain starts in state 0, by construction we have for $n \geq 1$

$$P(X_n = j) = \left(\tfrac{5}{6}\right)^j \tfrac{1}{6}, \qquad 0 \leq j \leq n-1,$$

(25) $$P(X_n = n) = \left(\tfrac{5}{6}\right)^n.$$

Notice that as $n \to \infty$ this converges to a probability distribution, which is denoted by π_j:

(27) $$\lim_{n \to \infty} P(X_n = j) = \left(\tfrac{5}{6}\right)^j \tfrac{1}{6} = \pi_j, \quad j \geq 0.$$

Now suppose we extend this chain by allowing a random initial state X_0. If you like, you can imagine taking the die over from someone else in the knowledge that they last rolled a six X_0 rolls ago. If X_0 has the distribution in (27), then

$$P(X_1 = j) = P(X_0 = j-1)\tfrac{5}{6} = \left(\tfrac{5}{6}\right)^j \tfrac{1}{6} = \pi_j, \quad j > 0$$

and
$$P(X_1 = 0) = \tfrac{1}{6} = \pi_0.$$

Thus in this case X_1 has the same distribution as X_0, and so does X_n for all n, after an easy induction. The chain is said to be in *equilibrium*, or *stationary*; $(\pi_j; j \geq 0)$ is called the *stationary distribution* and often denoted by $\boldsymbol{\pi}$.

It is an exercise for you to show that there is only one distribution that has this property for this chain. Furthermore, conditioning on the time Y at which a six is first rolled, for $n > Y$ we have

$$P(X_n = j \mid Y = y) = \left(\tfrac{5}{6}\right)^j \tfrac{1}{6}, \quad 0 \leq j \leq n-y-1.$$

As $n \to \infty$, $P(X_n = j \mid Y = y) \to \pi_j$ for all $j \geq 0$. That is to say, the distribution of X_n converges to $\boldsymbol{\pi}$ independently of the initial state X_0. ◯

Finally in this section, we note that the Markov property is preserved by some functions but not by others.

(28) **Sampling.** Suppose that X is a Markov chain. Show that $(Y_0, Y_1, \ldots)$, where $Y_n = X_{2n}$, is a Markov chain.

Solution. We must verify a suitable form of the Markov property. Thus

$$P(Y_{n+1} = k \mid Y_0, \ldots, Y_n = j) = P(X_{2n+2} = k \mid X_0, \ldots, X_{2n} = j)$$
$$= P(X_{2n+2} = k \mid X_{2n} = j), \quad \text{by (15)}$$
$$= P(Y_{n+1} = k \mid Y_n = j) = p_{jk}^{(2)},$$

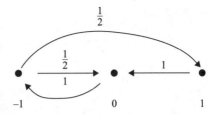

Figure 3.2 Diagram for the chain X.

where $p_{jk}^{(2)} = P(X_2 = k \mid X_0 = j)$ is the two-step transition probability for X.

On the other hand, if X is a Markov chain let us define $Y_n = |X_n|$ for all n. It is not hard to find a chain such that Y_n does not have the Markov property. For example, let X have the diagram in Figure 3.2.

Then
$$P(Y_{n+1} = 1 \mid Y_n = 1, Y_{n-1} = 1) = 0,$$

but
$$P(Y_{n+1} = 1 \mid Y_n = 1, Y_{n-1} = 0) = \tfrac{1}{2}.$$

So Y cannot be a Markov chain.

To sum up: a Markov chain is determined by its transition matrix **P**, and the initial distribution α of X_0.

We have given examples to show that chains and their states can exhibit markedly different behaviour:

- Some states are visited infinitely often, some are not.
- Some chains have a unique equilibrium distribution, some do not.
- Some chains converge in distribution to a probability distribution, some do not.

At first glance this all seems undesirably complicated, but in the next sections we shall see that there are in fact rather few different types of chains, and a chain's type is determined by a very few simply-stated properties of **P**.

Exercises
(a) A die is rolled repeatedly. Which of the following are Markov chains? For those that are, find the transition probabilities p_{jk}; and also find $\lim_{n\to\infty} P(X_n = k)$.
  (i) The largest number shown up to time n.
  (ii) The time until the next six after time n.
  (iii) The number of sixes shown by time n.

(b) If $(S_n; n \geq 0)$ is a simple random walk, which of the following are Markov chains? Justify your answer:
  (i) $Y_n = S_0 + S_1 + \cdots + S_n$
  (ii) $Z_n = (S_n, Y_n)$.

(c) **Discrete renewal (19) revisited.** Show that the excess life $S_{N(k)+1} - k = Y_k$ is a Markov chain, and write down its transition probabilities.

(d)* Show that the expected number of visits of a simple random walk to $j \neq 0$, before returning to 0, is 1. [Hint: Let θ be the probability that the walk visits j at all before returning to 0, and consider the distribution of the number of visits.]

3.2 Structure and *n*-step probabilities

The matrix **P** describes the short-term, one-step behaviour of the chain. A natural question is to ask if we can, in general, investigate the long-term behaviour of X. We define the *n*-step probabilities of the chain by

(1) $$P(X_{n+m} = k \mid X_m = j) = p_{jk}^{(n)}.$$

They are related to each other by a famous identity:

(2) **Chapman–Kolmogorov equations.** For all i and k in S, and any positive m and n

(3) $$p_{ik}^{(m+n)} = \sum_{j \in S} p_{ij}^{(m)} p_{jk}^{(n)}.$$

These equations arise from a very simple and natural idea: we condition on the state of the chain after m steps. The idea is obvious, the algebra is less so:

$$\begin{aligned} p_{ik}^{(m+n)} &= P(X_{m+n} = k \mid X_0 = i) \\ &= \sum_j P(X_{m+n} = k, X_m = j \mid X_0 = i) \end{aligned}$$

$$= \sum_j P(X_{m+n} = k \mid X_m = j, X_0 = i) P(X_m = j \mid X_0 = i),$$

by conditional probability

$$= \sum_j P(X_{m+n} = k \mid X_m = j) P(X_m = j \mid X_0 = i),$$

by the Markov property (3.1.15)

$$= \sum_j p_{ij}^{(m)} p_{jk}^{(n)}.$$

It follows immediately that if $P(X_0 = j) = \alpha_j$, then

(4) $$P(X_n = k) = \sum_j \alpha_j p_{jk}^{(n)}.$$

We may represent (3) and (4) more compactly in terms of the matrix $\mathbf{P} = (p_{jk})$ and the vector $\boldsymbol{\alpha} = (\alpha_j)$. Then (3) implies that the matrix $(p_{jk}^{(n)})$ is the nth power of $\mathbf{P}$, namely $\mathbf{P}^n$. And (4) tells us that the distribution of X_n is given by $\boldsymbol{\alpha} \mathbf{P}^n$.

Clearly a key distinction in these multistep transition probabilities $p_{ij}^{(n)}$ is between those that are zero, and those that are strictly positive. The preceding examples also illustrate that this depends crucially on the structure of the transition matrix $\mathbf{P}$. That is to say, which entries p_{jk} are actually greater than zero? (This aspect of such a diagram is sometimes called the *topology* of the chain. Note the convention that $p_{jj}^{(0)} = 1$.)

(5) **Chain structure.** We say that state k is *accessible* from state j (or j *communicates* with k) if the chain can ever reach k when started from j. That is, if $p_{jk}^{(m)} > 0$ for some $0 \leq m < \infty$. We denote this property by $j \to k$.

We say that j and k *intercommunicate* (or are mutually accessible) if $j \to k$ and $k \to j$, and we denote this property by $j \leftrightarrow k$. We interpret this condition $j \to k$ in terms of the diagram of the chain by observing that $p_{jk}^{(m)} > 0$ if and only if there is a path of steps from j to k, $j \to i_1 \to i_2 \to \cdots \to i_{m-1} \to k$ such that $p(i, i_1) p(i_1, i_2) \cdots p(i_{m-1}, k) > 0$.

It is an easy exercise to see that

(a) $j \leftrightarrow j$;
(b) $j \leftrightarrow k \Rightarrow k \leftrightarrow j$;
(c) $i \leftrightarrow j$ and $j \leftrightarrow k$ implies $i \leftrightarrow k$.

It follows that the state space S can be divided up into disjoint sets C_i such that all the states in any C_i are accessible to each other. These are called *equivalence classes*, and entail some jargon.

(6) Definition

(a) A state j with $p_{jj} = 1$ is called *closed* or *absorbing*.
(b) A set C is *closed* if $p_{jk} = 0$, for all $j \in C, k \notin C$.
(c) A set C is *irreducible* if $j \leftrightarrow k$, for all $j, k \in C$.

Once the chain enters a closed set of states, it never leaves subsequently; clearly the equivalence classes of $\leftrightarrow$ are irreducible.

However, $p_{jk}^{(m)} > 0$ for some m does not necessarily entail $p_{jk}^{(m+n)} > 0$ for all $n > 0$.

(7) Definition. The *period* $d(j)$ of a state j is the greatest common divisor (denoted by gcd) of the times at which return to j is possible. Formally

$$(8) \qquad d(j) = \gcd\{n : p_{jj}^{(n)} > 0\}.$$

If $d(j) = 1$ then j is *aperiodic*. $\triangle$

This entails that some chains exhibit a cyclic structure, such that there are d subclasses $C(1) \cdots C(d)$ of any irreducible class C, and the chain cycles through them in the same order. For example, in the simple random walk $\mathbb{Z} = C = C(1) \cup C(2)$, where $C(1) \equiv$ the odd integers, and $C(2) \equiv$ the even integers.

In fact, few real-life chains are periodic, and the simple random walk is almost the only periodic chain we consider here. This is fortunate, as the general theory of periodic chains requires ceaseless consideration of tiresome details, of technical but little other interest. So far as possible we exclude them, and confine our attention to aperiodic chains.

By contrast, there is one very attractive property that many finite chains have.

(9) Regularity. If for some $n_0 < \infty$ we have

$$p_{jk}^{(n_0)} > 0, \quad \text{for all } j \text{ and } k,$$

then the chain is said to be *regular*.

Roughly speaking, no matter where it started the chain could be anywhere in S at time n_0. This is stronger than irreducibility (defined in (6)), which asserts merely that any state can be reached from any other state eventually. $\triangle$

The practical upshot of these results is that we can concentrate our attention on irreducible chains in the long run, a point to which we return in more detail in Section 3.5. Let us look at some examples to illustrate these ideas.

(10) **Flip–flop chain: (3.1.17) revisited.** This runs on the state space $S = \{0, 1\}$, with transition matrix

$$\mathbf{P} = \begin{pmatrix} 1-a & a \\ b & 1-b \end{pmatrix}.$$

From (3), or simply by conditional probability,

(11) $$p_{00}^{(n+1)} = b p_{01}^{(n)} + (1-a) p_{00}^{(n)}.$$

Using the fact that $p_{00}^{(n)} + p_{01}^{(n)} = 1$, we obtain

(12) $$p_{00}^{(n+1)} = (1-a-b) p_{00}^{(n)} + b.$$

This is readily solved when $a + b > 0$ to give

(13) $$p_{00}^{(n)} = \frac{b}{a+b} + \frac{a}{a+b}(1-a-b)^n.$$

When $a + b = 0$, the chain is rather dull: it comprises the two irreducible subchains on $\{0\}$ and $\{1\}$, and simply stays at X_0 forever. On the other hand, if $a + b = 2$, then the chain is periodic with period 2:

(14) $$p_{00}^{(n)} = \begin{cases} 1, & n \text{ even,} \\ 0, & n \text{ odd.} \end{cases}$$

If either a or b is zero, but not both, then there is just one irreducible subchain, comprising an absorbing state. Wherever the chain starts, it is eventually absorbed there.

Excluding all the above cases, it is easy to see that the chain is then regular. The flip–flop chain is clearly trivial but friendly; we revisit it often in later sections.

Note that in view of (3) it is possible (at least in principle), to find n-step transition probabilities by using methods for matrix multiplication.

For example, looking at the flip–flop chain one can verify by induction that

$$(15) \qquad (a+b)\mathbf{P}^n = \begin{pmatrix} b & a \\ b & a \end{pmatrix} + (1-a-b)^n \begin{pmatrix} a & -a \\ -b & b \end{pmatrix}.$$

For larger state spaces, things get considerably more complicated.

Finally for the flip–flop chain, we record the interesting fact, important for later work, that the distribution of X_n converges if $|1 - a - b| < 1$. In fact, as $n \to \infty$,

$$(16) \qquad P(X_n = 0) \to b/(a+b), \qquad P(X_n = 1) \to a/(a+b).$$

Furthermore, if X_0 has this distribution, then so does X_n for all n. The chain is said to be *stationary*. ○

(17) **Branching process.** For the ordinary branching process with $\mathcal{G}(0) > 0$, the origin is an absorbing state; the chain ends up there with probability η, the extinction probability. For refreshed branching, by contrast, there is a single intercommunicating class; the origin is not absorbing. ○

(18) **Particles.** At time $n \geq 0$, a chamber contains a number X_n of particles; X_0 is Poisson parameter μ. In the interval $(n, n+1]$ each particle of X_n independently decays and disappears with probability q, independently of all other particles (present and past); otherwise it survives unaltered with probability $p = 1 - q$. At time n, $n \geq 1$, a number Y_n of fresh particles is injected into the chamber, where Y_n are Poisson (λ), and independent of each other and the past $(X_0, \ldots, X_n)$. Then we may write, to sum up all this information,

$$(19) \qquad X_{n+1} = f(X_n) + Y_{n+1}, \quad n \geq 0,$$

where, conditional on X_n, $f(X_n)$ is binomial $B(X_n, p)$. By the independence and (19), X_n is clearly a Markov chain. Now $Es^{X_0} = e^{\mu(s-1)}$, and

$$(20) \qquad \begin{aligned} Es^{X_1} &= E[E(s^{X_1} \mid X_0)], \quad \text{by conditional expectation} \\ &= E(q+ps)^{X_0} Es^{Y_1}, \quad \text{by independence} \\ &= \exp\{(\mu p + \lambda)(s-1)\}. \end{aligned}$$

Likewise

$$(21) \qquad Es^{X_2} = \exp\{(\mu p^2 + \lambda p + \lambda)(s-1)\}$$

and an easy induction gives

$$(22) \qquad Es^{X_n} = \exp\{(\mu p^n + \lambda p^{n-1} + \cdots + \lambda)(s-1)\}.$$

Note that this is yet another example of a chain such that the distribution of X_n converges as $n \to \infty$. The limiting distribution is Poisson $(\lambda/(1-p))$. Furthermore, if X_0 has this distribution, which is to say $\mu = \lambda/(1-p)$, then so does X_n for all n. ◯

Exercises
(a) Show that a chain with finite state space has at least one closed communicating set of states.

(b) Show that $\mathbf{P}^n$ is a stochastic matrix if $\mathbf{P}$ is.

(c) Show that if $j \leftrightarrow k$ then j and k have the same period.

(d)* On a table there are K plates of soup and a soup tureen. A fly, initially on the tureen, is equally likely to move from any plate or tureen to any of the other K vessels. Find the probability that it is revisiting the tureen after n flights.

(e) If X and Y are independent regular chains, show that $X_n = (X_n, Y_n)$, $n \geq 0$ is a regular chain.

3.3 First-step analysis and hitting times

If some state k of the chain X is accessible from j, that is: $j \to k$, then it is natural to ask for the probability that it in fact does travel from j to k, and if so when? We need some new notation for this.

(1) **Passage times.** The *first-passage time* from j to k is

(2) $$T_{jk} = \min\{n \geq 1 : X_n = k \mid X_0 = j\}.$$

In particular, when $X_0 = k$, then

(3) $$T_k = \min\{n \geq 1 : X_n = k \mid X_0 = k\}$$

is the first recurrence of k, or first return to k. We write

(4) $$ET_{jk} = \mu_{jk} \quad \text{and} \quad ET_k = \mu_k.$$

Commonly, μ_k is called the *mean recurrence time* of k.

In dealing with chains having absorbing states that stop further transitions, it is occasionally useful to work with a slightly different set of quantities.

(5) Hitting times. The *hitting time* of k from j is

(6) $$H_{jk} = \min\{n \geq 0 : X_n = k \mid X_0 = j\}.$$

Note that if $X_0 = k$, the hitting time of k from k is always zero, unlike the first recurrence time. (In particular, if k is absorbing then $T_k = 1$.)

More generally these ideas may be extended to include hitting subsets of the state space. For example, if A is such a subset, the hitting time of A from j is

$$H_{jA} = \min\{n \geq 0 : X_n \in A \mid X_0 = j\}$$

with expected value $K_{jA} = EH_{jA}$. In all these cases, we say that the passage time or hitting time is infinite if there is no n such that $X_n = k$ or $X_n \in A$. These events are important enough for more notation.

(7) Hitting probabilities. The probability that the chain ever hits k starting from j is

$$h_{jk} = P(H_{jk} < \infty).$$

By definition, $h_{kk} = 1$ for all k. Clearly $j \to k$ if and only if $h_{jk} > 0$.

(8) Passage probabilities. The probability that the chain first visits k on the nth step starting from j, $n \geq 1$, is

$$f_{jk}(n) = P(X_1 \neq k, \ldots, X_{n-1} \neq k, X_n = k \mid X_0 = j).$$

The probability that the chain ever returns to k is

$$f_{kk} = \sum_{n=1}^{\infty} f_{kk}(n)$$

and the mean recurrence time is

$$\mu_k = \sum_{n=1}^{\infty} n f_{kk}(n) = ET_k.$$

The distinction between certainty of return and no such certainty is also important enough to merit definitions and jargon.

(9) Definition. If $P(T_j < \infty) = 1$ then j is said to be *recurrent* (or *persistent*). Otherwise, if $P(T_j < \infty) < 1$, then j is said to be *transient*.

A recurrent state j such that $ET_j = \infty$ is said to be *null*. (We repeat the distinction between hitting times and passage times: in particular $EH_{kk} = 0$, in marked contrast to ET_k, which may be ∞.) △

We are already familiar with these concepts in special cases; for example,

(10) **Gambler's ruin.** This is a Markov chain on the integers $\{0, 1, \ldots, B\}$, where 0 and B are absorbing. The probability of ruin is the hitting probability of 0 from k, h_{k0}. By conditioning on the first step we showed (in the fair case $p = q = \frac{1}{2}$) that the hitting probabilities are

$$h_{k0} = 1 - k/B, \quad 0 \le k \le B$$

and the expected hitting times are

$$K_{k0} = k(B - k), \quad 0 \le k \le B.$$

Simple random walk. Again, by conditioning on the first step, we showed that $P(H_{01} < \infty) = p \wedge q/q$. By the independence of steps, $P(H_{0k} < \infty) = (p \wedge q/q)^k$, $k > 0$. This probability is 1 when $p = q$. If $p > q$, then $K_{0k} = EH_{0k} = k/(p-q)$, and if $p \le q$, then $K_{0k} = \infty$; by contrast $\mu_k = \infty$ in any case.

(11) **Waiting for sixes.** Here it is clear by construction, see (3.1.11), that $P(T_0 = r) = (\frac{5}{6})^{r-1}\frac{1}{6}$, and so $\mu_0 = 6$.

We have often used the idea of conditioning on the first step of a process in calculating probabilities and expectations. The same idea supplies a general result about hitting times. Note that a solution $\mathbf{x} = (x_0, x_1, \ldots,)$ of a set of equations is said to be *minimal* if, for any other solution $\mathbf{y} = (y_0, y_1, \ldots,)$, we have $y_r \ge x_r$ for all r.

(12) **Hitting times.**

(a) The probabilities $(h_{jk}; j \in S)$ are the minimal non-negative solution of

(13) $$h_{jk} = \sum_{i \in S} p_{ji} h_{ik}, \quad j \ne k$$

with $h_{kk} = 1$.

(b) The expected hitting times $(K_{jk}; j \in S)$ are the minimal non-negative solution of

(14) $$K_{jk} = 1 + \sum_{i \ne k} p_{ji} K_{ik}, \quad j \ne k$$

with $K_{kk} = 0$.

Proof of (a). It is easy to show that h_{jk} satisfies the equations; we condition on the first step to give

(15) $\quad h_{jk} = P(H_{jk} < \infty) = \sum_{i \in S} P(H_{ik} < \infty) p_{ji} = \sum_{i \in S} p_{ji} h_{ik}.$

Now let $(x_r; r \in S)$ be another solution. We have $x_k = 1$ and, for $j \neq k$, from (13)

$$x_j = p_{jk} + \sum_{i \neq k} p_{ji} x_i$$

$$= P(X_1 = k \mid X_0 = j) + \sum_{i \neq k} p_{ji} \left(p_{ik} + \sum_{\ell \neq k} p_{i\ell} x_\ell \right)$$

$$= P(X_1 = k \mid X_0 = j) + P(X_2 = k, X_1 \neq k \mid X_0 = j) + \sum_{i \neq k} \sum_{\ell \neq k} p_{ji} p_{i\ell} x_\ell.$$

Iterating this process reveals that for any n

$$x_j \geq P(H_{jk} \leq n) \to P(H_{jk} < \infty) = h_{jk}$$

as $n \to \infty$. Thus h_{jk} is the minimal solution.

Proof of (b). Condition on the first step to show that K_{jk} satisfies the equations. If $(x_r; r \in S)$ is another solution, then for $j \neq k$,

$$x_j = 1 + \sum_{i \neq k} p_{ji} x_i = P(H_{jk} \geq 1) + \sum_{i \neq k} p_{ji} + \sum_{i \neq k} \sum_{\ell \neq k} p_{ji} p_{i\ell} x_\ell.$$

Iterating shows that

(16)
$$x_j \geq P(H_{jk} \geq 1) + P(H_{jk} \geq 2) + \cdots + P(H_{jk} \geq n) \to EH_{jk} = K_{jk},$$

as $n \to \infty$. □

(17) Example. Non-negative simple random walk (Gambler's ruin with an infinitely rich casino). The transition probabilities are $p_{00} = 1$; $p_{j,j+1} = p$, $p_{j,j-1} = q = 1 - p$. So the probability of ever reaching 0 from $i > 0$, denoted by r_i, is the minimal non-negative solution of

(18) $\quad r_i = pr_{i+1} + qr_{i-1}, \quad i \geq 1, \text{ with } r_0 = 1.$

If $p \neq q$, the general solution of (18) is

(19) $\quad r_i = \alpha + \beta \left(\dfrac{q}{p} \right)^i.$

If $p > q$, then $r_0 = 1$ entails

$$r_i = \left(\frac{q}{p}\right)^i + \alpha\left(1 - \left(\frac{q}{p}\right)^i\right), \quad \text{for any } \alpha \geq 0,$$

$$\geq \left(\frac{q}{p}\right)^i, \quad \text{for all } i$$

and this is the required minimal solution. If $p \leq q$, the only bounded solution with $r_0 = 1$ is $r_i = 1$ for all i.

Likewise the expected hitting time $\kappa_i = K_{i0}$ satisfies

$$\kappa_0 = 0,$$
$$\kappa_i = 1 + q\kappa_{i-1} + p\kappa_{i+1}, \quad i \geq 1.$$

Hence, when $q \neq p$, for some constants α and β,

$$\kappa_i = \frac{i}{q-p} + \alpha + \beta\left(\frac{q}{p}\right)^i.$$

If $p < q$, then the minimal solution is

(20) $$\kappa_i = \frac{i}{q-p}.$$

If $p > q$, then the minimal non-negative solution is $\kappa_i = \infty$ for $i > 0$. Of course, this corresponds to the result of the first part, as there is a nonzero chance of never hitting 0 from i.

When $p = q$, $\kappa_i = -i^2 + \alpha + \beta i$, and in this case also the minimal non-negative solution is $\kappa_i = \infty$, $i > 0$. ○

Just as matrix methods can, in suitable cases, be used to find n-step transition probabilities, so too can they be used in connection with hitting times.

(21) Fundamental matrix. Suppose the state space S is the union of two sets A and T such that $p_{at} = 0$ for every $a \in A$ and $t \in T$, and for every state $t \in T$ there is at least one state a in A such that $t \to a$. Then for any such a and t

(22) $$h_{ta} = p_{ta} + \sum_{k \in T} p_{tk} h_{ka}.$$

If we label the states in T from 1 to $m = |T|$, and those in A from $m+1$ to $n = |S|$, then the transition matrix takes the partitioned form

$$\mathbf{P} = \begin{pmatrix} \mathbf{Q} & \mathbf{R} \\ \mathbf{0} & \mathbf{S} \end{pmatrix}.$$

Equations (22) may then be written in matrix form as

$$\mathbf{H} = \mathbf{R} + \mathbf{Q}\mathbf{H}.$$

Formally, we can solve this using the matrix $\mathbf{I}$, with ones on the diagonal and zeros elsewhere, as

$$\mathbf{H} = (\mathbf{I} - \mathbf{Q})^{-1}\mathbf{R},$$

provided that the inverse matrix $(\mathbf{I} - \mathbf{Q})^{-1}$ exists. In fact this does follow from our assumptions above; we omit the proof. The matrix $(\mathbf{I} - \mathbf{Q})^{-1}$ is called the *fundamental matrix* and may be written

$$(\mathbf{I} - \mathbf{Q})^{-1} = \lim_{n\to\infty} \sum_{r=0}^{n} \mathbf{Q}^r.$$

This approach can be developed much further, but not here.

Exercises
(a) A particle performs a symmetric random walk starting at O on the vertices {A,B,C,D,O} of this graph (Figure 3.3):
  (i) Show that the probability of visiting D before C is $\frac{2}{7}$.
  (ii) Show that the expected number of steps until hitting either of C or D is $\frac{15}{7}$.

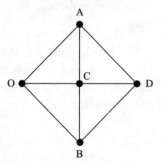

Figure 3.3 Diagram for Exercise 3.3(a).

(b)* Show that if X is a finite irreducible chain then $h_{jk} = 1$ and $K_{jk} < \infty$ for all j and k.

(c) Show that if X is a finite irreducible and aperiodic chain, then it is regular.

(d) In the set-up and notation of (21), let v_{jt} be the expected number of visits that the Markov chain X makes to the state t before it enters A, starting at $X_0 = j$. Show that the matrix $\mathbf{V} = (v_{jt})$ satisfies

$$\mathbf{V} = \mathbf{I} + \mathbf{QV},$$

where $\mathbf{I}$ is a matrix with ones on the diagonal and zeros elsewhere.

3.4 The Markov property revisited

In looking at simple random walks we often used the following type of argument: let T_{jk} be the first-passage time to k, starting from j. Then

(1)
$$\begin{aligned}P(T_{02} < \infty) &= P(T_{02} < \infty \mid T_{01} < \infty)P(T_{01} < \infty)\\ &= P(T_{12} < \infty)P(T_{01} < \infty),\end{aligned}$$

by the independence of steps. We would like to use the same idea in looking at Markov chains; for example, consider the probability that X revisits its starting point s at least twice. Intuitively, we expect this to be $\{P(T_s < \infty)\}^2$, but this requires the Markov property to hold at the random time T_s. In fact it does, as we now prove, following this definition.

(2) **Stopping times.** A non-negative integer-valued random variable T is called a *stopping time* with respect to $(X_r; r \geq 0)$ if, for all n, the event $\{T = n\}$ may depend only on $\{X_0, \ldots, X_n\}$, and does not depend on $(X_{n+m}; m > 0)$. To put this another way, the indicator $I(T = n)$, of the event that $T = n$, is a function only of $X_0, \ldots, X_n$. △

This includes any non-random time m, because the event $\{m = n\}$ does not depend on any X. Other most important examples of stopping times are:

Recurrence times, because

$$\{T_s = n\} = \{X_0 = s, X_1 \neq s, \ldots, X_{n-1} \neq s, X_n = s\}.$$

Hitting times, because

$$\{H_{jk} = n\} = \{X_0 \neq k, \ldots, X_{n-1} \neq k, X_n = k\}$$

and first-passage times, likewise.

It is easy to see, by contrast, that $T_s - 1$ is not a stopping time since the event $\{T_s - 1 = n\}$ depends on $\{X_{n+1} = s\}$. Stopping times are important because of this:

(3) **Strong Markov property.** Let T be a stopping time with respect to the chain $X = (X_r; r \geq 0)$. Then, conditional on $T < \infty$ and $X_T = s$, the sequence $(X_{T+n}; n \geq 0)$ is a Markov chain that behaves like X started at s, independently of $(X_0, \ldots, X_T)$.

Proof. Let A be any event depending only on $(X_0, \ldots, X_T)$, and let D be the event that $X_{T+1} = x_1, \ldots, X_{T+r} = x_r$; you may find it helpful to think of D as the development of the chain after T. Then on the event $\{T = t\}$, we can use the ordinary Markov property at the fixed time t, so that

(4) $P(D \cap A \mid T < \infty, X_T = s)$

$$= \sum_t P(D \cap A \cap \{T = t\} \cap \{X_t = s\})/P(T < \infty, X_T = s)$$

$$= P(X_r = x_r, \ldots, X_1 = x_1 \mid X_0 = s)$$

$$\times \sum_t \frac{P(A \cap \{T = t\} \cap \{X_t = s\})}{P(T < \infty, X_T = s)}$$

$$= P(X_r = x_r, \ldots, X_1 = x_1 \mid X_0 = s) P(A \mid T < \infty, X_T = s).$$

Thus, conditional on $\{T < \infty, X_T = s\}$, this shows A is independent of D, which has the same distributions as X started at s. □

Here is an example.

(5) **The moves chain.** Let X be a Markov chain with no absorbing states, and define $(Y_n; n \geq 0)$ to be the random process obtained by first recording X_0 and subsequently only those values of X_n such that $X_n \neq X_{n-1}$. Then by construction $Y_n = X_{M_n}$, where $M_0 = 0$, and, given $X_{M_n} = i$, $M_{n+1} - M_n$ is geometric with parameter $1 - p_{ii}$, for each $n \geq 0$.

The key point is that each M_n is a stopping time for X, so by the Markov property there

$$P(Y_{n+1} = k \mid Y_0, \ldots, Y_n = j) = P(X_{M_{n+1}} = k \mid X_0, X_{M_1}, \ldots, X_{M_n} = j)$$
$$= P(X_{M_{n+1}} = k \mid X_{M_n} = j)$$
$$= P(X_{M_1} = k \mid X_0 = j)$$
$$= \begin{cases} 0, & j = k, \\ p_{jk}/(1 - p_{jj}), & j \neq k. \end{cases}$$

Hence Y is a Markov chain with these transition probabilities. ○

(6) **First passages and recurrence times.** Recall that $f_{jk}(n) = P(X_1 \neq k, \ldots, X_{n-1} \neq k, X_n = k \mid X_0 = j)$ is the distribution of the first-passage time T_{jk}, from j to k, of the Markov chain X.

Let us consider the distribution of X_m, $P(X_m = k)$, conditional on the event $T_{jk} = r$, $1 \leq r \leq m$. We have, using the strong Markov property at T_{jk},

$$p_{jk}^{(m)} = P(X_m = k \mid X_0 = j)$$

(7)
$$= \sum_{r=1}^{m} P(X_m = k, T_{jk} = r \mid X_0 = j)$$

$$= \sum_{r=1}^{m} P(X_m = k \mid T_{jk} = r, X_0 = j) P(T_{jk} = r \mid X_0 = j)$$

$$= \sum_{r=1}^{m} P(X_m = k \mid X_r = k) P(T_{jk} = r),$$

by the strong Markov property,

$$= \sum_{r=1}^{m} p_{kk}^{(m-r)} f_{jk}(r), \quad m \geq 1$$

with the convention that

$$p_{jk}^{(0)} = \begin{cases} 1, & j = k, \\ 0, & j \neq k. \end{cases}$$

Denoting $E s^{T_{jk}}$ by $F_{jk}(s)$, and defining

$$P_{jk}(s) = \sum_{n=0}^{\infty} s^n p_{jk}^{(n)},$$

we obtain from the convolution in (7) that

(8) $$P_{jk}(s) = F_{jk}(s)P_{kk}(s), \quad j \neq k$$

and

(9) $$P_{jj}(s) = 1 + F_{jj}(s)P_{jj}(s).$$

Since $F_{jj}(1) = P(T_j < \infty)$, this has some interesting implications.

(10) **Recurrence and transience.** From (9) we have that

(11) $$P_{jj}(s) = \frac{1}{1 - F_{jj}(s)}, \quad \text{for } |s| < 1.$$

We recall from (3.3.9) that j is recurrent or transient accordingly as $F_{jj}(1) = 1$ or $F_{jj}(1) < 1$. It follows from (11) that j is recurrent or transient accordingly as $\sum_n p_{jj}^{(n)} = \infty$ or $\sum_n p_{jj}^{(n)} < \infty$. In this latter case we have also that $\lim_{n \to \infty} p_{jj}^{(n)} = 0$. Now using (8), we find that

(a) if k is recurrent then $\sum_n p_{jk}^{(n)} = \infty$, for all j such that $f_{jk} > 0$;

(b) if k is transient then $\sum_n p_{jk}^{(n)} < \infty$, for all j.

Of course, recalling (2.2.14), we see that these statements can be reinterpreted in terms of the expected number of visits made by the chain, thus:

(a)′ if k is recurrent then the chain expects to visit k infinitely often from j, if $f_{jk} > 0$;

(b)′ if k is transient, then the chain expects to visit k finitely often.

In fact a more extensive use of the strong Markov property can tell us even more about the visits of a Markov chain, because it enables us to use arguments similar to those used in analysing random walks; see (2.2.16) and subsequent paragraphs.

The key idea is this: let $T_{jk}^{(r)}$ be the time of the rth visit to k of the chain started in j. Thus $T_{jk}^{(1)}$ is simply the same as the first-passage time T_{jk}; then

(12) $$T_{jk}^{(2)} = \min\{n > T_{jk}^{(1)} : X_n = k\}$$

and so on. By the strong Markov property, $T_{jk}^{(2)} - T_{jk}^{(1)}$ is independent of $T_{jk}^{(1)}$, and has the same distribution as $T_k^{(1)}$. An obvious induction shows the equivalent statement to be true for all intervals between visits to k. This fact has a most important and useful interpretation that will be very useful later on; here it is:

(13) **Renewal and Markov chains.** The times of successive visits of a Markov chain to any state k form a renewal process. If $X_0 = k$, then it is an ordinary renewal process, otherwise it is a delayed renewal process. $\square$

For the moment, let us conclude with a brief look at another interesting quantity.

Visits. Let the total number of visits to the state k, by a chain started in j, be N_j^k, and denote its expected value by

$$
\begin{aligned}
v_j^k &= E N_j^k \\
&= \sum_{n=1}^{\infty} p_{jk}^{(n)},
\end{aligned}
\tag{14}
$$

where the last line follows easily by writing N_j^k as the sum of indicators of a visit to k at time n. Note that if $j = k$ we do not count its initial state j as a visit to j. We count only revisits in this case.

By definition then, we have for any j and k

$$
P(N_j^k > 0) = P(T_{jk} < \infty) = f_{jk}.
\tag{15}
$$

Next, we use the Markov property at $T_{jk}^{(r)}$ to find the distribution of N_j^k

$$
\begin{aligned}
P(N_j^k > r+1) &= P(T_{jk}^{(r+1)} < \infty) \\
&= P(T_{jk}^{(r+1)} - T_{jk}^{(r)} < \infty \mid T_{jk}^{(r)} < \infty) P(T_{jk}^{(r)} < \infty) \\
&= f_{kk} P(N_j^k > r), \quad \text{by (15)} \\
&= f_{kk}^r P(N_j^k > 0), \quad \text{by iterating} \\
&= f_{jk} f_{kk}^r, \quad \text{by (15) again.}
\end{aligned}
$$

Thus the distribution of N_j^k is geometric with parameter $1 - f_{kk}$, and mean $(1 - f_{kk})^{-1}$. Naturally this agrees with what we have already ascertained;

there are two cases to consider:

On the one hand, if $f_{kk} = 1$ then the state k is persistent and recurs infinitely often with probability 1. On the other hand, if $f_{kk} < 1$ then k is transient and recurs finitely often.

If $j \neq k$, then N_j^k has a distribution that is a mixture: it takes the value zero with probability $1 - f_{jk}$, or (with probability f_{jk}) it has the geometric distribution above.

> **Exercises**
> (a) Let X be a Markov chain, and define T_A to be the time that it first enters some set of states A, and U_A the last time at which it is ever in A. Show that T_A is a stopping time for X, but U_A is not.
>
> (b)* Let X be a Markov chain, and define Y to be the random process obtained by recording X only when it lies in some given subset A of its state space. Thus, for example, $Y_0 = X_{H_A}$, where H_A is the first hitting time of A by X.
> Show that Y is a Markov chain, and determine its transition probabilities in terms of the transition probabilities p_{jk} of X.
>
> (c)* Show that for any two states j and k
> $$\mathrm{E}T_{jk} + \mathrm{E}T_{kj} \geq \mathrm{E}T_j.$$
>
> (d) Show that for a chain with finite state space, if $j \to k$ then $p_{jk}^{(n)} > 0$ for some $n \leq |S|$.

3.5 Classes and decomposition

We have seen that any state of a Markov chain may be either transient or recurrent; and recurrent states may be either null or non-null. Furthermore, states may be periodic or aperiodic. This proliferation of types may seem a little complicated. However, recall from (3.2) that the state space of the chain can be partitioned into classes of intercommunicating states (by the equivalence relation $\leftrightarrow$).

It is gratifying that all states in such a class are of exactly the same type. Formally,

(1) **Theorem.** Suppose that the states j and k intercommunicate, $j \leftrightarrow k$. Then either both states are recurrent, or both are persistent. In either case they both have the same period.

Proof. Since $j \leftrightarrow k$, there exist finite m and n such that $p_{jk}^{(m)} > 0$, and $p_{kj}^{(n)} > 0$, so that for any integer r

(2) $$p_{jj}^{(m+r+n)} \geq p_{jk}^{(m)} p_{kk}^{(r)} p_{kj}^{(n)}.$$

Now summing over r, we see that $\sum_r p_{kk}^{(r)} < \infty$ if $\sum_r p_{jj}^{(r)} < \infty$. By (3.4.10), that is to say that k is transient if j is; the converse is similar. The statement asserting that periodicity is a class property is proved using the same inequality. For m, n such that $p_{jk}^{(m)} p_{kj}^{(n)} > 0$, the period of j must divide $m + n$, and hence also divides the period of k. The converse is similar. $\square$

In fact it is also true that in the recurrent case, j is null if and only if k is, but we defer the proof to a later section.

The most pleasant consequence of this theorem is that we need only find the type of one state in an irreducible chain; the rest are the same. The next result follows easily from the above theorem:

(3) **Decomposition.** The state space S of any chain can be written uniquely as a union of disjoint classes

$$S = T \cup C \cup C_2 \cup \cdots$$

where T is a set of transient states and $(C_n; n \geq 1)$ are closed irreducible sets of recurrent states.

Proof. Let $(C_n; n \geq 1)$ be the recurrent equivalence classes of intercommunicating states. We need to establish that they are closed. Suppose there exists j, and k, and a C_r, such that $j \in C_r$, $k \notin C_r$, and $p_{jk} > 0$. Then it is easy to show (Exercise (a)) that j is transient, which is a contradiction. $\square$

The point of this theorem is that if a chain starts in some $(C_n; n \geq 1)$ then it never leaves; you can take C_n to be the state space. If it starts in T, then either it stays there for ever, or it enters some C_r and never leaves.

Once the chain is in a recurrent class, the following result is useful.

(4) **Theorem.** In a recurrent irreducible chain $f_{jk} = f_{kj} = 1$ for all j and k.

Proof. We use the fact that a recurrent state k is revisited infinitely often with probability 1. Then for any i there is m such that $p_{ki}^{(m)} > 0$, and we can write

$$1 = P(X_n = k, \text{ for some } n > m \mid X_0 = k)$$

(5)
$$= \sum_j P(X_n = k, \text{ for some } n > m \mid X_m = j, X_0 = k)$$
$$\times P(X_m = j \mid X_0 = k)$$
$$= \sum_j f_{jk} p_{kj}^{(m)}, \quad \text{by the Markov property at } m.$$

Because $\sum_j p_{kj}^{(m)} = 1$, this implies that

$$\sum_j (1 - f_{jk}) p_{kj}^{(m)} = 0$$

and in turn this gives $f_{ik} = 1$, because $p_{ki}^{(m)} > 0$. Then we interchange the roles of i and k to show that $f_{ki} = 1$. □

For finite chains, things are even simpler; the option of transience is impossible.

(6) **Theorem.** If S is a finite state space, then at least one state is recurrent, and all the recurrent states are non-null.

Proof. For a transient state k, $p_{jk}^{(n)} \to 0$ as $n \to \infty$. If the states are all transient,

$$1 = \lim_{n \to \infty} \sum_{k \in S} p_{jk}^{(n)} = 0,$$

which is a contradiction. For the second part, we recall the renewal–reward theorems. If k is persistent null, then returns to k form a renewal process, and by the renewal–reward theorem the proportion of time spent in k is $(ET_k)^{-1} = 0$, since k is null. The same holds for any finite set of null states. But if C is the closed set of persistent null states, then the long-run proportion of time spent in C is 1, trivially. This contradiction proves that C is empty. □

Exercises

(a)* Show that if $j \to k$ but $k \not\to j$, then j is transient.

(b)* Show that if $j \to k$, and j is recurrent, then k is recurrent.

(c)* If the chain is irreducible, find the distribution of the number of visits to j before returning to k. Deduce that the expected number of visits to j before returning to k is finite and not zero.

3.6 Stationary distribution: the long run

Previous sections have supplied a number of methods for investigating the behaviour of Markov chains; for example: first step analysis, and use of the strong Markov property. This section introduces a fresh idea and technique.

We have seen many chains in which the probabilities converged as $n \to \infty$. These limits have some important and interesting properties. In the interests of simplicity, we display them first for chains on a finite state space.

(1) Equilibrium. Suppose that the transition probabilities of the finite chain X converge to a limit independent of X_0; that is to say for all j and k

$$\lim_{n \to \infty} p_{jk}^{(n)} = \pi_k.$$

Then $\pi = (\pi_1, \ldots, \pi_{|S|})$ has these key properties:

(a) π is a probability distribution, that is, $\sum_k \pi_k = 1$.
(b) The vector π is a solution of the equations

(2) $$\pi_k = \sum_j \pi_j p_{jk}.$$

Note that we may sometimes write this more compactly as $\pi = \pi P$, where P is the matrix of transition probabilities.

(c) If X_0 has the distribution π then, for all n, X_n also has distribution π.

Property (c) explains why π is called the *stationary distribution* of the chain, which is said to be in *equilibrium* in this case.

To see (a), we recall that P is a finite stochastic matrix so for all n

$$1 = \sum_k p_{jk}^{(n)}.$$

Since S is finite, taking the limit as $n \to \infty$ yields the result.

For (b), we use the recurrence

$$p_{ik}^{(n+1)} = \sum_j p_{ij}^{(n)} p_{jk}.$$

Let $n \to \infty$ to obtain the result.

Finally, for (c), we have

$$P(X_1 = k) = \sum_j P(X_0 = j) p_{jk} = \sum_j \pi_j p_{jk}$$

(3) $$= \pi_k.$$

Iteration, or a simple induction, gives what we want. □

Equally interesting and important is the following link to visits and mean recurrence times.

Pick any state k of the finite chain X, and let v_j^k be the expected number of visits to j between visits to k; note that $v_k^k = 1$. Denote the mean recurrence time of k by μ_k. Then we have

(4) **Representation of π**. If the transition probabilities of X converge to π, then $\pi_k = 1/\mu_k$, and

(5) $$\pi_j = v_j^k/\mu_k.$$

To see this, recall that visits of the chain to k constitute a renewal process (by the strong Markov property). Now use the renewal–reward theorem, (where a visit to j is a unit reward), to see that in the long run, the expected proportion of time the chain spends in j satisfies

(6) $$\pi_j = \lim_{n \to \infty} \frac{1}{n} \sum_{t=1}^{n} p_{kj}^{(t)} = \begin{cases} v_j^k/\mu_k, & j \neq k, \\ 1/\mu_k, & j = k. \end{cases}$$

Obviously $\sum_j v_j^k/\mu_k = 1$, as must be the case. □

All this is interesting and suggestive, but of course in practice we are more interested in the converse questions: given a chain X, is there a stationary distribution π, and does the distribution of X converge to it?

For finite aperiodic irreducible chains the answer is yes, and this follows from results in elementary linear algebra. First observe that because $\mathbf{P}$ is a stochastic matrix we have $\sum_k p_{jk} = 1$. This tells us that $\mathbf{P}$ has 1 as an eigenvalue, and hence there is a left eigenvector π such that $\pi = \pi \mathbf{P}$, and $\sum_k \pi_k = 1$. Routine methods of algebra now show that $\pi_k > 0$, and $\lim_{n \to \infty} p_{jk}(n) = \pi_k$. We omit this.

Unfortunately, these elementary methods do not readily handle chains with infinite state space. Such chains can behave very differently from finite chains; for example,

(7) **Simple random walk.** We have shown above that in this case $\lim_{n\to\infty} p_{jk}^{(n)} \to 0$ for all j and k. And even in the case $p = q = \tfrac{1}{2}$, when $P(T_k < \infty) = 1$, we have nevertheless that $\mu_k = \infty$.

Also it is easy to see that there is no stationary distribution, for such a distribution must have (by symmetry) the property that $\pi_j = \pi_k$ for all j, k, which is impossible when $\sum_k \pi_k = 1$. ○

Clearly we need to move more carefully in looking at infinite chains. Before we begin, it is convenient to insert this definition.

(8) **Invariant measure.** If $\mathbf{x} = (x_1, x_2, \ldots)$ is a non-negative vector such that $\mathbf{x} = \mathbf{x}\mathbf{P}$, then it is called an *invariant measure* for the stochastic matrix $\mathbf{P}$. If $\sum_k x_k = 1$, then $\mathbf{x}$ is a stationary distribution. △

(9) **Simple random walk revisited.** Here, even though there can be no stationary distribution, there is an invariant measure, viz: the vector of 1s, $\mathbf{x} = (\ldots 1, 1, 1, \ldots)$, as you can easily verify. With this extra idea, it turns out that we can carry out the program sketched above.

First, let us extend (1) to the infinite state space.

(10) **Equilibrium revisited.** Suppose that the transition probabilities of X converge to a limit independent of j,

$$\lim_{n\to\infty} p_{jk} = \pi_k,$$

where $\sum_k \pi_k = 1$. Then $\boldsymbol{\pi}$ is a solution of

(11) $$\pi_k = \sum_j \pi_j p_{jk}$$

and $\boldsymbol{\pi}$ is a stationary distribution.

To see (11), let F be any finite subset of the state space S. Then for any i and k

$$p_{ik}^{(n+1)} = \sum_{j \in S} p_{ij}^{(n)} p_{jk}$$

(12) $$\geq \sum_{j \in F} p_{ij}^{(n)} p_{jk}.$$

Since F is finite we can let $n \to \infty$ in this to find

(13) $$\pi_k \geq \sum_{j \in F} \pi_j p_{jk}.$$

Since F was arbitrary we therefore have that

(14) $$\pi_k \geq \sum_{j \in S} \pi_j p_{jk}.$$

But if the inequality were strict for any k, it would follow that

$$1 = \sum_k \pi_k > \sum_k \sum_j \pi_j p_{jk} = \sum_j \pi_j = 1,$$

which is a contradiction. Thus equality holds in (14), establishing (11). The second part now follows trivially, as in the finite case. Furthermore, the representation given in (4) is easily seen to be still valid. □

The converse question, the existence of π, is slightly trickier because of the examples above. However, it turns out that a more general form of the representation given in (4) will do it for us. Because of the decomposition theorem, it is natural for us to confine our attention to irreducible chains; this is the key result:

(15) Existence of invariant measure. Let X be a recurrent and irreducible chain with transition matrix **P**, and denote the expected number of visits to j before returning to k by v_j^k, where $v_k^k = 1$. Then

(a) $0 < v_j^k < \infty$;
(b) $v_j^k = \sum_i v_i^k p_{ij}$,

which is to say that $\mathbf{v}^k = (v_1^k, v_2^k, \ldots)$ is an invariant measure.

Part (a) was Exercise (3.4(c)). To see (b) we denote the indicator of a visit to j by $I(X_n = j)$, recall that T_k is the time of first return to k, and write

$$v_j^k = E\left\{\sum_{n=1}^{T_k} I(X_n = j) \mid X_0 = k\right\}$$

(16) $$= \sum_{n=1}^{\infty} P(X_n = j, T_k \geq n \mid X_0 = k)$$

$$= \sum_{i \in S} \sum_{n=1}^{\infty} P(X_{n-1} = i, X_n = j, T_k \geq n \mid X_0 = k)$$

$$= \sum_{i \in S} p_{ij} \sum_{n=1}^{\infty} P(X_{n-1} = i, T_k \geq n \mid X_0 = k),$$

using the Markov property,

$$= \sum_{i \in S} p_{ij} E\left(\sum_{r=0}^{T_k - 1} I(X_r = i) \mid X_0 = k\right)$$

$$= \sum_{i \in S} p_{ij} v_i^k. \qquad \square$$

(17) **Corollary.** v^k is in fact the minimal invariant measure. That is, if $\mathbf{x}$ is any other invariant measure with $x_k = 1$, then $x_j \geq v_j^k$, for all j.

The proof follows the same lines as those in (3.3.12). That is, we write

$$x_j = p_{kj} + \sum_{i \neq k} x_i p_{ij}$$

(18)
$$= P(X_1 = j, T_k \geq 1) + \sum_{i \neq k} p_{ki} p_{ij} + \sum_{\substack{i \neq k \\ r \neq k}} x_r p_{ri} p_{ij}$$

$$\geq P(X_1 = j, T_k \geq 1) + P(X_2 = j, T_k \geq 2)$$
$$+ \cdots + P(X_n = j, T_n \geq n)$$

on iterating n times, and noting that the remainder is always non-negative. Now let $n \to \infty$ to find that $x_j \geq v_j^k$. $\qquad \square$

It is most important to note that if k is a non-null recurrent state, so that $\mu_k < \infty$, then the vector

(19)
$$\mathbf{v}^k / \mu_k = (v_1^k / \mu_k, v_2^k / \mu_k, \ldots, v_{|S|}^k / \mu_k)$$

is a stationary distribution. This observation enables us to prove the remark following Theorem (3.4.1) about the class properties of irreducible chains, and add a useful criterion for the nature of a chain.

(20) **Theorem.** Let X be irreducible with transition matrix $\mathbf{P}$.

(a) If any state is non-null recurrent, then all states are.
(b) The chain is non-null recurrent if and only if there is a stationary distribution π. In this case, for every k, $\pi_k = 1/ET_k$.

Proof. Suppose that k is non-null recurrent. By Theorem (3.4.1) it follows that the chain is recurrent. Hence there is an invariant measure, by (15), and therefore the vector in (19) is a stationary distribution in which $\pi_k = 1/ET_k$.

Conversely, suppose there is a stationary distribution $\boldsymbol{\pi}$. For any state k we have for all n

$$\pi_k = \sum_j \pi_j p_{jk}^{(n)}, \tag{21}$$

which must be >0 for some n, since the chain is irreducible. So $\pi_k > 0$ for all k. Now, again for any k, the vector

$$\mathbf{x} = \left(\frac{\pi_1}{\pi_k}, \ldots, \frac{\pi_k}{\pi_k}, \ldots \right) \tag{22}$$

is an invariant measure with $x_k = 1$. By the result (17) proved above, it follows that $x_j \geq v_j^k$ for all j. Hence

$$ET_k = \sum_{j \in S} v_j^k \leq \sum_j x_j = \sum_j \frac{\pi_j}{\pi_k} = \frac{1}{\pi_k} < \infty. \tag{23}$$

Therefore k is non-null recurrent, for all $k \in S$.

Finally, because the chain is recurrent, $\mathbf{v}^k$ is also an invariant measure and it is minimal (by Corollary (17)). Hence $\mathbf{x} - \mathbf{v}^k$ is invariant, with

$$\pi_j/\pi_k - v_j^k \geq 0, \quad j \in S.$$

Now, for any j there exists n such that $p_{jk}^{(n)} > 0$. Therefore, we can write

$$0 = 1 - v_k^k = \sum_i (\pi_i/\pi_k - v_i^k) p_{ik}^{(n)}$$

$$\geq (\pi_j/\pi_k - v_j^k) p_{jk}^{(n)}. \tag{24}$$

Hence $v_j^k = \pi_j/\pi_k$, and using (23) gives

$$ET_k = \frac{1}{\pi_k}, \quad k \in S. \tag{25}$$

Note that because of this the stationary distribution must be unique in this case, and null-recurrence is indeed a class property for irreducible chains as we claimed above. You can show by essentially the same argument that if the chain is null-recurrent, then any invariant measure $\mathbf{x}$ is unique. $\square$

The main practical application of this theorem is that if we find a stationary distribution π for an irreducible chain, we know the chain is non-null recurrent, and π also supplies all the mean recurrence times. Furthermore, π_i gives the proportion of time that the chain spends visiting the state i, in the long run.

The final step in our program is to establish when the distribution of the chain converges to the stationary distribution that we have shown exists when the chain is recurrent and non-null. A brief consideration of the flip–flop chain with $p_{01} = 1 = p_{10}$ shows that we cannot expect the distribution of periodic chains to converge. The key result is therefore this:

(26) **Convergence theorem.** Let X be aperiodic and irreducible with transition matrix $\mathbf{P}$, having a stationary distribution π. Then for all j and k, as $n \to \infty$,

(27) $$p_{jk}^{(n)} \to \pi_k.$$

Proof. We use a simple form of coupling (discussed above in connexion with Poisson approximation). Let Y be a Markov chain independent of X with the same transition matrix $\mathbf{P}$, and define $Z = (X, Y)$ by $Z_n = (X_n, Y_n)$, $n \geq 0$. Then Z is a Markov chain with transition probabilities

$$P(Z_{n+1} = (k, \beta) \mid Z_n = (j, \alpha)) = p_{jk} p_{\alpha\beta}$$

and stationary distribution

(28) $$\pi(j, k) = \pi_j \pi_k.$$

Therefore, Z is non-null recurrent. Because X and Y are aperiodic there exists an n_0 such that $p_{jk}^{(n)} p_{\alpha\beta}^{(n)} > 0$ for all $n > n_0$. Hence Z is also irreducible.

Now pick a state s, and define the first-passage time of Z to (s, s)

$$T_s = \min\{n \geq 1 : Z_n = (s, s)\}.$$

Define the meeting time of X and Y to be

$$T = \min\{n \geq 1 : X_n = Y_n\}.$$

We have $P(T < \infty) = 1$, because Z is recurrent, and $T \leq T_s$.

The core of the proof is the fact that on the event $\{T \leq n\}$, X_n and Y_n have the same distribution. This follows immediately from an appeal to the strong Markov property; but we may also prove it directly.

$$P(X_n = x, T \leq n)$$

(29)
$$= \sum_j \sum_{r=1}^n P(X_n = x, X_r = j, T = r)$$

$$= \sum_j \sum_r P(X_n = x \mid X_r = j) P(T = r, X_r = j),$$

by the Markov property at r

$$= \sum_j \sum_r P(Y_n = x \mid Y_r = j) P(T = r, Y_r = j)$$

$$= P(Y_n = x, T \leq n).$$

Now, using this,

$$P(X_n = k) = P(X_n = k, T \leq n) + P(X_n = k, T > n)$$
$$= P(Y_n = k, T \leq n) + P(X_n = k, T > n)$$
$$\leq P(Y_n = k) + P(T > n).$$

Likewise
$$P(Y_n = k) \leq P(X_n = k) + P(T > n).$$

Hence,

$$|P(X_n = k) - P(Y_n = k)| \leq P(T > n)$$

(30)
$$\to 0, \quad \text{as } n \to \infty.$$

If we let $X_0 = j$, and give Y_0 the stationary distribution π, then we have shown

$$p_{jk}(n) \to \pi_k, \quad \text{as } n \to \infty. \qquad \square$$

(31) **Corollary.** For any irreducible aperiodic chain,

$$p_{jk}(n) \to \frac{1}{\mu_k}, \quad \text{as } n \to \infty.$$

If the chain is transient $\mu_k = \infty$ and $p_{jk}(n) \to 0$. If the chain is non-null recurrent, then it has a stationary distribution π, where $\pi_k = 1/\mu_k$, and $p_{jk}(n) \to \pi_k$.

It only remains to consider the case when the chain is null-recurrent, and prove that $p_{jk}(n) \to 0$. This is a little trickier, and we omit it. However, we can easily establish an interesting but weaker result as a corollary of the following:

(32) **Ergodic theorem.** Let X be an irreducible Markov chain, and define $N_j^k(n)$ to be the number of visits to k, starting at j, before time n. Then, as $n \to \infty$, with probability 1,

(33) $$\frac{1}{n} N_j^k(n) \to \frac{1}{\mu_k}.$$

Proof. If the chain is transient, the result is trivial. Suppose it is not transient. Since $P(T_{jk} < \infty) = 1$, and $N_j^k(n + T_{jk}) = 1 + N_k^k(n)$, it is sufficient to consider $(1/n)N_k^k(n)$. Let $Y(r + 1)$ be the time between the rth and the $(r + 1)$th visit to k. By the strong Markov property these are independent and identically distributed, and $EY(r) = \mu_k$.

We use the same line of argument as we did for renewal and renewal–reward theorems; here $N_k^k(n)$ plays the same role as the renewal function $N(t)$ did there. We have

(34) $$\frac{Y(1) + \cdots + Y(N_k^k(n) - 1)}{N_k^k(n)} \leq \frac{n}{N_k^k(n)} \leq \frac{Y(1) + \cdots + Y(N_k^k(n))}{N_k^k(n)}.$$

Since k is recurrent $N_k^k(n) \to \infty$, and by the strong law of large numbers, each bound in (34) converges to μ_k. Hence,

$$\frac{N_k^k(n)}{n} \xrightarrow{\text{a.s.}} \frac{1}{\mu_k},$$

where the right side is zero if $\mu_k = \infty$. Now $0 \leq N_k^k(n)/n \leq 1$, so by dominated convergence we therefore have that

(35) $$\frac{1}{n} EN_j^k(n) = \frac{1}{n} \sum_{r=1}^{n} p_{jk}^{(r)} \to \frac{1}{\mu_k},$$

where the right side is zero if the chain is null-recurrent. □

We can also derive an ergodic reward theorem in the non-null case. Let $g(k)$ be the gain arising from a single visit to the state k. The total reward gained by the chain up to time n is then $R_n = \sum_{r=0}^{n} g(X_r)$.

(36) **Ergodic reward theorem.** Let $g(\cdot)$ be bounded; then

$$\frac{1}{n} R_n \xrightarrow{a.s.} \sum_{k \in S} \pi_k g(k),$$

where π_k is the stationary distribution.

Proof. Without loss of generality we can assume $|g| \leq 1$. Let B be any finite subset of the state space S. Then

$$\left| \frac{1}{n} R_n - \sum_k \pi_k g(k) \right| = \left| \sum_{k \in S} \left(\frac{1}{n} N_j^k(n) - \pi_k \right) g(k) \right|$$

$$\leq \sum_{k \in B} \left| \frac{1}{n} N_j^k(n) - \pi_k \right| + \sum_{k \notin B} \left(\frac{1}{n} N_j^k(n) + \pi_k \right)$$

$$\leq 2 \sum_{k \in B} \left| \frac{1}{n} N_j^k(n) - \pi_k \right| + 2 \sum_{k \notin B} \pi_k.$$

Now as $n \to \infty$ the first term on the right-hand side converges to zero with probability 1, and the second may be chosen as small as we please by letting $B \to S$. The result follows.

With suitable conditions, an expected-value version of this also holds; we omit the details. □

We conclude with some examples.

(37) **Flip–flop chain (3.1.17) revisited.** Recall from (3.2.10) that for $P = \begin{pmatrix} 1-a & a \\ b & 1-b \end{pmatrix}$, we find

$$p_{00}^{(n)} = \frac{b}{a+b} + \frac{a}{a+b}(1-a-b)^n \to \frac{b}{a+b} = \pi_0,$$

whenever $|1 - a - b| < 1$.

For the recurrence time T_0 of 0, we have easily that $P(T_0 = 1) = 1 - a$, and for $n > 1$,

$$P(T_0 = n) = a(1-b)^{n-2}b.$$

Thus

$$\mu_0 = ET_0 = \sum_{n=1}^{\infty} nP(T_0 = n) = (a+b)/b = \pi_0^{-1},$$

as required. Hence, because all the time away from 0 is spent at 1, we find that for v_j^k, as defined in (15),

$$v_1^0 = \mu_0 - 1 = \frac{a}{b}$$

and of course

$$(\pi_0, \pi_1) = \left(\frac{b}{a+b}, \frac{a}{a+b}\right) = \left(\frac{v_0^0}{\mu_0}, \frac{v_1^0}{\mu_0}\right).$$

The case when $a = b = 1$ is interesting because, for example,

$$p_{00}(n) = \frac{1}{2}(1 + (-1)^n) = \begin{cases} 1, & n \text{ even}, \\ 0, & n \text{ odd}. \end{cases}$$

This chain is periodic.

There is no convergence to a limit as $n \to \infty$, but there is a stationary distribution $\pi = (\frac{1}{2}, \frac{1}{2})$. And furthermore this stationary distribution is still given by $(v_0^0/\mu_0, v_1^0/\mu_0)$. ○

Finally we consider these results in an example with infinite state space.

(38) Survival chain. This is a Markov chain on the non-negative integers with transition probabilities $p_{j,j+1} = p_j$ and $p_{j0} = q_j$, $j \geq 0$, where $p_j + q_j = 1$, and $p_j > 0$ for all j. You should write down the diagram and transition matrix for the chain. The name arises because this may be interpreted as a model for survival. Suppose we have any number of copies of some component whose lifetimes in use are independent and identically distributed. In detail, if the first is installed at time 0 it survives to 1 with probability p_0, or fails and is replaced by a fresh one with probability q_0. If it has survived until time j, then it survives to $j+1$ with probability p_j, or fails and is replaced by a fresh one with probability q_j. Then the state of the chain X_n is the length of time since the most recent replacement. The chain is clearly irreducible and aperiodic because $p_j > 0$. We cannot easily calculate $p_{jk}^{(n)}$ explicitly, but we can gain a lot of insight by using the above results. Let us focus on the state 0, and assume that $p_0 = 1$; this rules out the trivial case when the component never works at all. For $r \geq 0$, we have for the recurrence time T_0 that

$$P(T_0 > r) = p_0 p_1 \cdots p_r = \prod_{j=0}^{r} p_j.$$

Therefore,

$$\mu_0 = ET_0 = \sum_{r=0}^{\infty} P(T_0 > r)$$

(39)
$$= \sum_{r=0}^{\infty} \prod_{j=0}^{r} p_j.$$

Hence the chain is recurrent if $\prod_{r=0}^{\infty} p_r = 0$, and also non-null if

(40)
$$\sum_{r} \prod_{j=0}^{r} p_j = \mu_0 < \infty.$$

Any stationary vector π must satisfy $\pi = \pi P$, which amounts to

$$\pi_{k+1} = p_k \pi_k, \quad \text{for } k \geq 0$$

and

$$\pi_0 = \sum_{j=0}^{\infty} \pi_j (1 - p_j).$$

Hence $\pi_{r+1} = p_0 \cdots p_r \pi_0$, and if the condition (40) holds, so the chain is non-null,

$$\pi_0 = \left[\sum_{r} \prod_{j=0}^{r} p_j \right]^{-1} = \mu_0^{-1}.$$

In this case π is a stationary distribution.

We may also calculate, for $r \geq 1$,

$$v_r^0 = P(T_0 > r) = p_0 \cdots p_r,$$

so that if $\mu_0 < \infty$

$$v_r^0 / \mu_0 = \frac{p_0 \cdots p_r}{\sum_r \prod_{j=0}^{r} p_j} = \pi_r.$$

The vector $\mu_0^{-1} v^0$ provides a stationary distribution of the chain in the non-null case. (Remember that $v_0^0 = 1$.)

By contrast, if $\prod_{r=0}^{\infty} p_r = c > 0$, then the chain is transient. Explicitly, there must be some $m < \infty$ and $0 < \epsilon < 1$ such that

(41)
$$\prod_{r=0}^{n} p_r > (1 - \epsilon)c, \quad \text{for all } m > m.$$

Then, if we seek a stationary distribution, imposing the condition

$$1 = \sum_{r=0}^{\infty} \pi_r = \pi_0 \sum_{r=0}^{\infty} \prod_{j=0}^{r} p_j$$

$$> \pi_0 \sum_{r=m}^{\infty} c(1-\epsilon), \quad \text{by (41)}$$

gives $\pi_0 = 0$. There is, as expected, no stationary distribution. □

Exercises

(a) Let X be an irreducible Markov chain. For any two states labelled 0 and k, define

$$\theta_k = P(\text{the chain visits } k \text{ before returning to } 0),$$
$$\phi_k = P(\text{the chain visits } 0 \text{ before returning to } k).$$

(i) Show that v_k^0, the expected number of visits to k before returning to 0, is given by $v_k^0 = \theta_k/\phi_k$.

(ii) Find θ_k and ϕ_k for the simple random walk, and verify that $v_k^0 = (p/q)^k$.

(iii) Show that v_k^0, $k \in \mathbb{Z}$, is an invariant measure for simple random walk on the integers.

(iv) For simple random walk on the non-negative integers, find the stationary distribution (when $p < q$) in the two cases

$$(\alpha) \quad p_{01} = 1, \qquad (\beta) \quad p_{01} = p, \qquad p_{00} = q.$$

(b)* Verify that the measure $x_k = (p/q)^k$ is the minimal invariant measure with $x_0 = 1$ for simple random walk on the non-negative integers when $p < q$.

3.7 Reversible chains

Suppose you were to record the progress of a Markov chain, and then play your recording backwards. Would it look the same? Would it even be a Markov chain? The answers to these questions are maybe, and yes, respectively; as we now show.

The essence of the Markov property is that, conditional on its present value, its future and its past are independent. This statement is symmetrical, so we expect reversed Markov chains to retain the Markov property. But because chains converge to equilibrium (non-null recurrent chains, anyway), we must expect to start it in equilibrium to get the same process when reversed.

The key results are these.

(1) Reversed chain. Let X be a Markov chain, with transition probabilities p_{jk}, and define $Y_n = X_{m-n}$ for $0 \leq n \leq m$. Then Y_n has the Markov property, but is not necessarily homogeneous. (Recall (3.1.5).)

Proof. We use the form of the Markov property displayed in (3.1.16). Then for $n \leq m$

$$P(Y_1 = y_1, \ldots, Y_{r-1} = y_{r-1}, Y_{r+1} = y_{r+1}, \ldots, Y_n = y_n \mid Y_r)$$
$$= P(X_{m-1} = y_1, \ldots, X_{m-r+1} = y_{r-1},$$
$$X_{m-r-1} = y_{r+1}, \ldots, X_{m-n} = y_n \mid X_{m-r})$$
$$= P(Y_1 = y_1, \ldots, Y_{r-1} = y_{r-1} \mid Y_r) P(Y_{r+1} = y_{r+1}, \ldots, Y_n = y_n \mid Y_r),$$

by (3.1.16). Hence Y has the Markov property. However, a simple calculation gives

(2) $$P(Y_{n+1} = k \mid Y_n = j) = \frac{p_{kj} P(Y_{n+1} = k)}{P(Y_n = j)}.$$

This will be independent of n only if $P(Y_{n+1} = k)/P(Y_n = j)$ is independent of n for each pair of values (Y_n, Y_{n+1}).

For Y_n to be homogeneous we must therefore require both the chains X and Y to have the equilibrium distribution π, and be irreducible. In this case Y has transition probabilities

(3) $$P(Y_{n+1} = k \mid Y_n = j) = \frac{\pi_k p_{kj}}{\pi_j}.$$

It is now obvious that the reversed chain Y looks the same as X only if their transition probabilities are the same, which is to say that

(4) $$\pi_j p_{jk} = \pi_k p_{kj}, \quad \text{for all } j \text{ and } k.$$

An irreducible chain X with stationary distribution π, satisfying (4), is called *reversible* (in equilibrium). The equations (4) are called the *detailed balance equations*, and the following theorem is most useful.

(5) Reversible chains. Let X be an irreducible Markov chain, and suppose there is a distribution π such that (4) is true. Then π is the stationary distribution of X, and X is reversible.

Proof. If π satisfies (4) then

$$\sum_k \pi_k p_{kj} = \sum_k \pi_j p_{jk} = \pi_j,$$

because $\mathbf{P}$ is stochastic. Thus π is the stationary distribution, and X is reversible by definition. $\square$

This result is useful because if a solution to (4) exists it is usually quicker to find than solving the full equations $\pi = \pi \mathbf{P}$. Here is an important example.

(6) Birth and death process. This may also be thought of as a non-homogeneous simple random walk on the non-negative integers; the transition probabilities are, for $k \geq 0$,

$$p_{k,k+1} = \lambda_k, \qquad p_{k,k-1} = \mu_k, \qquad p_{kk} = 1 - \lambda_k - \mu_k,$$

where $\mu_0 = 0$.

The equilibrium distribution (if any) satisfies

(7) $$\pi_k = \pi_{k+1}\mu_{k+1} + \pi_k(1 - \lambda_k - \mu_k) + \pi_{k-1}\lambda_{k-1}.$$

However, it is easy to see that any solution of the detailed balance equations

(8) $$\pi_k \lambda_k = \pi_{k+1} \mu_{k+1}$$

also satisfies (7) and is also the minimal invariant measure. Thus birth and death processes are reversible in equilibrium, and (8) supplies the stationary distribution easily, whenever there is one. $\circ$

Another important example is this:

(9) Random walk on a graph. Recall that a graph is a collection of vertices, where the ith vertex v_i is directly joined (by at most one edge each) to d_i other vertices (called its neighbours). If $\sigma = \sum_i d_i < \infty$, then a random walk on the vertices of the graph simply goes to any one of the neighbours of v_i with equal probability $1/d_i$. Such a walk is reversible. To see this we note that the detailed balance equations take the form

$$\frac{\pi_j}{d_j} = \frac{\pi_k}{d_k}$$

for neighbours j and k, and the solution such that $\sum_i \pi_i = 1$ is obviously $\pi_j = d_j/\sigma$.

There are of course plenty of non-reversible chains; see the exercises for one of them.

Exercises

(a) **Kolmogorov's criterion.** Let X be an irreducible ergodic chain. Show that X is reversible if and only if

$$p(j_1, j_2)p(j_2, j_3) \cdots p(j_n, j_1)$$
$$= p(j_1, j_n)p(j_n, j_{n-1}) \cdots p(j_2, j_1),$$

for all finite sequences $j_1, \ldots, j_n$. Deduce that birth and death processes are reversible in equilibrium.

(b)* **Ehrenfest model.** Show that the Markov chain with transition probabilities

$$p_{j,j+1} = 1 - \frac{j}{m}, \quad p_{j,j-1} = \frac{j}{m}, \quad 0 \le j \le m$$

is reversible with stationary distribution $\pi_j = \binom{m}{j} 2^{-m}$.

(c) Suppose that the Markov chain X has the diagram on the left (Figure 3.4):

Show that the reversed chain has the diagram on the right. Is X reversible?

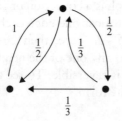 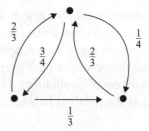

Figure 3.4 Diagram for X (left), and X reversed (right).

3.8 Simulation and Monte Carlo

A large and increasing number of applications of probability in statistics, physics, and computing, require the calculation of many sums having the form $\sum_j g(j)\pi(j)$, where $\pi(j)$ is a probability distribution. This may not seem problematical, but typically in practical applications, the number of terms in the sum is far too large for even the fastest computer to sum them one by one in a reasonable time.

However, a glance at the ergodic theorem at the end of (3.6) shows us another way to do it. If we could run a Markov chain X whose unique stationary distribution is π, then the easily calculated sequence $S_n = (1/n) \sum_{r=1}^{n} g(X_r)$ converges with probability 1 to exactly the sum required.

Greatly simplified, our task thus has three parts: we must devise methods for

(A) simulating a given chain;
(B) specifying a chain with given stationary distribution;
(C) estimating when the sum is close to the desired limit.

We tackle these tasks in the order given:

(A) **Simulating** $(X_n; n \geq 1)$. For the first, it is customary to use the fact that most computers can be persuaded to disgorge an endless sequence of numbers that are evenly and uniformly scattered on the unit interval. These cannot in fact be uniform random variables, both because they are rational and also because the computer was programmed to produce them. However, they behave for many purposes as though they were approximately uniform and random, so we call them *pseudo-random* and denote them by $(U_r; r \geq 1)$.

Now, given a transition matrix $\mathbf{P} = (p_{jk})$, for any j we can divide the interval $[0, 1]$ into disjoint intervals D_{jk} such that

$$p_{jk} = |D_{jk}| = \text{the length of } D_{jk}$$

and

$$\sum_k |D_{jk}| = 1.$$

Next, we define a function $\psi(\cdot, \cdot)$, taking values in S, such that for $j \in S$ and $u \in [0, 1]$,

(1) $$\psi(j, u) = k, \quad \text{whenever } u \in D_{jk}.$$

Finally, we define a sequence of random variables (actually pseudo-random, but never mind) like this: $X_0 = x_0 \in S$,

$$X_{n+1} = \psi(X_n, U_{n+1}), \quad \text{for } n \geq 0.$$

Trivially, by our construction,

$$P(X_{n+1} = k \mid X_0, \ldots, X_n = j) = P(U_{n+1} \in D_{jk})$$
(2)
$$= p_{jk}.$$

Hence X_n is a simulation of the Markov chain X.

(B) Specifying P. Given that we can simulate any chain with transition probabilities p_{jk}, the second task is to ensure that **P** has the required stationary distribution π. It turns out to be most convenient to restrict our attention to reversible chains. Therefore, we must have

(3) $$\pi_k p_{jk} = \pi_j p_{jk}, \quad j \text{ and } k \in S$$

and we require a procedure to determine X_{n+1}, given that $X_n = j$, for any $j \in S$.

The procedure is surprisingly simple; here are the rules:

Rule (i). Let $\mathbf{H} = (h_{jk}; j, k \in S)$ be a stochastic matrix, called the *proposal matrix*. We begin by choosing a random variable $W \in S$ with distribution

$$P(W = k \mid X_n = j) = h_{jk}.$$

This choice W is sometimes called the *candidate variable*. We choose **H** so that this W is easy to simulate.

Rule (ii). Next let $\mathbf{A} = (a_{jk}; j, j \in S)$ be an arbitrary matrix of probabilities, called the *acceptance matrix*. Conditional on $W = k$ we let

$$X_{n+1} = \begin{cases} k, & \text{with probability } a_{jk}, \\ X_n, & \text{with probability } 1 - a_{jk}. \end{cases}$$

Then unconditionally, $P(X_{n+1} = k \mid X_n = j) = p_{jk}$, where

(4) $$p_{jk} = \begin{cases} 1 - \sum_{s \neq j} h_{js} a_{js}, & \text{when } j = k, \\ h_{jk} a_{jk}, & \text{when } j \neq k. \end{cases}$$

The key point is that we can indeed now fix **A** so that the detailed balance equations (3) hold. One possible choice is

$$a_{jk} = \min\left\{1, \frac{\pi_k h_{kj}}{\pi_j h_{jk}}\right\}, \tag{5}$$

as you can readily verify by simple substitution in (4) and then (3).

Our account so far has been very general, but it is often the case in practice that the state space of the chain has a special structure which simplifies things. Most commonly S is a product space, which is to say that $S = L^V$. Here L is some finite set that we think of as the possible states of any locality, and V is the index set of possible localities. Often this arises quite naturally as a process defined on a graph with vertex set V, whose state at any vertex is an element of L. Then we can simplify the task of running a Markov chain on L^V by updating its value at only one of the V vertices at each step of the chain (where the vertex to be updated may be selected at random, or by some other method that runs through all of V eventually). More generally we may allow different local spaces at each vertex, so $S = \prod_{v \in V} L_v$; we do not pursue this here.

Here are two particularly famous examples of this type.

(6) Metropolis method. A very simple choice for the proposal matrix is to choose any one of the $|L| - 1$ states other than j at random. Hence,

$$h_{jk} = \frac{1}{|L|-1}, \quad j \neq k,$$

which is symmetric, so that the acceptance probabilities are

$$a_{jk} = \min\left\{1, \frac{\pi_k}{\pi_j}\right\}. \tag{7}$$

○

(8) Gibbs sampler. In this case we restrict our choice of new state to those that leave everything unchanged except possibly the state of the vertex v in question. Call this set of states $U_j(v)$. Then we make a weighted choice thus: the new state k is chosen with probability

$$h_{jk} = \frac{\pi_k}{\sum_{r \in U_j(v)} \pi_r}, \quad k \in U_j(v), \tag{9}$$

and it is easy to see that $a_{jk} = 1$. ○

Here is an example of the Gibbs sampler at work.

(10) Repulsive model. This model has various applications, for example, in modelling the behaviour of objects that cannot overlap; in this context it is often called the hard-core model. The state space has a product form $S = \{0, 1\}^V$, where for any vertex v in the graph V the states 0 and 1 correspond to unoccupied and occupied, respectively. We are interested in configurations in which no two adjacent vertices are occupied; these are called feasible. Let the total number of feasible configurations be ϕ, and select one at random. What is the expected number v of occupied vertices? It is practically impossible to calculate this except in trivially small graphs. But it is simple to estimate v by running a Markov chain with uniform stationary distribution $\pi_j = \phi^{-1}$. The Gibbs sampler gives us this one:

The state space S is the set of feasible configurations. Then the procedure is this:

(i) Pick a vertex at random; call it v. We may change only this.
(ii) We can only let the state of v be 1 if all its neighbours are 0. In this case (9) tells us to let it be 1 with probability $\frac{1}{2}$, or let it be 0 with probability $\frac{1}{2}$.

The chain is obviously aperiodic, because it may stay unchanged at any step. It is also easy to see that it is irreducible (an exercise for you). From the above, the chain converges to the uniform stationary distribution, as required; but it is instructive to verify explicitly that the detailed balance equations are indeed satisfied. The only non-trivial case is when j and k differ at one vertex, and then by the above construction

$$(11) \qquad \pi_j p_{jk} = \frac{1}{\phi} \frac{1}{|V|} \frac{1}{2} = \pi_k p_{kj}.$$

○

(C) Precision. Our final task, mentioned early in this section, is to investigate when the sum is close to the desired limit. For well-behaved functions one may look at the variance $E(S_n - n^{-1} \sum_{k=1}^n g(k)\pi_k)^2$. Another approach is to find bounds on the rate at which X approaches its stationary distribution. In general, this is a hard problem, but in simple cases a number of results from the classical theory of matrices are useful here, and we state two typical results without proof.

(12) Theorem. Let X be an ergodic reversible chain on S whose transition matrix $\mathbf{P}$ has right eigenvectors $\mathbf{v}_r = (v_r(1), v_r(2), \ldots)$, and λ_2 is the eigenvalue with second largest modulus. Then

$$\sum_{k \in S} |p_{jk}(n) - \pi_k| \leq |S||\lambda_2|^n \sup_{r \in S} |v_r(j)|$$

and
$$\sum_{k \in S} |p_{jk}(n) - \pi_k| \leq |\lambda_2|^n \pi_j^{-1/2}.$$

The second bound has the advantage of not depending on the eigenvectors, which are often difficult to calculate. But the dependence on $|\lambda_2|$ has entailed a substantial effort to find bounds for $|\lambda_2|$ that are relatively easy to calculate. We omit this; it is remarkably difficult to obtain (in general) bounds of practical use. And note that even tight bounds do not imply that the chain ever actually attains its stationary distribution.

Some recently developed methods avoid these difficulties, by delivering a simulated random variable that has exactly the stationary distribution required. Furthermore it is immediately known that it has done so, thus removing all uncertainty. We give a brief account of the first of these techniques, known as *exact simulations*.

(13) **Propp–Wilson method.** Let π be the distribution that we wish to simulate; that is to take a sample W having distribution π. Just as described above in the first part of this section, we construct an ergodic reversible Markov chain X having π as its stationary distribution. The function $\psi(j, u)$ defined in (1) is used to simulate X, using the recurrence given in (2) and a collection of independent Uniform (0,1) random variables $(U(r); r \geq 0)$.

The crux of the Propp–Wilson technique is to simulate X (with $X_0 = j$), for all $j \in S$, over an increasing sequence of time periods. This family of simulations stops when (for the first time) all the sequences arrive at a common value, which will be W. Note that all realizations use the same sequence $U(r)$ until a value of W is determined.

This clever and complex idea is best grasped by seeing its definition as an iteration, or inductive sequence:

First we fix a decreasing sequence of negative integers $(n_0, n_1, n_2, \ldots)$; a popular and easily-remembered choice is $n_r = -2^r$.

(i) At the rth step, begin by simulating values for

(14) $$U(n_r + 1), U(n_r + 2), \ldots, U(n_{r-1}).$$

(ii) Then, for all $j \in S$, simulate a realization of X, starting at time $n_r + 1$, with initial state j, running until time 0 (using the same values of U from (14) for each simulation).

(iii) If all $|S|$ simulations have the same value w when stopped at time 0, then we record this as a value of W and quit. Otherwise we increase r by 1, and return to step (i) to begin the $(r+1)$th iteration, reusing the values of U in (14) where required.

(iv) After recording a value of W, we repeat the iteration starting at $r = 0$, and generating a fresh independent sequence of values of U.

You can now see why Propp and Wilson called this idea *coupling from the past*. When the iteration stops, all the simulations are coupled at W. It is a remarkable fact that if the iteration stops, then the recorded result W has distribution π.

Formally, if the probability that the iteration never stops is zero, then

(15) $$P(W = j) = \pi_j.$$

Proof. Let $a > 0$ be arbitrary. Since the iteration certainly stops, there is some finite m such that

$$P(\text{at most } m \text{ iterations are necessary}) \geq 1 - a.$$

Now simulate a chain Z, using the same sequence of Us as yielded W, but starting Z at time n_m with the stationary distribution π, explicitly

(16) $$P(Z(n_m) = j) = \pi_j.$$

Hence, as always, $P(Z(0) = j) = \pi_j$. When at most m iterations are necessary for coupling, $Z(0) = W$. Therefore, from (8), $P(Z(0) \neq W) \leq a$.

Now, using the type of argument familiar from the convergence theorem for ergodic Markov chains,

(17)
$$P(W = j) - \pi_j = P(W = j) - P(Z(0) = j)$$
$$\leq P(\{W = j\} \cap \{Z(0) \neq j\})$$
$$\leq P(W \neq Z(0)) \leq a.$$

Likewise $\pi_j - P(W = j) \leq a$, and hence

$$|P(W = j) - \pi_j| \leq a.$$

Since a is arbitrary, we have proved (15). □

Similar ideas can be made to work in continuous state spaces, and to solve a variety of optimization problems, but this is beyond our scope.

> **Exercises**
> (a) **Barker's method.** Show that a feasible choice for the acceptance probabilities in (4) is
>
> $$a_{jk} = \pi_k h_{kj} / \{\pi_j h_{jk} + \pi_k h_{kj}\}.$$

(b)* Let X be an ergodic Markov chain with finite state space and stationary distribution π. Suppose you run the Propp–Wilson iteration into the future, starting at time 0, until all the $|S|$ chains (started in each of the $|S|$ states) first coincide in a value W. Show, by constructing a counterexample, that W need not have the stationary distribution π. [Hint: Set $|S| = 2$.]

(c) **Simulating a random permutation.** Consider a pack of n cards labelled $1, \ldots, n$. The top card (#1) is taken and reinserted at random in any one of the n possible positions. Show that the ordering of the cards is an aperiodic irreducible Markov chain. Show that all the cards underneath the original bottom card (#n) are equally likely to be in any order. Deduce that when #n is first reinserted, the pack is in completely random order. Show that the expected number of insertions to reach this state is $n \log n$, approximately.

Finally, show that the chain is doubly stochastic.

3.9 Applications

We conclude this chapter with a look at some practical and amusing applications of Markov chain theory.

(1) **Patterns.** A fair coin is flipped repeatedly, and we record the sequence of heads and tails (H and T) until we first see some given sequence, or pattern, such as HH, or HT, or HTTH, etc. What is the expected number of flips? For any pattern π, we denote this expectation by m_π. Now adopt the obvious strategy of looking at short patterns first:

1-patterns. Trivially, the expected number m_H of flips until we see the 'pattern' H is 2, and the same is true for T, so $m_T = 2 = m_H$.

2-patterns. These are less trivial; we display a number of methods for deriving m_{HH}. The most obvious approach is by conditioning on the first few flips. A natural partition of the sample space is $\{T, HT, HH\}$. If either of the first two events occurs at the start of our sequence, then we are in effect starting from scratch in our quest for HH. If the third event occurs, we are done. Hence, by conditional expectation,

$$m_{HH} = \tfrac{1}{2}(1 + m_{HH}) + \tfrac{1}{4}(2 + m_{HH}) + \tfrac{1}{4} \cdot 2,$$

which gives $m_{HH} = 6$.

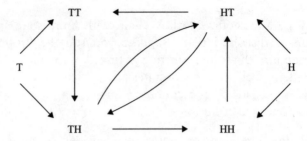

Figure 3.5 Diagram for 2-pattern chain. All arrows carry probability $\frac{1}{2}$.

For m_{HT}, we simply condition on whether the flip first shows T or H, respectively, to obtain

$$m_{HT} = \tfrac{1}{2}(1 + m_{HT}) + \tfrac{1}{2}(1 + m_T).$$

Hence $m_{HT} = 4$.

This is quick, but a few moments' thought should convince you that it may not be easy to extend this elementary approach to sequences such as HTHTHTH. We turn to Markov chains for further and better methods:

In this framework, the Markov chain of interest for 2-patterns has the diagram above (Figure 3.5); we argue as follows:

The initial distribution is $P(H) = P(T) = \frac{1}{2}$, and the chain immediately leaves the transient states T and H to enter the closed set of non-null recurrent states {HH, HT, TH, TT}. Trivially, the stationary distribution is uniform, and the mean recurrence times are all the same: $\mu_{HH} = 4$, and so on.

Suppose we start the chain in HH, and condition on the next step, either to the state HH or to the state HT. By conditional expectation

$$4 = \mu_{HH} = \tfrac{1}{2} \cdot 1 + \tfrac{1}{2}(1 + m_{HH}),$$

where the final term arises from the observation that the appearance of a tail puts us in the same position as starting from T, if we are waiting for HH. Hence $m_{HH} = 6$.

Alternatively, we can observe that starting from H is also essentially the same as starting from HH, if we are waiting for a recurrence of HH. That is to say, the expected number of steps to go from H to HH equals μ_{HH}. Hence

$$m_{HH} = m_H + \mu_{HH} = 6.$$

3-patterns. Let us try to apply each of the above methods to find m_{HTH}.

First method. Consider the partition of the sample space given by

$$\Omega = \{T, HT, HHT, HHHT, \ldots\}.$$

Given any of the events (H^rT, $r \geq 1$), if the next flip is H then we are home, if it is T we are back to our initial task. Hence, by conditional expectation

$$m_{HTH} = \tfrac{1}{2}(1 + m_{HTH}) + \tfrac{1}{8} \cdot 3 + \tfrac{1}{8}(3 + m_{HTH})$$
$$+ \tfrac{1}{16} \cdot 4 + \tfrac{1}{16}(4 + m_{HTH}) + \cdots$$
$$= \tfrac{1}{2} + 2 + \tfrac{3}{4} m_{HTH}, \quad \text{on summing the series.}$$

Hence $m_{HTH} = 10$.

Second method. Suppose you wait until the 2-pattern Markov chain enters HT. Then either the next flip is H, and we are home, or it is T and our task begins afresh from the beginning. Hence, as usual,

$$m_{HTH} = m_{HT} + \tfrac{1}{2} \cdot 1 + \tfrac{1}{2}(1 + m_{HTH}).$$

Hence $m_{HTH} = 2(1 + m_{HT}) = 10$.

Third method. The expected number of flips to go from H to HTH is the same as the mean recurrence time of HTH. Hence

$$m_{HTH} = m_H + \mu_{HTH} = 2 + 8 = 10.$$

You should now need no convincing that the third method is the neatest and quickest. Let us look at the rest of the 3-patterns this way.

Looking at the Markov chain for 3-patterns we find easily that $\mu_{HHH} = 8$. (You could draw the diagram if you wish.) Because of the overlap of HH at the start of HHH with HH at the end of HHH, we see that this is the same as the expected number of flips needed to go from HH to HHH. But $m_{HH} = 6$, so

$$m_{HHH} = 6 + 8 = 14.$$

Since TTH, THH, HHT, and HTT have no such overlaps $m_{TTH} = \mu_{TTH} = 8$, and likewise for the rest.

Now we can easily find the expected number of flips to see any pattern of length n, with mean recurrence time 2^n.

Example. For THTH there is one overlap TH, so

$$m_{THTH} = m_{TH} + \mu_{THTH} = 4 + 16 = 20.$$

But for THTT the overlap is only T, so

$$m_{\text{THTT}} = m_{\text{T}} + \mu_{\text{THTT}} = 18.$$

Exactly the same idea works for biased coins and deMoivre trials:

(2) **Biased coins.** Here we consider flipping a coin that shows H with probability $p = 1 - q = 1 - P(T)$. Then, in the Markov chain of 3-patterns, the stationary probability of (say) HHH is p^3, so μ_{HHH} is $1/p^3$. Thus, since the overlap is HH,

$$\begin{aligned}
m_{\text{HHH}} &= m_{\text{HH}} + \frac{1}{p^3} \\
&= m_{\text{H}} + \frac{1}{p^2} + \frac{1}{p^3}, \quad \text{by the same argument for } m_{HH}, \\
&= \frac{1}{p} + \frac{1}{p^2} + \frac{1}{p^3}.
\end{aligned}$$

By contrast, HHHT (say) has no overlap so, (using an obvious notation for the stationary probability),

$$m_{\text{HHHT}} = \mu_{\text{HHHT}} = \frac{1}{\pi_{\text{HHHT}}} = \frac{1}{p^3 q}.$$

(3) **DeMoivre trials.** The same argument works for dice (say), so the expected number of rolls of a fair die needed to observe the sequence 34534 is

$$m_{34} + \mu_{34534} = \mu_{34} + 6^{+5} = 6^2 + 6^5.$$

Another natural question is to ask whether one sequence is more likely to be seen before another. This is of course a particular application of our general results about hitting probabilities of Markov chains, so we sketch the details for some special cases.

(4) **MOTTO.** Player A picks a pattern of three coins, for example THT; this is his motto. Player B then picks her motto, a pattern of length three which is different from that of A.

A coin is flipped repeatedly, and the player whose motto first appears wins $10 from the other. Would you rather be A or B?

Solution. We shall show that it is far preferable to make the second choice; to be B is best. The demonstration is by exhaustion; we run through all the possibilities. Let A's motto be α, and B's β.

(i) Suppose $\alpha = $ HHH and $\beta = $ THH. Let p_A be the probability that A's motto α is seen first. By inspection, α is first if and only if the first three flips are HHH, and therefore $p_A = \frac{1}{8}$. Note that it is the overlap of B's motto β with the beginning of α that gives B this advantage.

(ii) Let $\alpha = $ HHT and $\beta = $ THH. Once again, A wins if the first two flips are HH, but loses otherwise. Hence $p_A = \frac{1}{4}$.

(iii) If $\alpha = $ HTT and $\beta = $ HHT, nobody can think of winning until there is a head H. If the next is H, then B must win; if the next two are TT then A wins; if the next two are TH then the game is back to the situation with a single head H. Hence

$$p_A = \tfrac{1}{4} + \tfrac{1}{4} p_A, \quad \text{and so} \quad p_A = \tfrac{1}{3}.$$

(iv) Likewise if $\alpha = $ HTH, $\beta = $ HHT, then $p_A = \frac{1}{3}$.

The rest of the cases are exercises for you:

(v) P(HHH before THT) $= \frac{5}{12}$.

(vi) P(HHH before HTH) $= \frac{3}{10}$.

(vii) P(HHH before TTH) $= \frac{3}{10}$.

(viii) P(HTH before THT) = P(HHH before TTT) $= \frac{1}{2}$, by symmetry.

All the other comparisons are obtained by interchanging the roles of H and T. Thus, even if A makes the best possible choice, B has twice A's chance of winning the game, if B chooses wisely.

Even more interesting, we may note that

$$P(\text{HHT before THH}) = \tfrac{1}{4}$$
$$P(\text{THH before TTH}) = \tfrac{1}{3}$$
$$P(\text{TTH before HTT}) = \tfrac{1}{4}$$
$$P(\text{HTT before HHT}) = \tfrac{1}{3}.$$

So if we use the symbol '>' to mean 'more likely to occur before', we have shown

$$\text{HHT} > \text{HTT} > \text{TTH} > \text{THH} > \text{HHT},$$

which is a little eerie.

Finally, we note another odd feature of waiting for patterns. We showed that $m_{\text{THTT}} = 18$, and $m_{\text{HTHT}} = 20$. But the probability that HTHT occurs before THTT $= \frac{9}{14} > \frac{1}{2}$.

To see this is a simple, but tedious, exercise. One can also generalize the idea of overlaps to obtain the so-called 'Conway magic formula', to give the required probabilities but that would take us too far afield. ○

We turn to another popular area for applications of Markov chains: the theory of queues. Here is a classic illustrative example.

(5) Queueing. Customers arrive at a counter and form a single line for service. During the unit time-interval $(t, t+1)$ the number of arrivals is Poisson with parameter λ, and independent of all other arrivals and departures. The time taken for C_n the nth customer to be served is X_n, where the $(X_n; n \geq 1)$ are integer-valued and independent of each other and all arrivals.

Let Q_n be the length of the queue at the instant when C_n departs. By construction,

$$(6) \qquad Q_{n+1} = \begin{cases} Q_n + A_{n+1} - 1, & \text{if } Q_n > 0, \\ A_{n+1}, & \text{if } Q_n = 0, \end{cases}$$

where A_{n+1} is the number of arrivals during the service time X_{n+1}, and we denote $P(A_{n+1} = j) = a_j$. By the independence, Q_n is an irreducible Markov chain with transition matrix

$$(7) \qquad \mathbf{P} = \begin{pmatrix} a_0 & a_1 & a_2 & a_3 & \cdots \\ a_0 & a_1 & a_2 & a_3 & \cdots \\ 0 & a_0 & a_1 & a_2 & \cdots \\ 0 & 0 & a_0 & a_1 & \cdots \\ \vdots & \vdots & \vdots & & \ddots \end{pmatrix}.$$

Here a natural question is whether Q_n is recurrent or transient, so we seek a stationary distribution π. This must satisfy

$$(8) \qquad \begin{aligned} \pi_0 &= a_0 \pi_0 + a_0 \pi_1 \\ \pi_1 &= a_1 \pi_0 + a_1 \pi_1 + a_0 \pi_2 \\ &\vdots \\ \pi_j &= a_j \pi_0 + \sum_{i=1}^{j+1} a_{j-i+1} \pi_i. \end{aligned}$$

Multiplying the jth equation above by s^j, $j \geq 0$, and summing over all j yields

$$\Pi(s) = \pi_0 A(s) + s^{-1}\{\Pi(s) - \pi_0\} A(s),$$

where $\Pi(s) = \sum_{r=0}^{\infty} s^r \pi_r$ and $A(s) = \sum_{r=0}^{\infty} s^r a_r$. Hence

$$(9) \qquad \Pi(s) = \frac{\{\pi_0 (s-1) A(s)\}}{\{s - A(s)\}}.$$

It is an exercise for you to show that $\pi_r \geq 0$ for all $r > 0$. This is therefore indeed a stationary distribution if $\Pi(1) = \sum_r \pi_r = 1$. By L'Hopital's rule in (9), this occurs if and only if $\pi_0 = 1 - A'(1) > 0$, and therefore the chain Q_n is non-null recurrent if and only if

(10) $$A'(1) = \mathrm{E}A_1 < 1.$$

However, conditional on $X_1 = x$, A_1 is Poisson (λx), and so we may write

$$\begin{aligned} A(s) = \mathrm{E}s^{A_1} &= \mathrm{E}\{\mathrm{E}(s^{A_1} \mid X_1)\} \\ &= \mathrm{E}\exp\{\lambda X_1(s-1)\} \\ &= M_X(\lambda(s-1)), \end{aligned}$$

where $M_X(\theta)$ is the moment-generating function (mgf) of the Xs. Hence the condition (10) may be written as

$$\lambda M'_X(0) = \lambda \mathrm{E}X_1 < 1.$$

That is to say, the arrival rate λ is less than the service rate $(\mathrm{E}X_1)^{-1}$.

Finally, we note an entertaining connexion with another class of Markov chains, namely branching processes. Let B be the length of time for which the server is continuously busy. That is to say, if the first customer arrives at 0, then B is the first time at which the queue is empty. We shall show that the mgf $M_B(\theta) = \mathrm{E}e^{\theta B}$ satisfies

(11) $$M_B(\theta) = M_X(\theta - \lambda + \lambda M_B(\theta)).$$

To see this, call any customer C_a a scion of customer C if C_a arrives while C is being served. Since arrivals during different service intervals are independent, the scions of different customers are independent. Let T be the total number of customers arriving during the service period X_1. If $\lambda \mathrm{E}X_1 < 1$ then the chain is recurrent and $\mathrm{P}(B < \infty) = \mathrm{P}(T < \infty) = 1$. Let B_j be the sum of the service times of the jth customer $C_j, 1 \leq j \leq T$, and C_j's scions; conditional on T the B_j are independent and have the same distribution as B. But

$$B = X_1 + \sum_{j=1}^{T} B_j$$

and, conditional on X_1, T is Poisson (λX_1). Hence

$$E\left\{\exp\left(\theta\sum_{j=1}^{T} B_j\right)\bigg| X_1\right\} = E\{M_B(\theta)^T \mid X_1\}$$

$$= \exp\{\lambda X_1(M_B(\theta) - 1)\}.$$

Hence,

$$M_B(\theta) = E\exp\{\theta X_1 + \lambda X_1(M_B(\theta) - 1)\} = M_X\{\theta - \lambda + \lambda M_B(\theta)\}.$$

Exercises
(a) Show that the solution π of (8) is indeed non-negative.

(b)* If a fair coin is flipped repeatedly, show that the expected number of flips to get THTHT is 42.

(c) Verify the assertion that in repeated flips of a fair coin

$$P(\text{HTHT occurs before THTT}) = \tfrac{9}{14}.$$

(d) Suppose that the service time X in (5) is in fact a constant d with probability 1. Show that in equilibrium the expected queue length is

$$\tfrac{1}{2}\lambda d(2 - \lambda d)/(1 - \lambda d).$$

(e) Let W be the time that an arriving customer waits to begin being served. In equilibrium, show that for the queue specified in (5)

$$M_W(\theta) = \frac{(1 - \lambda E X_1)\theta}{\lambda + \theta - \lambda M_X(\theta)},$$

where M_W is the mgf of W.

Problems

1.* Let the transition matrix **P** be doubly stochastic; $\sum_i p_{ij} = \sum_j p_{ij} = 1$. Show that
   (a) if **P** is finite, then all states are non-null recurrent;
   (b) if **P** is infinite and irreducible then all states are transient or null recurrent.

2. **Excess life.** Let X_n be a Markov chain with transition probabilities $p_{r,r-1} = 1$ for $r \geq 1$, and $p_{0r} = f_r$, for $r \geq 0$ where $f_0 = 0$ and $\sum_{r \geq 1} f_r = 1$.

   Find conditions on these probabilities for the chain to be (a) irreducible, (b) recurrent, (c) non-null. In case (c), find the stationary distribution.

3.* Let X_n be a Markov chain with transition matrix **P**. Show that

$$Y_n = (X_n, X_{n+1}), \quad n \geq 0$$

is a Markov chain, and write down its transition probabilities. If X_n has unique stationary distribution π, does Y have a unique stationary distribution?

4.* **Left-continuous random walk.** Let $(X_r, r \geq 1)$ be independent and identically distributed on $\{-1, 0, 1, \ldots\}$ with probability-distributing function (pgf) $G(s) = Es^{X_1}$. Define the random walk $S_n = 1 + \sum_{r=1}^{n} X_r, n \geq 0$. Let e_0 be the probability that S_n ever reaches 0. Show that e_0 satisfies

$$e_0 = 1, \quad \text{if } EX_1 \leq 0,$$
$$e_0 < 1, \quad \text{if } EX_1 > 0.$$

Calculate e_0 explicitly when $G(s) = qs^{-1} + ps^2$, where $p + q = 1$.

5. Let j and k be two states of a finite irreducible Markov chain X_n with transition probabilities p_{ij}. Let r_i be the probability that the chain visits j before k, starting at i. Show that

$$r_i = \sum_{x \in S} p_{ix} r_x, \quad i \neq j \text{ or } k,$$

with $r_j = 1$ and $r_k = 0$.

6. **Chess.** A knight moves at random on a conventional chessboard. At each jump it makes any one of the moves allowed to it. Show that
   (a) if it starts in a corner, the expected number of moves to return to that corner is 168;
   (b) if it starts on one of the four central squares, the expected number of moves to return to it is 42.

7.* A spider moves at random between the vertices of a symmetrical web with r rays and l loops. Here is a four-ray two-loop web

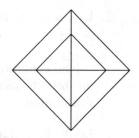

Figure 3.6 A four-ray, two-loop spider's web.

(Figure 3.6):

If it is at the centre it is equally likely to move out along any ray. If it is at any other internal vertex it is equally likely to pick any one of the four adjacent possibilities; on the perimeter of its web it picks any one of three adjacent possibilities. Find
(a) the expected number of moves to reach the perimeter, starting at the middle;
(b) the expected number of moves to reach the middle, starting on the perimeter.

8. **Bookshelf.** You have b books on your shelf, that we may identify with the integers $1, \ldots, b$. On any occasion when you take a book, you select the book numbered n with probability $s(n) > 0$, $1 \leq n \leq b$, independently of all other selections. When you are done with a book, you replace it on the left of all the others on the shelf. Let $X_r = (j_1, \ldots, j_b)$ be the order of the books after the rth such selection. Show that X is an irreducible aperiodic Markov chain with stationary distribution

$$\pi(j_1, \ldots, j_b) = s(j_1) \frac{s(j_2)}{1 - s(j_1)} \cdots \frac{s(j_b)}{1 - s(j_1) - \cdots - s(j_{b-1})}.$$

9. **Another bookshelf.** You have another bookshelf containing a books, that we label $1, \ldots, a$. Each selection is independent, with book number n selected with probability $s(n) > 0$. When you are done with the book, you replace it one place to the left. Show that this is an irreducible reversible chain, and deduce that the stationary distribution is

$$\pi(j_1, \ldots, j_a) = cs(j_1)^a s(j_2)^{a-1}, \ldots, s(j_a),$$

where c is chosen so that $\Sigma \pi = 1$.

10.* Two adjacent containers C_i and C_2 each contain m particles. Of these particles, m are type A and m are type B. At $t = 1, 2, \ldots$ a particle is selected at random from each container and switched to the other. Let X_n be the number of type A particles in C_1 after n such exchanges. Show that X_n is a Markov chain and find its stationary distribution.

11.* With the set-up of problem (10), suppose that if the particles are of different types, they are switched with probability α if the type A particle is in C_1, or switched with probability β if the type A particle is in C_2. Find the stationary distribution of the chain.

12* A Markov chain X with state space $\{1, 2, 3, 4\}$ has this transition matrix:
$$\mathbf{P} = \begin{pmatrix} 0 & a & 1-a & 0 \\ 1-b & 0 & b & 0 \\ 0 & 0 & 0 & 1 \\ 0 & 0 & 1 & 0 \end{pmatrix}.$$

Given b, show that there is exactly one value of a such that $c = \lim_{n\to\infty} p_{14}(n)$ exists. Find this value of a as a function of b, and the limit c.

13.* A Markov chain Y has the diagram below (Figure 3.7):

Show that $c = \lim_{n\to\infty} p_{16}(n)$ exists if either of a or α takes the value $\frac{2}{3}$. Does the limit exist in any other case?

14. Find the mgf of the busy period of the server in (3.9.5) when the service periods X_i are Exponential (μ). [Hint: Use (3.9.11).]

15. Show that if a Markov chain X has two distinct stationary distributions, then it has infinitely many distinct stationary distributions.

16. Let S_n, $n \geq 0$, be a simple random walk started at the origin. Show that $|S_n|$ is a Markov chain.

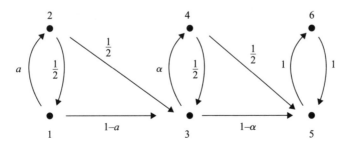

Figure 3.7 Diagram for the chain Y.

4 Markov chains in continuous time

> We may describe a Markov process picturesquely by the statement that it represents the gradual unfolding of a transition probability.
>
> S. Chandrasekhar (1943)

4.1 Introduction and examples

Of course many random processes in the real world do not change state only at integer times. Your telephone can ring; you can inhale a rhinovirus; meteorites can fall; your cam-belt can fail, at any time. But many of these processes still have the key property of Markov chains discussed in Chapter 3:

(1) **Markov property.** Given a complete description of the state of the process at time t, its future development is independent of its track record up to t.

The difference is that we now allow such a Markov chain to change its state at any time $t \in \mathbb{R}$. Let us look at some illustrative instances.

Our first example is trivial but important.

(2) **One-way Markov switch.** Your computer may be protected against power surges by a cut-out switch; let $X(t) \in \{0, 1\}$ be the state of the switch at time $t \geq 0$. (There are two states that we may call 'open' and 'closed'.) Suppose that the first surge that triggers the switch from 1 to 0 occurs at T, where T is Exponential (λ). Then recalling the lack of memory of the exponential density, and using the obvious notation parallelling discrete-time Markov chains, we have

$$P(X(t+s) = k \mid X(u), 0 \leq u \leq s; X(s) = j)$$
(3) $$= P(X(t+s) = k \mid X(s) = j) = p_{jk}(t).$$

To see this explicitly, just check it for all four possible values of (j, k). These calculations yield

$$\mathbf{P} = (p_{jk}(t)) = \begin{pmatrix} p_{00}(t) & p_{01}(t) \\ p_{10}(t) & p_{11}(t) \end{pmatrix} = \begin{pmatrix} 1 & 0 \\ 1 - e^{-\lambda t} & e^{-\lambda t} \end{pmatrix}.$$

Of course (3) is just a symbolic statement of the verbal form of the Markov property in (1).

It is important to note that if the distribution of T is anything other than Exponential (λ), for any value of λ, then $X(t)$ does not satisfy the Markov property (except in trivial cases, which we exclude). ○

Extending this example, it is natural to consider a switch that resets itself.

(4) **Exponential flip–flop circuit.** Let $X(t) \in \{0, 1\}$. When $X(t)$ enters 1 it remains there for a time that is Exponential (λ), and then goes to 0; it remains there for a time that is Exponential (μ), and then goes to 1. The process repeats indefinitely, and all these exponential random variables are independent.

As for the one-way switch, we consider transitions conditional on the past of $X(t)$; for example, the probability that $X(t + s)$ is in state 1, given the past of the process up to time s, with $X(s) = 1$.

Now, the conditioning event includes the most recent time at which the chain entered state 1 prior to s, at v (say). Hence, by the lack of memory of the exponential density, (and the independence of the times between switches), the first time after s at which the chain leaves 1 is an Exponential (λ) random variable that is independent of events prior to s, and all subsequent switches. But this is simply a flip–flop process as defined above, started in state 1. Formally, this construction shows that

(5) $$P(X(t+s) = 1 \mid X(u), 0 \leq u \leq s; X(s) = 1)$$
$$= p_{11}(t) = P(X(t) = 1 \mid X(0) = 1).$$

This is again the Markov property, but it is not so easy to calculate $p_{jk}(t)$ in this case. (We return to find these transition probabilities later.) Note that the argument above can succeed only if the times between switches have an exponential density, and are independent. ○

With these examples, and Chapter 3, in mind, it is natural to make the following formal definition.

(6) **Markov chain in continuous time.** Let $X(t), t \geq 0$, be a family of random variables taking values in a countable set S, called the state space. Then $X(t)$ is a Markov chain if for any collection $k, j_1, \ldots, j_{n-2}, j$ of states, and $t_1 < t_2 < \cdots < t_n$ of times, we have

(7) $$P(X(t_n) = k \mid X(t_1) = j_1, \ldots, X(t_{n-1}) = j)$$
$$= P(X(t_n) = k \mid X(t_{n-1}) = j)$$
$$= p_{jk}(t_n - t_{n-1}).$$

The collection (as a matrix), $\mathbf{P} = (p_{jk}(t))$, comprises the transition probabilities of the chain, which we have required to be time-homogeneous. △

Note that we have abandoned the ideal of conditioning on the entire past of the process up to t_{n-1}, $\{X(u); 0 \leq u \leq t_{n-1}\}$, because this is an uncountable collection of random variables. Trying to condition on this leads to interesting technical difficulties, that are beyond our scope to resolve here.

One classic example of a Markov chain where we do have all the transition probabilities $p_{jk}(t)$ is this:

(8) **Poisson process.** We showed in (2.5.8) that the Poisson process $N(t)$ has independent increments. It follows immediately that for $t_1 < t_2 < \cdots < s < t+s$ and $j_1 \leq j_2 \leq \cdots \leq j \leq k$

$$P(N(t+s) = k \mid N(t_1) = j_1, \ldots, N(s) = j) = P(N(t+s) - N(s) = k - j)$$
(9)
$$= e^{-\lambda t}(\lambda t)^{k-j}/(k-j)!.$$

We shall meet one more very tractable example in Section 4.3. ○

Note that, as in discrete time, $\mathbf{P}$ is a stochastic matrix and in general, also as in the discrete case, the Markov property (7) immediately gives the following:

(10) **Chapman–Kolmogorov equations.** Considering the state of the chain at time s, we use conditional probability, and the Markov property (7), to give

$$p_{ik}(s+t) = \sum_{j \in S} P(X(t+s) = k \mid X(s) = j, X(0) = i) P(X(s) = j \mid X(0) = i)$$

(11)
$$= \sum_{j \in S} p_{jk}(t) p_{ij}(s).$$

In matrix form this may be written as

(12) $$\mathbf{P}_{s+t} = \mathbf{P}_s \mathbf{P}_t, \quad \text{for } s, t \geq 0.$$

It follows from this, that, if we know the distribution of $X(t)$ for $0 \leq t \leq a$, where $a > 0$, then we know the distribution of $X(t)$ for all $t \geq 0$. This resembles the discrete case, see (3.2), but the analogy is not as close as we would like. For discrete-time Markov chains X_n, all the joint distributions may be written in terms of the one-step transition probabilities p_{jk}; and furthermore these supply a construction of the chain as displayed in the first part of (3.8).

But in continuous time this does not follow, because there is no implicit unit of time, and the probabilities $p_{jk}(t)$ do not immediately yield a way of constructing the chain $X(t)$. However, the examples and discussions above lead us to define naturally a class of random processes with the following structure.

(13) **Jump chain and holding times.** Let a random process $X(t) \in S$, $0 \leq t < \infty$, be constructed according to these rules:

At time zero it enters the state $X(0) = Z_0$.

(i) On entering any state j, the process remains there for a random interval U that is Exponential (q_j), called its *holding time*. In general, we require $0 < q_j < \infty$ to exclude trivial cases. (If $q_j = 0$ it never leaves j, and if $q_j = \infty$ it leaves immediately.)

(ii) At the conclusion of its holding time, the chain leaves j and enters k with probability r_{jk}. We generally assume that $r_{jj} = 0$ for all j; this is not essential, but of course in continuous time a jump from j to itself is not observable. (However, see (4.2.41) for an instance when it is convenient to take $r_{jj} > 0$.)

This ensures that all transitions are indeed jumps in the state space, and the sequence of distinct states thus visited is a Markov chain which we call the *jump chain*. Its transition matrix is $\mathbf{R} = (r_{jk})$.

(iii) Given the sequence $Z_0 = z_0, Z_1 = z_1, \ldots, Z_{n-1} = z_{n-1}, \ldots$, the corresponding sequence of holding times $U_0, U_1, \ldots, U_{n-1}, \ldots$, are independent exponential random variables with respective parameters $q_{z_0}, q_{z_1}, \ldots, q_{z_{n-1}}, \ldots$.

We define the jump times

(14) $$J_n = \sum_{r=0}^{n-1} U_r, \quad n \geq 1$$

and this construction defines $X(t)$ for all $t \geq 0$, provided that $\lim_{n \to \infty} J_n = \infty$. This final condition will always be required to hold, except when explicitly noted otherwise: for example, when we discuss birth processes in (4.3).

Note that by convention our processes are defined to be right-continuous; that is to say a jump from j to k at t means that $X(t) = k$, and indeed $X(s) = X(t) = k$ for $t \leq s \leq t + a$, for some $a > 0$. This choice has important and helpful theoretical consequences, but it is beyond our scope to explore the details.

The importance of processes constructed in this way depends on the following key property (which should come as absolutely no surprise to you).

(15) **Theorem.** The process $X(t)$ defined in (13) has the Markov property (7), where $p_{jk}(t) = P(X(t) = k \mid X(0) = j)$.

Sketch proof. Let A be the event that $X(t_r) = x_r$ for $0 \leq r \leq n$ and $0 \leq t_0 \leq \cdots \leq t_n \leq s$. We seek to show that

(16) $$P(X(t+s) = k \mid X(s) = j, A) = p_{jk}(t).$$

Let B be any event that gives a complete description of the process $X(u), 0 \leq u \leq s$. (It is here that our sketch proof omits a number of technical details about the nature of events like these.) Then B certainly includes the length of time for which the process has been in j prior to s.

Hence (by the lack-of-memory property of the exponential density), conditional on B the further holding time after s is Exponential (q_j). This does not depend on anything in B, and this and subsequent holding times and jumps are independent of the process prior to s. Furthermore, the construction of the process is exactly that given in (13) with $Z_0 = j$. Formally, then,

(17) $$P(X(t+s) = k, X(s) = j, A \mid B) = p_{jk}(t) P(X(s) = j, A \mid B).$$

Now removing the conditioning on B gives, by conditional expectation,

(18) $$P(X(t+s) = k \mid X(s) = j, A) = p_{jk}(t).$$

Conversely, given $p_{jk}(t)$, the Chapman–Kolmogorov equations yield all the joint distributions of $X(t)$ including its holding times and the jump chain. Thus, (13) and (18) are equivalent definitions of $X(t)$. ○

These ideas can be most easily seen (free of technical details) in the context of the Poisson process.

(19) **The Poisson process.** We defined this in several ways in (2.5), but most importantly in (2.5.7) we defined it as a counting process with independent identically distributed Exponential (λ) intervals between events. That is to say, in the terminology of (13) above, the holding times all have the same parameter $q_j = \lambda$, and the jump chain has transition matrix $r_{jj+1} = 1, j \geq 0$.

Then, in (2.5.8), we showed that the Poisson process has independent increments. The Markov property follows immediately from that, and it is useful and instructive to re-examine the argument in (2.5.8), which will be seen as a special case of the sketch proof in (15) above. ○

Here is another interesting example.

(20) **Poisson-driven Markov chain.** Let $N(t)$ be a Poisson process of rate λ, and let X_n be a Markov chain with transition matrix **P**. Let $X(t)$ be a process

that takes a step, whose distribution is given by **P**, at each event of the Poisson process $N(t)$. We can write

(21) $$X(t) = X_{N(t)}, \quad t \geq 0.$$

In the terminology of (15), the holding times all have the same parameter $q_j = \lambda$, and the jump chain has transition matrix $\mathbf{R} = \mathbf{P}$. Therefore, $X(t)$ is a Markov chain in continuous time.

A Markov process constructed in this way is sometimes called a *uniform*, or *subordinated*, chain; we return to this idea in (4.2.41). ○

A classic example of this type is the

(22) **Telegraph process.** Let $N(t)$ be a Poisson process, and define

(23) $$Z(t) = (-1)^{N(t)}, \quad t \geq 0.$$

It is an exercise for you to verify explicity that this is a Markov process.
○

Another important elementary chain is this:

(24) **Yule process.** This is a process $X(t)$ taking values in the positive integers. In terms of the holding time and jump chain it is defined by $r_{k,k+1} = 1, k \geq 1$, and $q_k = \lambda k$. It arises as a population model in which any individual enjoys a lifetime that is Exponential (λ) before splitting into two fresh individuals; all individuals live and split independently. Thus, if $X(t) = n$, the time until one of the n individuals alive splits into two is Exponential (λn). Assume $X(0) = 1$ and, without loss of generality, we let $\lambda = 1$. By construction,

(25) $$P(X(t) = n) = P(J_n > t) - P(J_{n-1} > t),$$

where the jump times J_n were defined in (14).

Now we claim that, for $t \geq 0$, and $n \geq 1$,

(26) $$P(X(t) = n) = e^{-t}(1 - e^{-t})^{n-1}$$

and we prove it by showing that both the above expressions have the same Laplace transform. (This is an exercise for you.) We develop an altogether easier method (for the transform-averse reader) in the next section. ○

Finally, in this section we recall that Markov processes in discrete time satisfied the strong Markov property, and note that continuous-time Markov processes also have this property. We sketch the basics, omitting technical details. As in the discrete case, we require this definition:

(27) **Stopping time.** A non-negative random variable T is a stopping time for the Markov process $X(t)$ if the indicator of the event $\{T \leq t\}$ is a function only of values of $\{X(s) : s \leq t\}$, and does not depend on values of $\{X(s) : s > t\}$. △

Certainly the most important examples of stopping times are first-passage times such as,

(28) $$T_{jk} = \inf\{t > 0 : X(t) = k; X(0) = j \neq k\}.$$

For such times, just as in the discrete case, we need and use this theorem:

(29) **Strong Markov property.** Let $X(t)$ be a Markov process and T a stopping time for X. Let D be any event depending on $\{X(s) : s > T\}$ and B any event depending on $\{X(s) : s \leq T\}$. Then

(30) $$P(D \mid X(T) = j, B) = P(D \mid X(T) = j).$$

Sketch proof. Let C be the event that provides a complete description of $\{X(u) : 0 \leq u \leq T\}$. (We omit the technicalities needed to construct events like C.) Then, C includes the determination of the value of T, and we can use the ordinary Markov property at this T fixed on C to write

$$P(D \mid X(T) = j, B, C) = P(D \mid X(T) = j)$$

and the required result follows by conditional expectation:

$$P(D \mid X(T) = j, B) = E(E(I(D) \mid X(t) = j, B, C) \mid B) = P(D \mid X(T) = j),$$

where $I(D)$ is the indicator of D. More strongly, by setting $D = \{X(T + t) = k\}$, we see that

$$P(X(T + t) = k \mid X(T) = j, B) = p_{jk}(t).$$

Thus, given $X(T)$, the Markov process $X(T + t)$ after T is constructed in the same way as the original $X(t)$, started at time zero, and is independent of $X(u)$, $0 \leq u \leq T$.

> **Exercises**
> (a) Verify that the transition probabilities of the Poisson process (8) satisfy the Chapman–Kolmogorov equations (10).

(b)* **Yule process.**
   (i) Let $p_{1k}^*(\theta) = \int_0^\infty e^{-\theta t} e^{-t}(1-e^{-t})^{k-1} \, dt$. Show that $p_{1k}^*(\theta) = p_{1k-1}^*(\theta) - p_{1k-1}^*(1+\theta)$, $k \geq 2$, and deduce that

$$p_{1k}^*(\theta) = \frac{1}{1+\theta} \cdot \frac{2}{2+\theta} \cdots \frac{k-1}{k-1+\theta} \cdot \frac{1}{k+\theta}.$$

   (ii) Let $J_k = \sum_{r=1}^k X_r$, where X_r is Exponential (r), and the Xs are independent. Show that

$$\int_0^\infty e^{-\theta t} P(J_k > t) \, dt = \frac{1}{\theta}\left\{1 - \frac{1}{1+\theta} \cdots \frac{k-1}{k-1+\theta} \cdot \frac{k}{k+\theta}\right\}$$

$k \geq 1$.

   (iii) Hence establish equation (26).

(c)* **Yule process continued.** Show that the general transition probabilities of the Yule process are

$$p_{jk}(t) = \binom{k-1}{j-1} e^{-\lambda j t}(1-e^{-\lambda t})^{k-j}, \quad 1 \leq j \leq k.$$

(d) **Telegraph process.** Verify explicitly that the Telegraph process defined in (23) has the Markov property.

(e) Show that the transition probabilities $p_{jk}(t)$, defined in (6), are continuous, in that, for $s, t \geq 0$,

$$|p_{ik}(s+t) - p_{ik}(t)| \leq 1 - p_{ii}(s) \leq 1 - \exp(-q_i s).$$

[Hint: Use (10).] Note that the continuity is uniform in k.

4.2 Forward and backward equations

The holding-time and jump-chain characterization of a Markov chain $X(t)$ has several good points: it shows that such processes exist, gives a conceptually easy method for simulating $X(t)$, and supplies a nice picture of what it does. However, it does not in general (very simple processes excepted) immediately supply a tractable way of answering several basic questions: for example, the value of $P(X(t) = k)$ for all t, or the joint distribution of $X(t_1), \ldots, X(t_n)$.

4.2 Forward and backward equations

In discrete time, use of the transition matrix **P** answers all such questions. In continuous time we shall now discover that there is an analogous matrix **Q** that plays a similar role. We generate this **Q**-matrix by looking at the distribution of increments in the chain over arbitrarily small time-intervals.

Explicitly, for $X(0) = j$, as $h \downarrow 0$, we write

$$p_{jj}(h) = P(X(h) = j \mid X(0) = j)$$
(1) $$\geq P(J_1 > h \mid X(0) = j), \quad \text{where } J_1 \text{ is the time of the first jump,}$$
$$= e^{-q_j h} = 1 - q_j h + o(h).$$

Similarly, for $k \neq j$, the chain may go there (to k) during $[0, h)$ and not leave. Hence,

$$p_{jk}(h) = P(X(h) = k \mid X(0) = j)$$
(2) $$\geq (1 - e^{-q_j h}) r_{jk} e^{-q_k h}$$
$$= q_j r_{jk} h + o(h).$$

Now suppose the state space is finite. Summing (1) and (2) gives

$$1 = \sum_{k \in S} p_{jk}(h) \geq 1 + \left(\sum_{k \in S} r_{jk} - 1 \right) q_j + o(h)$$
(3) $$\geq 1 + o(h), \quad \text{since } \mathbf{R} \text{ is a stochastic matrix.}$$

Hence equality holds in (1) and (2), because if any inequality were strict, this would establish that $1 > 1$.

Writing $q_{jk} = q_j r_{jk}$, we therefore have the formulas

(4) $$p_{jj}(t) = 1 - q_j t + o(t)$$

and

(5) $$p_{jk}(t) = q_{jk} t + o(t), \quad j \neq k, \quad \text{for small } t.$$

This representation explains why q_{jk} is called the *transition rate* from j to k; it may also be called the jump rate or instantaneous transition rate. The quantities $q_{jk} t$ and $1 - q_j t$ are sometimes called the infinitesimal transition probabilities. In the interests of uniformity, it is conventional to set

(6) $$q_j = -q_{jj}$$

and then the array

(7) $$\mathbf{Q} = (q_{jk})$$

is called the **Q-matrix**. By construction its row sums are zero. (Note that some books write $g_{jk} = q_{jk}$, and refer to the resulting **G-matrix** as the *generator* of the chain.) These preliminaries lead us to a result of great importance.

(8) **Forward equations.** Because we are still holding the state space finite, we may use the Chapman–Kolmogorov equations, together with (4), (5), and (6) to give, for any j and k

(9)
$$\lim_{h \to 0} \frac{1}{h} [p_{jk}(t+h) - p_{jk}(t)] = \lim_{h \to 0} \frac{1}{h} \left[\sum_{r \in S} p_{jr}(t) p_{rk}(h) - p_{jk}(t) \right]$$
$$= \lim_{h \to 0} \frac{1}{h} \left[\sum_{r \in S} p_{jr}(t) q_{rk} h + o(h) \right]$$
$$= \sum_{r \in S} p_{jr}(t) q_{rk}.$$

Therefore, $p_{jk}(t)$ is differentiable, and in matrix form the equations (9) may be written concisely as

(10) $$\mathbf{P}' = \mathbf{PQ},$$

where the superscript prime denotes differentiation with respect to t, and these are called the *forward equations*. ○

In many classic cases of interest, these equations may be solved to yield any required joint distributions of $X(t)$. Here are some first examples; we see more later.

(11) **Flip–flop process (4.1.4) revisited.** Because the chain is held in states 0 and 1 for exponential times with parameters λ and μ, respectively, we have immediately that $q_0 = \lambda$, $q_1 = \mu$, and the jump chain has $r_{01} = r_{10} = 1$. Hence, when $X(0) = 0$, the forward equations are

(12) $$p'_{00}(t) = -\lambda p_{00}(t) + \mu p_{01}(t),$$
(13) $$p'_{01}(t) = \lambda p_{00}(t) - \mu p_{01}(t).$$

We have $p_{00}(t) + p_{01}(t) = 1$ for all t, and the initial conditions $p_{00}(0) = 1$; $p_{01}(0) = 0$. It is now quick and easy to obtain

(14) $$p_{00}(t) = \{\mu + \lambda e^{-(\lambda+\mu)t}\}/(\lambda+\mu),$$

(15) $$p_{01}(t) = \{\lambda - \lambda e^{-(\lambda+\mu)t}\}/(\lambda+\mu).$$

By interchanging the roles of λ and μ we readily find, when $X(0) = 1$, that

(16) $$p_{11}(t) = \{\lambda + \mu e^{-(\lambda+\mu)t}\}/(\lambda+\mu),$$

(17) $$p_{10}(t) = \{\mu - \mu e^{-(\lambda+\mu)t}\}/(\lambda+\mu).$$

Such simple expressions for $p_{jk}(t)$ are rare. ○

When the state space S is infinite, many of the preceding arguments are invalid without a suitable extra condition. For a first example, it is no longer necessarily true that $\mathbf{P}(t)$ is a stochastic matrix for all t. It may be that

$$\sum_{k \in S} p_{jk}(t) < 1$$

and we give an example in Section 4.3. One can exclude this possibility by imposing conditions on $\mathbf{Q}$; for example, if $|q_{jj}| < c < \infty$, for all j, then $\sum_k p_{jk}(t) = 1$ for all t.

However, this condition is not necessary, and is not satisfied by even quite ordinary and well-behaved chains such as the simple birth process above.

Second, the interchange of limit and summation in (9) is no longer necessarily justified. One can guarantee that it is by imposing an extra condition on $\mathbf{P}(t)$, for example, if for given k, convergence to the limit

(18) $$\frac{1}{h} p_{jk}(h) \to q_{jk}$$

is uniform in j, then the forward equations follow.

We can avoid having to confront this difficulty by deriving a different set of equations satisfied by $p_{jk}(t)$.

Essentially, we obtained the forward equations for $p_{jk}(t)$ by conditioning on the state of the process $X(t)$ at time t, and considering all the possible eventualities in $(t, t+h)$ for small h. By contrast, we now obtain equations for $p_{jk}(t)$ by conditioning on the first jump of $X(t)$ from j. For this reason they are called the

(19) **Backward equations.** Let $X(0) = j$, and suppose that the first jump is from j to i at time $J_1 = s < t$. Conditional on this event, by the Markov property at s, the probability of the event $\{X(t) = k\}$ is $p_{ik}(t - s)$.

If the first jump is at time $J_1 = s > t$, then the chain remains at j. Now recall that J_1 has density $q_j e^{-q_j s}$, and the chain jumps from j to i with probability r_{ji}. Therefore, unconditionally,

$$(20) \quad p_{jk}(t) = \delta_{jk} e^{-q_j t} + \sum_{i \neq j} r_{ji} \int_0^t q_j e^{-q_j s} p_{ik}(t - s) \, ds,$$

where

$$\delta_{jk} = \begin{cases} 1, & \text{if } j = k, \\ 0, & \text{if } j \neq k. \end{cases}$$

We can interchange the order of summation and integration in (20), because the summands are non-negative. If we also multiply by $e^{q_j t}$, and set $t - s = v$ in the integrals, then the equations take the form

$$(21) \quad e^{q_j t} p_{jk}(t) = \delta_{jk} + \int_0^t q_j e^{q_j v} \sum_{i \neq j} r_{ji} p_{ik}(v) \, dv.$$

It follows that $p_{jk}(t)$ is differentiable for all $t \geq 0$. To see this, recall that $|p_{ik}(v)| \leq 1$, and $\sum_{i \neq j} r_{ji} = 1$. Therefore, the integral and hence $p_{jk}(t)$, are both continuous in t. Hence, more strongly, the integrand $q_j e^{q_j v} \sum_{i \neq j} r_{ji} p_{ik}(v)$ is continuous, because it is a uniformly convergent sum of continuous functions. Thus (by the fundamental theorem of calculus) we can differentiate (21) to obtain

$$(22) \quad q_j p_{jk}(t) + \frac{d}{dt} p_{jk}(t) = q_j \sum_{i \neq j} r_{ji} p_{ik}(t).$$

Finally, using $q_j = -q_{jj}$ and $q_j r_{ji} = q_{ji}$, these backward equations take the simple and memorable form

$$(23) \quad \frac{d}{dt} p_{jk}(t) = \sum_i q_{ji} p_{ik}(t).$$

In matrix form this is even more concise:

$$(24) \quad \mathbf{P}' = \mathbf{QP}.$$

Contrast this with the forward equations $\mathbf{P}' = \mathbf{PQ}$.

Some remarks are in order here: first we stress that (unlike the forward equations) we needed no extra technical assumptions to derive the backward equations. Second, we note that (10) and (24) yield

(25) $$\mathbf{QP} = \mathbf{PQ},$$

which is remarkable, as matrices do not in general commute.

Third, it is natural to wonder if we can solve the forward and backward equations using this compact matrix notation. If **P** and **Q** were scalars, we would have easily that $\mathbf{P}(t) = \exp(t\mathbf{Q})$, with the natural initial condition $\mathbf{P}(0) = 1$. This observation makes it appropriate to introduce the series

(26) $$\sum_{r=0}^{\infty} \mathbf{Q}^r t^r / r!,$$

whose sum is denoted by $\exp(t\mathbf{Q})$. As usual, matters are much more straightforward for chains with finite state space, and in that case we have this:

(27) **Theorem.** For a chain with finite state space, $\mathbf{P}(t) = \exp(t\mathbf{Q})$ is the unique solution of the forward and backward equations, with initial condition $\mathbf{P}(0) = \mathbf{I}$. (Here, **I** is the matrix with 1s in the diagonal and 0s elsewhere.) Furthermore, for all $s, t \geq 0$,

(28) $$\mathbf{P}(s+t) = \mathbf{P}(s)\mathbf{P}(t).$$

Proof. The power series defined in (26) converges for all finite t, and is therefore differentiable term by term in each component. Hence,

(29) $$\mathbf{P}'(t) = \sum_{r=1}^{\infty} \mathbf{Q}^r t^{r-1}/(r-1)! = \mathbf{QP}(t) = \mathbf{P}(t)\mathbf{Q}.$$

Trivially $\mathbf{P}(0) = \mathbf{I}$, and (28) is immediate because $s\mathbf{Q}$ and $t\mathbf{Q}$ commute for any s and t. Finally, suppose $\mathbf{S}(t)$ is any solution of the backward equations and consider

$$\frac{d}{dt}[\exp(-t\mathbf{Q})\mathbf{S}] = \exp(-t\mathbf{Q})(-\mathbf{Q})\mathbf{S} + \exp(-t\mathbf{Q})\mathbf{S}'$$

(30) $$= \exp(-t\mathbf{Q})[-\mathbf{QS} + \mathbf{QS}] = 0.$$

Hence $\exp(-t\mathbf{Q})\mathbf{S} = \text{constant} = \mathbf{I}$, and thus $\mathbf{S} = \exp(t\mathbf{Q}) = \mathbf{P}$, establishing uniqueness. The forward equations are treated likewise. □

These techniques are sometimes useful but, before proceeding any further down this theoretical highway, let us illustrate the forward and backward equations in a familiar environment.

(31) **Yule process.** We derived the transition probabilities for this process in (4.1.24) using the holding-time jump-chain construction. Forward and backward equations offer a different approach.

Forward equations. Assume that $X(0) = 1$. The Q-matrix is easily seen to be

(32) $$\begin{pmatrix} -\lambda & \lambda & 0 & 0 & \cdots \\ 0 & -2\lambda & 2\lambda & 0 & \cdots \\ 0 & 0 & -3\lambda & 3\lambda & \cdots \\ \vdots & \vdots & \vdots & \vdots & \ddots \end{pmatrix},$$

so that $\mathbf{P}' = \mathbf{PQ}$ yields

(33) $$p'_{1k}(t) = (k-1)\lambda p_{1,k-1}(t) - k\lambda p_{1,k}(t), \quad k \geq 1.$$

Alternatively, you may derive (33) from first principles by using conditional probability to write

$$p_{1k}(t+h) = p_{1k}(t)(1 - k\lambda h) + p_{1,k-1}(t)(k-1)\lambda h + o(h),$$

then rearranging and taking the limit as $h \to 0$.

We seek the solution of (33) having initial condition $p_{1k}(0) = \delta_{1k}$, corresponding to $X(0) = 1$. It is easy to check that the solution is

(34) $$p_{1k}(t) = e^{-\lambda t}(1 - e^{-\lambda t})^{k-1}.$$

Alternatively, one may introduce the probability-generating function (pgf) $G(z) = \mathrm{E}z^{X(t)} = \sum_k z^k p_{1k}(t)$, which enables you to write the set of equations (33) as the single relation

(35) $$\frac{\partial}{\partial t}G = \lambda(z^2 - z)\frac{\partial G}{\partial z}.$$

To show this, multiply (33) by z^k, and sum over k to yield (35).

This can be solved by standard methods, or you can check that

(36) $$G(z,t) = z/\{z + (1-z)e^{\lambda t}\}$$

is the solution satisfying $G(z,0) = z$. Either way, $X(t)$ is Geometric $(e^{-\lambda t})$.

Backward equations. Conditioning on the time of the first jump in $(0, t)$ yields, following (20),

$$(37) \qquad p_{1k}(t) = \delta_{1k} e^{-\lambda t} + \int_0^t \lambda e^{-\lambda s} p_{2,k}(t-s) \, ds.$$

Proceeding as we did in the general case, we rewrite this as

$$(38) \qquad \frac{d}{dt}(e^{\lambda t} p_{1k}(t)) = \lambda e^{\lambda t} p_{2,k}(t)$$

and it is routine to check that the solution is as given in Exercise (4.1(c)). Alternatively, we may more easily seek a backward equation for $G(z, t) = \sum_k z^k p_{1k}(t)$. By conditioning on the first jump at J_1 we obtain,

$$G = E(E(z^{X(t)} \mid J_1))$$
$$= z e^{-\lambda t} + \int_0^t \lambda e^{-\lambda s} [G(z, t-s)]^2 \, ds.$$

To see why G^2 appears in this integral, recall from (4.1.24) that the process arises as a population model; the two individuals present at J_1 give rise to a sum of two independent Yule processes.

Rearranging in the usual way yields

$$(39) \qquad \frac{\partial G}{\partial t} = \lambda(G^2 - G),$$

which is easily solved to give (36) as the unique solution such that $G(z, 0) = z$. See Problem (10) for more details. ○

In this case, the Yule process, we have shown explicitly that the solution of the forward equations is unique, and satisfies the backward equations (and, of course, vice versa). This is not always true, because of difficulties arising when $\sum_k p_{jk}(t) < 1$. Such problems will not arise in the models and processes we consider, so we relegate their resolution to the next section which deals only with that special case.

A natural question is to wonder whether these transition rates q_{jk} are sufficient to define the Markov process $X(t)$ directly, without developing the holding-time jump-chain construction. That is to say, can we use this as a definition?

(40) **Infinitesimal construction.** Let $X(t)$ take values in the state space S, and let q_{jk} satisfy, for all j,k in S,

(a) $q_{jk} \geq 0$, $j \neq k$;
(b) $0 \leq -q_{jj} < \infty$, $j \in S$;
(c) $\sum_{k \in S} q_{jk} = 0$, $j \in S$.

The distribution of $X(t)$ is defined for all $t \geq 0$ by

(i) Conditional on $X(t) = j$, $X(t+s)$ is independent of $X(u)$, ($0 \leq u \leq t$), where $s \geq 0$.
(ii) As $h \downarrow 0$, for all k, uniformly in t,

$$P(X(t+h) = k \mid X(t) = j) = \delta_{jk} + q_{jk}h + o(h).$$

It turns out that if S is finite this does define a Markov process whose transition probabilities satisfy the forward and backward equations. Unfortunately, if S is infinite then complications may ensue, and it is beyond our scope to explore all the difficulties. In practice, and in all the examples we consider, definition (40) does in fact characterize the Markov chain with holding times that are exponential with parameter $-q_{jj} = q_j < \infty$, and jump chain given by

$$r_{jj} = \begin{cases} 1, & q_j = 0, \\ 0, & q_j \neq 0 \end{cases}$$

and for $j \neq k$,

$$r_{jk} = \begin{cases} 0, & q_j = 0, \\ q_{jk}/q_j, & q_j \neq 0. \end{cases}$$

It is, therefore, common practice to specify models and processes in terms of **Q**, while remaining alert for possible complications. △

Finally in this section we consider a useful idea that aids simulation. It is known as:

(41) **Uniformization.** In our general approach to Markov processes we did not include the possibility that the chain might 'jump' to its current position at the end of its holding time there. This is natural, because of course such jumps to the same place are invisible.

However (for reasons that will emerge), let us extend our model to permit such non-moving jumps in the flip–flop model.

Specifically, let $Y(t) \in \{0, 1\}$ behave like this:

Whichever state it is in, after a holding time that is exponential with parameter $\lambda + \mu$, it seeks to move. If it is in 0, it goes to 1 with probability

$\lambda/(\lambda+\mu)$, or remains in 0 with probability $\mu/(\lambda+\mu)$ and begins another holding time. If it is in 1, it moves to 0 with probability $\mu/(\lambda+\mu)$, or remains in 1 to begin another holding time.

Then (see Exercise (e)) it actually succeeds in moving from 0 to 1 (or 1 to 0), after a total sojourn that is Exponential (λ) (or Exponential (μ)), respectively. This new process is essentially the same as the original flip–flop process, but all the 'jumps' (including the non-moving jumps) take place at the instants of a Poisson process having intensity $\lambda + \mu$.

This fact greatly aids simulation, which is now easily executed by first simulating a Poisson process, and then (in effect) flipping a sequence of biased coins to decide which jumps are movers and which are stayers.

The entire procedure is called *uniformization*, and can be readily extended to any Markov process that has uniformly bounded rates q_{jj}, $j \in S$.

In detail, consider a Markov chain $X(t)$ with jump-chain transition probabilities r_{jk}, and such that

(42) $$q_j \leq \nu < \infty, \quad j \in S.$$

Following the idea introduced above for the flip–flop chain, define $U(t)$ with the following structure:

(a) The holding times are independent and Exponential (ν). They thus form a Poisson process.
(b) At the end of each holding time the chain makes a transition according to this stochastic matrix of probabilities

(43) $$u_{jk} = q_j r_{jk}/\nu, \quad j \neq k,$$
(44) $$u_{jj} = 1 - q_j/\nu.$$

Just as for the flip–flop chain, it is easy to see that the effective holding time in j before a jump to a different state is Exponential (q_j), and the jump-chain probabilities (conditional on leaving j), are r_{jk}. This $U(t)$ is easily simulated, and is said to be the uniformized version of $X(t)$. Furthermore, this supplies a tractable expression for the transition probabilities $p_{jk}(t)$ of $X(t)$ in terms of the transition probabilities $u_{jk}(n)$ of the jump chain U_n of $U(t)$. We display it as follows:

(45) **Lemma.** We have

$$p_{jk}(t) = \sum_{n=0}^{\infty} u_{jk}(n) \, e^{-\nu t} (\nu t)^n / n!.$$

Proof. Let $N(t)$ be the number of 'jumps' of $U(t)$; these include those which do not change the state. We know that $N(t)$ is Poisson (vt), and conditional on $N(t)$

$$P(X(t) = k \mid X(0) = j) = P(U(t) = k \mid U(0) = j)$$

$$= \sum_{n=0}^{\infty} P(U(t) = k \mid U(0) = j, N(t) = n) \, e^{-vt} (vt)^n / n!$$

$$= \sum_{n=0}^{\infty} P(U_n = k \mid U_0 = j; N(t) = n) \, e^{-vt} (vt)^n / n!$$

$$= \sum_{n=0}^{\infty} u_{jk}(n) \, e^{-vt} (vt)^n / n!$$

□

Exercises

(a)* Write down the forward and backward equations for the Poisson process, and solve them.

(b) Find the forward and backward equations for the flip–flop chain (4.1.11), and solve them. (Note that one set is easier to solve than the other.)

(c) Show that transition probabilities of the process $X(t)$ defined in (40) satisfy the forward and backward equations when the state space is finite.

(d) Let $\mathbf{Q}$ be a finite matrix. Show that it is the Q-matrix of a Markov chain $X(t)$ if and only if $\mathbf{P}(t) = \exp(t\mathbf{Q})$ is stochastic for all $t \geq 0$.

(e)* Let N be Geometric (p) and independent of the independent family $(X_r; r \geq 1)$ of Exponential (λ) random variables. Show that $Y = \sum_{r=1}^{N} X_r$ has an exponential density, and find its parameter.

4.3 Birth processes: explosions and minimality

Deriving the forward and backward equations for finite chains has been straightforward and fruitful. By contrast, our treatment of chains with infinite state space has been guarded and incomplete. In this section we look at an example to illustrate the difficulty, and then outline its resolution.

(1) Birth process. This is a Markov counting process $X(t)$ in which the rate at which events occur depends on the number counted already. It arises naturally as a model for the growth of populations, for example. It may be defined in terms of holding times thus:

(2) **Definition A.** Conditional on $X(0) = j$, the successive holding times are independent exponential random variables with parameters $\lambda_j, \lambda_{j+1}, \ldots$. The jump chain is trivial: $p_{j,j+1} = 1$, $j \geq 0$. The parameters $(\lambda_j; j \geq 0)$ are called the *birth rates* of the process.

Alternatively, it may be defined in terms of its increments thus:

(3) **Definition B.** If $s < t$ then, conditional on the value of $X(s)$, the increment $X(t) - X(s)$ is non-negative and independent of $(X(u); 0 \leq u \leq s)$; furthermore as $h \downarrow 0$, uniformly in $t \geq 0$,

$$P(X(t+h) = j+m \mid X(t) = j) = \begin{cases} \lambda_j h + o(h), & \text{if } m = 1, \\ o(h), & \text{if } m > 1, \\ 1 - \lambda_j h + o(h), & \text{if } m = 0, \end{cases}$$

for all $j \geq 0$.

By conditioning on $X(t)$, either of these characterizations yields

$$p_{jk}(t+h) - p_{jk}(t) = p_{j,k-1}(t)\lambda_{k-1}h - p_{jk}(t)\lambda_k h + o(h), \quad 0 \leq j \leq k < \infty,$$

where $p_{j,j-1}(t) = 0$. Hence $p_{jk}(t)$ is differentiable, and we have the forward equations

(4) $\qquad p'_{jk} = \lambda_{k-1} p_{j,k-1} - \lambda_k p_{jk}, \quad k \geq j, \text{ where } p_{j,j-1} = 0.$

Alternatively, conditioning on the time of the first jump yields

(5) $\qquad p_{jk}(t) = \delta_{jk} e^{-\lambda_j t} + \int_0^t \lambda_j e^{-\lambda_j s} p_{j+1,k}(t-s) \, ds,$

where δ_{jk} is the usual delta-function. It follows that p_{jk} is differentiable, and we have the backward equations

(6) $\qquad p'_{jk} = \lambda_j p_{j+1,k} - \lambda_j p_{jk}, \quad k \geq j.$

The key result is the following theorem.

(7) **Theorem.** With the initial (boundary) condition $p_{jk}(0) = \delta_{jk}$, the forward equations (4) have a unique solution which satisfies the backward equations (6).

Proof. Taking the forward equations in natural order, we find trivially that $p_{jk}(t) = 0$ for $k < j$. Then

$$p_{jj}(t) = e^{-\lambda_j t},$$

(8) $$p_{j,j+1}(t) = e^{-\lambda_{j+1} t} \int_0^t \lambda_j e^{-(\lambda_j - \lambda_{j+1})s} \, ds$$

(9) $$= \frac{\lambda_j}{\lambda_j - \lambda_{j+1}} \{e^{-\lambda_{j+1} t} - e^{-\lambda_j t}\}.$$

This procedure may be iterated to find $\lambda_{jk}(t)$ for any $k \geq j$ using the sequence of integrals

(10) $$p_{jk}(t) = \lambda_{k-1} \int_0^t e^{-\lambda_k s} p_{j,k-1}(t-s) \, ds,$$

obtained from (4).

This solution is unique, by construction. To show that it also satisfies the backward equations (6), we may differentiate (10) with respect to t, and use induction to verify that $p_{jk}(t)$ is a solution of (6) for all j, and fixed k.

Alternatively, it is neater and quicker to use the family of Laplace transforms

(11) $$p_{jk}^*(\theta) = \int_0^\infty e^{-\theta t} p_{jk}(t) \, dt.$$

The transformed version of the forward equations (4) is

$$(\theta + \lambda_k) p_{jk}^* = \delta_{jk} + \lambda_{k-1} p_{j,k-1}^*$$

with solution

(12) $$p_{jk}^* = \frac{\lambda_j}{\lambda_j + \theta} \cdot \frac{\lambda_{j+1}}{\lambda_{j+1} + \theta} \cdots \frac{\lambda_{k-1}}{\lambda_{k-1} + \theta} \cdot \frac{1}{\lambda_k + \theta}, \quad k \geq j.$$

This determines $p_{jk}(t)$ uniquely, by Laplace transform theory. The transformed version of the backward equations is

(13) $$(\theta + \lambda_j) p_{jk}^* = \delta_{jk} + \lambda_j p_{j+1,k}^*$$

and it is easy to check that (12) satisfies these equations, yielding a solution $p_{jk}(t)$ of the backward equations. It does *not* follow that this solution is unique, and indeed it may not be. The best we can do is this:

(14) **Theorem.** If $p_{jk}(t)$ is the unique solution of the forward equations, and $b_{jk}(t)$ is any solution of the backward equations, then

(15) $$p_{jk}(t) \leq b_{jk}(t), \quad \text{for all } j, k, t.$$

Proof. Given $b_{j+1,k}(t)$, we have from the backward equations (5) that $b_{jk}(t)$, and also in particular $p_{jk}(t)$, satisfy the equations

(16) $$y_{jk}(t) = e^{-\lambda_j t}\delta_{jk} + \int_0^t \lambda_j e^{-\lambda_j s} y_{j+1,k}(t-s)\,ds.$$

Hence, if $p_{j+1,k}(t) \leq b_{j+1,k}(t)$ for any j, it follows that $p_{jk}(t) \leq b_{jk}(t)$. But it is trivial to see that $0 = p_{k+1,k}(t) \leq b_{k+1,k}(t) = 0$. The required result now follows by induction on the sequence $j = k, k-1, k-2, \ldots$. □

This result has the following important consequences, which explains all our earlier caution.

(17) **Corollary.** If for all t we have $\sum_k p_{jk}(t) = 1$, then the solution of the backward equations is the unique solution $p_{jk}(t)$ of the forward equations.

Proof. This is immediate, because we then have

(18) $$1 = \sum_k p_{jk}(t) \leq \sum_k b_{jk}(t) \leq 1.$$

If the first inequality were strict for any k, then $1 < 1$, which is a contradiction. ○

The possibility that $\sum_k p_{jk}(t) < 1$ is formalized in a definition.

(19) **Honesty and explosions.** Let J_n be the jump times of the Markov process $X(t)$, and define

$$J_\infty = \lim_{n\to\infty} J_n = \sup_n J_n.$$

If $P(J_\infty = \infty) = 1$, then the process is said to be *honest*. In this case, $\sum_k p_{jk}(t) = 1$ for all t. If $P(J_\infty < \infty) > 0$, then the process is said to be *dishonest*. In this latter case the process may make infinitely many jumps in a finite time; this is called explosion, and J_∞ is the explosion time. △

In the case of the birth process, there is a simple criterion for honesty:

(20) Theorem. Let $X(t)$ be the birth process with birth rates $(\lambda_j; j \geq 0)$. Then $X(t)$ is honest if and only if $\sum_{j=0}^{\infty} 1/\lambda_j = \infty$.

Proof. Let the holding times be denoted by $(Z_j; j \geq 0)$. Then, because the terms of the sum are non-negative, we can calculate

$$(21) \qquad \mathrm{E} J_\infty = \mathrm{E} \sum_{n=0}^{\infty} Z_n = \sum_{n=0}^{\infty} \mathrm{E} Z_n = \sum_n \lambda_n^{-1}.$$

Therefore, if $\sum_n \lambda_n^{-1} < \infty$, $\mathrm{P}(J_\infty < \infty) = 1$.

For the converse, consider $\mathrm{E} e^{-J_\infty}$. By monotone convergence (and independence) we calculate

$$\mathrm{E} e^{-J_\infty} = \mathrm{E} \left(\prod_{n=0}^{\infty} e^{-Z_n} \right) = \lim_{m \to \infty} \prod_{n=0}^{m} \mathrm{E}(e^{-Z_n})$$

$$(22) \qquad\qquad = \prod_{n=0}^{\infty} (1 + \lambda_n^{-1})^{-1},$$

when we recall that $\mathrm{E} e^{-Z_n} = \lambda_n/(1 + \lambda_n)$.

Now the product in (22) is zero if and only if $\sum_{n=0}^{\infty} \lambda_n^{-1} = \infty$, and in this case $\mathrm{E} e^{-J_\infty} = 0$. Because $e^{-J_\infty} \geq 0$, this occurs if and only if $\mathrm{P}(J_\infty < \infty) = 0$, as required. □

The ideas we have introduced in the context of the basic birth process apply equally to continuous-time Markov chains in general. However, the details and proofs are not short, so we content ourselves with this summary. Recall that a solution $(x_0, x_1, \ldots)$ is said to be minimal if, for any other solution $(y_0, y_1, \ldots)$, we have $y_r \geq x_r$, for all r.

(23) Minimality. Let $X(t)$ be a Markov process with transition matrix $\mathbf{P}(t)$, and generator $\mathbf{Q}$. The backward equations $\mathbf{P}' = \mathbf{QP}$ have a minimal non-negative solution $\mathbf{P}$ satisfying $\mathbf{P}(0) = \mathbf{I}$. This solution is also the minimal non-negative solution of the forward equations $\mathbf{P}' = \mathbf{PQ}$ that satisfies $\mathbf{P}(0) = \mathbf{I}$.

The solution $\mathbf{P}$ is unique if the process is honest, having no explosion.

We do not pursue the details of this any further. In practice, models that correspond to real applications usually lead to honest processes. Explosive and non-minimal Markov chains are chiefly of interest to mathematicians having a keen eye for those bizarre examples that run counter to our intuition. One constructs the more freakish examples by extending the

Markov process after its explosion. Such chains satisfy the backward equations but not the forward equations, and a great deal of harmless fun is available constructing variously atrocious processes, if you have the time and energy. But all the chains in the rest of this book are honest, and the following theorem is useful.

(24) Honesty. Let $X(t)$ be a Markov process with generator $\mathbf{Q}$, and jump chain $(Z_n; n \geq 0)$. Then $X(t)$ does not explode if any one of the following is true:

(a) $\sup_j q_j < \infty$;
(b) the state space S is finite;
(c) $X(0) = j$, where j is a recurrent state for the jump chain Z.

The proof is Exercise (c).

Exercises
(a) **Yule process.** Let J_n be the time of the nth jump in the Yule process. Show that with probability 1, $J_n \to \infty$ as $n \to \infty$. (The Yule process cannot explode.)
(b) Show that the two definitions (2) and (3) of the birth process in (1) are equivalent.
(c)* **Honesty.** Prove the final result (24) of this section.

4.4 Recurrence and transience

Just as for chains in discrete time, we identify certain important types of essentially distinct behaviour; fortunately, in continuous time this is more straightforward.

The following summary echoes (3.2) and (3.5) with X_n replaced by $X(t)$.

(1) Chain structure. The state k is said to be *accessible* from j if $p_{jk}(t) > 0$ for some $t \geq 0$; we write $j \to k$. If also $k \to j$, then j and k are *mutually accessible* (or intercommunicating), and we write $j \leftrightarrow k$.

This relationship gives rise to communicating classes of mutually accessible states as in (3.2.5) and (3.2.6). If C is a class that the chain cannot leave then C is closed and if S comprises exactly one communicating class then the chain is said to be *irreducible*.

An important and useful property of chains in continuous time is this: for any pair of states either $p_{jk}(t) > 0$ for all $t > 0$, or $p_{jk}(t) = 0$ for all $t > 0$. Clearly this must be linked to the generator matrix $\mathbf{Q}$ of the chain, and we display the connexion formally.

(2) **Accessibility**. The state k is accessible from j if and only if either of the following is true:

(a) There exist $j_1, j_2, \ldots, j_n$ such that

$$q(j, j_1)q(j_1, j_2), \ldots, q(j_n, k) > 0,$$

where $(q(j, k)) = \mathbf{Q}$.

(b) $p_{jk}(t) > 0$, for all $t > 0$.

Proof. If we can show that (a) implies (b), the rest is easy. Note that from (4.2.40) for any $j \neq k$, $q_{jk} > 0$ if and only if $r_{jk} > 0$. For any states j and k such that $q(j, k) = q_{jk} > 0$ we have, by considering that the first jump may be to k at time $J_1 < t$, and the second jump at $J_2 > t$,

$$p_{jk}(s) \geq P(J_1 \leq s, Z_1 = k, J_2 > s \mid Z_0 = j)$$

(3)
$$\geq (1 - e^{-q_j s}) r_{jk} e^{-q_k s}$$

$$> 0, \quad \text{for all } s > 0.$$

Therefore, assuming (a) is true, and using $s = t/n$,

$$p_{jk}(t) \geq p_{j,j_1}(s), \ldots, p_{j_n, k}(s) > 0, \quad \text{for all } t > 0.$$

□

If the chain can reach k from j, the next natural questions are to ask if, when, and how often it does so. This leads to fresh definitions and notation similar to that for discrete chains.

(4) **First-passage times**. For any two states j and k we define for $X(0) = j$,

$$T_{jk} = \inf(t > J_1 : X(t) = k).$$

This is the first-passage time from j to k, and in particular

$$\mu_j = ET_{jj}$$

is called the *mean recurrence time* of j.

(5) **Transience**. The state j is *transient* if

$$P(\text{the set } \{t : X(t) = j\} \text{ is unbounded} \mid X(0) = j) = 0.$$

(6) **Recurrence**. The state j is *recurrent* if

$$P(\text{the set } \{t : X(t) = j\} \text{ is unbounded} \mid X(0) = j) = 1.$$

If $ET_{jj} = \infty$, then j is *null-recurrent*; if $ET_{jj} < \infty$, then j is *non-null recurrent*. (Recurrent states may be called *persistent*.) △

Fortunately, in working out the consequences of these definitions we can use both the close link between $X(t)$ and its jump chain, and also the interplay between $X(t)$ and $\mathbf{P}(t)$. Here are the essential results:

(7) **Theorem.**

(i) If $q_j = 0$, then j is recurrent.
(ii) If $q_j > 0$, then j is recurrent for $X(t)$ if and only if it is recurrent for the jump chain Z_n.

Furthermore, j is recurrent if
$$\int_0^\infty p_{jj}(t)\,dt = \infty$$
and is transient if this integral is finite.

Proof. Part (i) is obvious; we turn to part (ii). If j is transient for Z_n, then the chain visits j only finitely often; hence the set $\{t : X(t) = j\}$ is finite with probability 1 and j is transient for $X(t)$. Conversely, if j is recurrent for Z_n and $X(0) = j$ then $X(t)$ does not explode; see (4.3.24). Hence the total duration of visits to j is unbounded and j is recurrent for $X(t)$.

Finally, since the integrand is positive, we have

$$\int_0^\infty p_{jj}(t)\,dt = E\left\{\int_0^\infty I(X(t) = j)\,dt \mid X(0) = j\right\}$$

(8)
$$= E\left(\sum_{r=0}^\infty U_r I(Z_r = j) \mid Z_0 = j\right),$$

where $(U_r; r \geq 0)$ are the holding times of $X(t)$.

Denoting the n-step transition probabilities of Z by $r_{jk}(n)$, the right-hand side of (8) is

(9)
$$\sum_{n=0}^\infty E(U_0 \mid X(0) = j) r_{jj}(n) = q_j^{-1} \sum_{n=0}^\infty r_{jj}(n),$$

which diverges if and only if j is recurrent for Z_n.

Two obvious and useful corollaries are exercises for you:

(a) Every state is either recurrent or transient.
(b) The states in an irreducible class are either all transient or all recurrent. □

Exercises
(a)* Fix $h > 0$ and let $X_n = X(nh)$, where $X(t)$ is a Markov process. Show that
  (i) if j is transient for $X(t)$, then it is transient for X_n;
  (ii) if j is recurrent for $X(t)$, then it is recurrent for X_n.

Remark: The chain X_n is called the *skeleton* of $X(t)$.

(b) Let $X(t)$ be a continuous-time simple symmetric random walk on the integers, which is to say that $p_{j,j+1}(h) = p_{j,j-1}(h) = \lambda h + o(h)$. Show that the walk is recurrent.

4.5 Hitting and visiting

Let us consider an example. Suppose that $X(t)$ is a simple birth–death process, which is to say that it has Q-matrix given by

(1) $q_{j,j+1} = \lambda j, \quad q_{j,j-1} = \mu j, \quad q_j = \lambda j + \mu j, \quad$ for $j \geq 0$.

Further suppose that $X(0) = b > 0$; what is the probability that $X(t)$ reaches $m > b$ before it hits 0 and is absorbed?

The point is that $X(t)$ hits m before 0 if and only if the jump chain does. But the jump chain has transition probabilities

$$r_{j,j+1} = \lambda/(\lambda + \mu), \quad r_{j,j-1} = \mu/(\lambda + \mu), \quad j \geq 1,$$

which is a simple random walk. Therefore, we find the required probability by reading off the appropriate result in the gambler's ruin problem (2.2.7).

Obviously, for any chain the hitting probabilities are obtained from those of the jump chain. However, if we seek the expected time to hit some fixed set, this must also depend on the holding times encountered on the way. Here is the general result (recall (3.3.5) for some background):

(2) **Hitting times.** For the Markov chain $X(t)$, with $X(0) = j$, we define the hitting time of any subset A of S by

$$H_{jA} = \inf\{t \geq 0 : X(t) \in A; X(0) = j\}$$

and denote its expectation by k_{jA}. We assume that $q_j > 0, j \notin A$. Then the collection $(k_{jA}; j \in S)$ is the minimal non-negative solution of

(3) $$\sum_{i \in S} q_{ji} k_{iA} + 1 = 0, \quad j \notin A$$

with $k_{jA} = 0$ for $j \in A$.

The proof uses the same ideas and methods as (3.3.12), and is left as an exercise for you. Here is an example.

(4) Birth–death chain. Let $X(t) \geq 0$ have **Q**-matrix

$$q_{j,j+1} = \lambda_j, \quad q_{j,j-1} = \mu_j, \quad q_j = \lambda_j + \mu_j$$

with $\mu_0 = 0$. We seek $k_{j,j+1}$, for $j \geq 0$; that is, the expected time it takes to reach $j+1$ starting from j.

The holding time in $j > 0$ has mean $(\lambda_j + \mu_j)^{-1}$; therefore, conditioning on whether the first jump is to $j+1$ or $j-1$, we have

$$k_{j,j+1} = (\lambda_j + \mu_j)^{-1} + \mu_j/(\lambda_j + \mu_j)\{k_{j-1,j} + k_{j,j+1}\}.$$

Rearranging this yields

(5) $$k_{j,j+1} = \lambda_j^{-1} + (\mu_j/\lambda_j)k_{j-1,j}, \quad j \geq 1$$

and for $j = 0$ the first jump is surely to 1, so

(6) $$k_{0,1} = \lambda_0^{-1}.$$

The solution follows by iterating (5) with the initial condition (6). ○

Finally, in this section we consider the question of the total time spent visiting some state of interest. Here is a simple example.

(7) Visits. Consider the birth–death process $X(t)$ with $\lambda_j = \mu_j = j\lambda$, and $X(0) = 1$. The chain is transient, and is absorbed at 0 with probability 1. But it is natural to ask (for example) how much time the chain spends at its initial state before absorption.

Since the jump chain is simple symmetric random walk, it is easy to see that the number of separate occasions when the chain is held in state 1 is Geometric $(\frac{1}{2})$. And these holding times are independent and Exponential (2λ). Thus the total time T for which $X(t) = 1$ is Exponential (λ).

The basic idea of this example is readily extended to similar problems.
○

Exercises
(a) **Birth–death chain.**
  (i) In (4.5.4) write down an expression for the expected time to reach j starting from 0, denoted by e_j.

(ii) When $\lambda_j = \mu_j = \lambda_0$, $j > 0$, and $\mu_0 = 0$, show that
$$e_j = \tfrac{1}{2}j(j+1)\lambda_0^{-1}.$$

(This is a random walk, with reflection at 0.)

(iii) When $\lambda_j = \mu_j = \lambda_0(j+1)$, $j > 0$, and $\mu_0 = 0$, show that
$$e_j = [j + \tfrac{1}{2}(j-1) + \tfrac{1}{3}(j-2) + \cdots + j^{-1}]\lambda_0^{-1}.$$

(This is simple birth–death, with reflection at 0.)

(b)* **Visits.** Let $X(t)$ be a continuous-time simple random walk on the integers, which is to say that $p_{n,n+1}(h) = \lambda h + o(h)$, $p_{n,n-1}(h) = \mu h + o(h)$. Find the distribution of the time V spent visiting $r > 0$ on an excursion from 0.

4.6 Stationary distributions and the long run

We have seen several examples in which transition probabilities $p_{jk}(t)$ converge as $t \to \infty$, of which the simplest is the flip–flop process. Here, from (4.2.11), $q_0 = \lambda$, $q_1 = \mu$, and as $t \to \infty$, for $j \in \{0, 1\}$,

(1) $$p_{j1} \to \frac{\lambda}{\lambda + \mu}, \quad p_{j0} \to \frac{\mu}{\lambda + \mu}.$$

We consider the general consequences of convergence, but for simplicity we first confine ourselves to a finite state space.

(2) **Equilibrium.** Suppose that, for all j and k in the finite set S,
$$\lim_{t \to \infty} p_{jk}(t) = \pi_k.$$

Then π has a number of key properties.

(i) Allowing $t \to \infty$ in the Chapman–Kolmogorov identity $\mathbf{P}(t + s) = \mathbf{P}(t)\mathbf{P}(s)$ we find from any row (all being the same) that

(3) $$\pi = \pi \mathbf{P}(s),$$

for all s. That is to say, π is an *equilibrium* or *stationary* distribution.

(ii) Allowing $t \to \infty$ in the forward equations

$$\mathbf{P}'(t) = \mathbf{P}(t)\mathbf{Q},$$

we find from any row that

(4) $$\pi \mathbf{Q} = 0.$$

(iii) Recalling from (4.2.20) that the transition probabilities r_{jk} for the jump chain satisfy

$$q_{jk} = q_j r_{jk} - q_j \delta_{jk},$$

we have

(5) $$\sum_{j \in S} \pi_j q_j (r_{jk} - \delta_{jk}) = \sum_{j \in S} \pi_j q_{jk} = 0.$$

If we denote

$$\mathbf{v} = (v_1, \ldots, v_{|S|}) = (\pi_1 q_1, \ldots, \pi_{|S|} q_{|S|}),$$

then (5) may be written as

(6) $$\mathbf{v} = \mathbf{v}\mathbf{R}.$$

Therefore, if we write $c = \sum_i \pi_i q_i$, then $\mathbf{v}/c$ is a stationary distribution for the jump chain.

In fact these three conditions on π ((3), (4), and (6)) are equivalent. The equivalence of (4) and (6) is easy; for the equivalence of (3) and (4) we have that

$$\pi \mathbf{Q} = 0 \quad \text{if and only if} \quad \sum_{n=1}^{\infty} t^n \pi \mathbf{Q}^n / n! = 0$$

for all t. Since $\mathbf{Q}^0 = \mathbf{I}$, this is equivalent to $\pi \sum_{n=0}^{\infty} t^n \mathbf{Q}^n / n! = \pi$, and the result follows from (4.2.27).

(iv) Finally, if $p_{jk}(t) \to \pi_k > 0$, we have the important result that

(7) $$\pi_k = (q_k \mathsf{E} T_{kk})^{-1}.$$

To see this, recall that visits to k form a renewal process (by the strong Markov property). If we accumulate a reward equal to the holding time Z in k at each visit, then the renewal–reward theorem (2.6.38) yields

$$\pi_k = \lim_{t\to\infty} p_{jk}(t) = \lim_{t\to\infty} \frac{1}{t}\int_0^t E(I(X(s)=k))\,ds = \frac{EZ}{ET_{kk}}$$

(8) $\qquad\qquad\qquad = (q_k ET_{kk})^{-1}.$

These facts are illuminating, but we are of course really interested in a more general version of the converse question: given a chain (on a possibly infinite state space), when does it have a stationary distribution, and when does $p_{jk}(t)$ converge to it? We must be careful here, for all the usual reasons involving explosions. However, note that the arguments of (4)–(6) remain true, even when the state space is infinite. That is to say, π is an invariant measure for $\mathbf{Q}$ if and only if $\boldsymbol{v}$ is an invariant measure for $\mathbf{R}$ where $v_k = \pi_k q_k$. It follows (recall Theorem 3.6.20) that if $\mathbf{Q}$ is irreducible and recurrent and $\pi\mathbf{Q} = 0$, then π is unique up to scalar multiples. Now we have this:

(9) **Recurrence.** Let $X(t)$ be irreducible with transition matrix $\mathbf{P}$ and generator $\mathbf{Q}$. Then the following are equivalent:

(a) The chain is non-null recurrent.
(b) Some state j is non-null recurrent.
(c) $X(t)$ is honest and $\pi\mathbf{Q} = 0$ for some distribution π.
(d) $X(t)$ is honest and $\pi\mathbf{P}(t) = \pi$ for all t, and for some distribution π.

Furthermore, $\pi_j = 1/q_j ET_{jj}$ for all j when the chain is non-null.

This theorem is simply the continuous-time version of what we have already proved in (3.6.20) for discrete-time chains. It is therefore natural to seek to use the jump chain of $X(t)$, which has transition matrix $\mathbf{R} = (r_{jk})$. With this in mind, let v_k^j be the expected number of visits to k by the jump chain, between visits to j.

It is trivial that (b) follows from (a); to show that (c) and (d) follow from (b), we construct a stationary distribution π in terms of the expected time spent in k, by $X(t)$, in between visits to j, that is

(10) $\qquad v_k = E\left\{\int_0^{T_{jj}} I(X(t)=k)\,dt \mid X(0)=j\right\} = v_k^j/q_j.$

First, the jump chain is recurrent and, from (3.6.15),

$$v_k^j = \sum_i v_i^j p_{ik}.$$

It now follows from (4) to (6) (see the remark before (9)), that $\pi Q = 0$, where
$$\pi_k = v_k^j/(q_j ET_{jj}).$$

Second, we invoke the Markov property at T_{jj} to see that
$$E\left(\int_0^t I(X(s) = k)\, ds \mid X(0) = j\right) = E\left(\int_{T_{jj}}^{T_{jj}+t} I(X(s) = k)\, ds \mid X(0) = j\right).$$

Hence
$$v_k = E\left\{\int_t^{t+T_{jj}} I(X(s) = k)\, ds \mid X(0) = j\right\}$$
$$= \int_0^\infty P(X(t+v) = k, v < T_{jj} \mid X(0) = j)\, dv$$
$$= \int_0^\infty \sum_{i \in S} P(X(v) = i, v < T_{jj} \mid X(0) = j) p_{ik}(t)\, dv$$

by conditioning on $X(v)$
$$= \sum_{i \in S} v_i p_{ik}(t).$$

Conversely, if $\pi Q = 0$ for some distribution π, then (4)–(6) show that, for any j, $v_k = \pi_k q_k/(\pi_j q_j)$ is an invariant measure for $\mathbf{R}$, with $v_j = 1$. But we know from (3.6.17) that v_k^j is a minimal invariant measure for $\mathbf{R}$. Therefore, as in the proof of (3.6.20),
$$ET_{jj} = \sum_k v_k = \sum_k v_k^j/q_j \leq \sum_k v_k/q_k = (\pi_j q_j)^{-1} < \infty.$$

Thus, j is recurrent and non-null. It follows that $\mathbf{R}$ is recurrent, so that $v_k = v_k^j$, and thus $ET_{jj} = (\pi_j q_j)^{-1}$, as required.

Finally, as $X(t)$ is honest, we can use (3.6.20) to show that π is unique such that $\pi = \pi P(t)$. Hence, by the above, $\pi Q = 0$, completing the proof.

Thus, if $X(t)$ is irreducible and honest and has a stationary distribution π, then π is unique and the chain is non-null recurrent. In this case we have,

(11) **Convergence.** For all j and k in S,
$$p_{jk}(t) \to \pi_k, \quad \text{as } t \to \infty.$$

This is proved by considering the discrete chain $Y_n = X(nh)$ for fixed $h > 0$; the chain Y is called a *skeleton* of $X(t)$. From (3.6.26) we have that

$p_{jk}(nh) \to \pi_k$, as $n \to \infty$, for any $h > 0$. The result follows by using the continuity of $p_{jk}(t)$, as expressed in Exercise (4.1(e)). We omit the details.

Note that, also as in the discrete case, for an irreducible non-null chain $X(t)$, with stationary distribution π, we have this:

(12) **Ergodic theorem.** Let $g(\cdot)$ be any bounded real-valued function defined on S. Then, with probability 1,

$$\frac{1}{t} \int_0^t g(X(u)) \, du \to \sum_{j \in S} \pi_j g(j), \quad \text{as } t \to \infty.$$

This is easily proved using the same techniques as in discrete time, and is left as an exercise.

These results may seem a little complicated, but they are frequently applied in practice in a very straightforward way. Here is an important and useful example.

(13) **M/M/1 queue.** This is the basic model that we have met before. Customers arrive at the instants of a Poisson process of rate λ. There is a single server, so they form a line as necessary if the server is busy; the rule is first-come first-served (often abbreviated to FIFO). The server works continuously if any customers are waiting, and service times are Exponential (μ) random variables that are independent of each other and arrivals.

Let $X(t)$ be the length of the queue, including the customer being served (if any) at time t. By construction, $X(t)$ is a Markov process, which is clearly honest (non-explosive), having transition rates

$$p_{j,j+1}(h) = \lambda h + o(h), \quad j \geq 0$$
$$p_{j,j-1}(h) = \mu h + o(h), \quad j \geq 1.$$

In the context of our previous examples, this is a birth–death process on the non-negative integers, with a reflecting barrier at the origin. We may therefore seek a stationary solution by solving the equations $\pi Q = 0$, that is,

$$\mu \pi_{n+1} - \lambda \pi_n = \mu \pi_n - \lambda \pi_{n-1}, \quad n \geq 1,$$
$$\mu \pi_1 - \lambda \pi_0 = 0.$$

Thus, there is such a distribution π if and only if $\lambda < \mu$, and then you can verify that

(14) $$\pi_n = \left(1 - \frac{\lambda}{\mu}\right)\left(\frac{\lambda}{\mu}\right)^n, \quad n \geq 0.$$

In this case the queue is non-null recurrent, and as $t \to \infty$,

$$P(X(t) = n) \to \pi_n, \quad n \geq 0.$$

If $\lambda = \mu$, then we glance at the jump chain to see that $X(t)$ is null-recurrent, and if $\lambda > \mu$ we find likewise that $X(t)$ is transient. In either case, as $t \to \infty$,

$$P(X(t) = n) \to 0, \quad n \geq 0.$$

Because we know that $q_0 = \lambda$ and $q_n = (\lambda + \mu)$, $n \geq 1$, (14) also gives us the mean recurrence times $\mu_0 = 1/(\pi_0 q_0) = \mu/(\lambda(\mu - \lambda))$, and $\mu_n = (\mu/(\mu^2 - \lambda^2))(\mu/\lambda)^n$, when $\lambda < \mu$.

Of course the jump chain has a different stationary distribution, v, given for $\lambda < \mu$ by

$$v_n = \frac{\mu^2 - \lambda^2}{2\mu} \left(\frac{\lambda}{\mu}\right)^{n-1}, \quad n \geq 1,$$

$$v_0 = (\mu - \lambda)/(2\mu).$$

A natural question is to ask how long an arriving customer can expect to wait, when the queue is in equilibrium. It is left as an exercise for you to find the distribution of the time such a customer spends in the queue, and to show that the expected time spent in the queue is $(\mu - \lambda)^{-1}$. Of course the time spent waiting to be served is therefore,

$$\frac{1}{\mu - \lambda} - \frac{1}{\mu} = \frac{\lambda}{\mu(\mu - \lambda)},$$

because the arriving customer's expected service time is μ^{-1}. ○

Exercises

(a)* Let $X(t)$ be a simple birth–death process, with $\lambda_j = j\lambda$, $\mu_j = j\mu$, for $j \geq 1$, and $\lambda_0 = \lambda$. Find the stationary distribution of $X(t)$, and also of the jump chain, when they exist. Explain the difference in your answers.

(b) **Erlang's loss formula.** Suppose there are s servers, each of whose service times are independent Exponential (μ). Customers arrive at the instants of a Poisson process of intensity λ, but leave immediately if all servers are busy. Show that, in equilibrium, the probability that an arriving customer is not served is

$$\{(\lambda/\mu)^s/s!\} \bigg/ \left\{\sum_{r=0}^{s}(\lambda/\mu)^r/r!\right\}.$$

Remark: This queue may be denoted by M/M/s.

(c) **M/M/1.** Customers arrive at the instants of a Poisson process of rate λ; service times are independent Exponential (μ); the number in the queue at time t is $X(t)$, including the customer being served. In equilibrium show that the distribution of the time spent in the queue by an arriving customer (including service time) is Exponential $(\mu - \lambda)$, where $\lambda < \mu$. [Hint: You may recall Exercise 1.8(d).]

What is the distribution of the time spent waiting for service to begin?

4.7 Reversibility

The ideas and techniques of reversibility are very much the same for continuous-time chains as they were discovered in (3.7) for discrete chains. First, if we consider the chain $X(t)$ in reversed time, that is $Y(t) = X(T-t)$, $0 \le t \le T$, we have as in (3.7.1) that $Y(t)$ is a Markov process, and by conditional probability

(1) $\quad P(Y(s+t) = k \mid Y(s) = j) = p_{kj}(t) P(Y(s+t) = k)/P(Y(s) = j),$

where $p_{jk}(t)$ are the transition probabilities for $X(t)$. For $Y(t)$ to be time-homogeneous it is therefore necessary that the chain is in equilibrium, so that (1) becomes

$$\pi_j \tilde{p}_{jk}(t) = \pi_k p_{kj}(t),$$

where $\tilde{p}_{jk}(t)$ are the transition probabilities for $Y(t)$, and π is the equilibrium distribution. As in the discrete case, the reversed chain has the same transition probabilities as $X(t)$ if

(2) $\quad\quad\quad\quad\quad\quad \pi_j p_{jk}(t) = \pi_k p_{kj}(t)$

and such a chain is said to be reversible (in equilibrium).

It is natural and obvious to express the above detailed balance conditions in terms of the transition rates, so dividing (2) by t, and letting $t \downarrow 0$, gives

(3) $$\pi_j q_{jk} = \pi_k q_{kj}, \quad \text{for all } j \text{ and } k \text{ in } S.$$

Conversely, summing the equations (3) over all $k \neq j$ yields

$$0 = \sum_{k \neq j} \pi_k q_{kj} - \pi_j \sum_{k \neq j} q_{jk} = \sum_{k \neq j} \pi_k q_{kj} + \pi_j q_{jj}$$

(4) $$= \sum_k \pi_k q_{kj}$$

and so π is the equilibrium distribution; recall (4.6.4). Now recalling that the q_{jk} characterize the distributions of $X(t)$ and $Y(t)$ we have proved this:

(5) Theorem. A stationary Markov process $X(t)$ is reversible if and only if there exists π satisfying (3), and in this case π is the equilibrium distribution of $X(t)$.

Many other results carry over from discrete-time reversibility; for example, Kolmogorov's criterion, Exercise 3.7(a), remains valid, and birth–death processes are reversible in equilibrium. These are exercises for you; here is a remarkable application.

(6) Multiserver queue. Customers arrive as a Poisson process of rate λ, and are served by any one of s servers; service times are independent Exponential (μ) random variables. If no server is free, customers form a line as usual, with the FIFO discipline ≈ 'first come, first served' (this queue is often denoted by the symbol M/M/s).

Let $X(t)$ be the total number of customers in the system at time t, including both those waiting and those being served. This is clearly a birth–death process so it is reversible in equilibrium by the remark above. You showed in Exercise (4.6(b)) that a stationary distribution exists if $\lambda < \mu s$, but it is also easy to show directly from the detailed balance equations. For $k \geq s$ these are seen to be

(7) $$\pi_k \lambda = \pi_{k+1} \mu s$$

and our claim follows.

When we consider the chain running forwards from time t in equilibrium, jumps upwards after t are a Poisson process of arrivals which are independent of $X(t)$, by construction. If we now consider the same chain run in reverse, these are seen as jumps down, which are departures before t. But of course they are still a Poisson process, and up to time t these departures are still independent of $X(t)$. This establishes the following surprising result.

(8) **Departures from M/M/s.** If $X(t)$ is the number of customers present at time t for the M/M/s queue in equilibrium, then the departures up to t form a Poisson process of rate λ independent of $X(t)$.

This has some even more interesting consequences that we sketch in the next section. ○

We conclude here with another example of the amenability of reversible chains.

(9) **Censored state space.** Suppose a reversible chain $X(t)$ with equilibrium distribution π is running on a state space S. Now it is decided that the chain is forbidden to enter a subset C of S. That is to say, any attempted move to $k \in C^c$ is allowed to succeed, but if the chain attempts to move to a state in C, then it simply remains where it is. The effect of this is that for $j \in C^c = S \setminus C$,

$$q_{jk} = 0, \quad k \in C.$$

Then the new chain $Y(t)$ is still reversible, with stationary distribution $\eta_j \propto \pi_j$, $j \in C^c$. In detail we have

(10) $$\eta_j = \pi_j / \left(\sum_{j \in S \setminus C} \pi_j \right).$$

Proof. It is trivial to check that η satisfies the detailed balance equations, so the result follows from (5). □

(11) **Shared waiting room.** Suppose that two independent queues are forced to share a waiting room with capacity w, where the first queue is M/M/s and the second is M/M/r. A customer who arrives to find the waiting room full (containing w customers waiting for service) leaves immediately never to return.

By Exercise (4.7(d)) the queues are jointly reversible in equilibrium, with stationary distributions π and ν, say. The finite shared waiting room truncates the state space to the set R of states such that $x \leq s + w$, $y \leq r + w$, and $x + y \leq s + r + w$.

By the previous example, the equilibrium distribution is

$$\pi_{jk} = \pi_j \nu_k \Big/ \left\{ \sum_{(j,k) \in C} \pi_j \nu_k \right\}.$$

○

Exercises
(a)* Let $X(t)$ be an irreducible non-null recurrent birth–death process. Show that $X(t)$ is reversible in equilibrium.

(b) **Kolmogorov's criterion.** Show that $X(t)$ is reversible if and only if for any sequence $i_1, \ldots, i_n, j$, we have

$$q(j, i_1) q(i_1, i_2), \ldots, q(i_n, j) = q(j, i_n), \ldots, q(i_1, j).$$

(c)* Show that an irreducible Markov chain $X(t)$ with two states is reversible in equilibrium.

(d)* If $X(t)$ and $Y(t)$ are independent reversible processes, prove that $(X(t), Y(t))$ is a reversible process.

4.8 Queues

We have seen several individual queues in previous sections, but in the real world we are often faced with the prospect of moving through several queues in succession. For example, if you telephone a faceless (or indeed, any) corporation seeking to speak to someone who has some relevant knowledge, you will first spend a random time conversing with a computer, then join a queue on hold to speak to a switchboard operator, and then at least one further queue on hold before succeeding in your quest.

We first consider a simple version of such nightmares.

(1) **Tandem queue.** There are two single-server queues, whose service times are independent and exponentially distributed with parameters μ_1 and μ_2, respectively. Customers arrive at the first queue as a Poisson process of rate λ, where $\lambda < \mu_1 \wedge \mu_2$. On completing that service, they move immediately to the second queue. On completing the second service, they leave the system forever.

Let the total number in each queue be $X_1(t)$ and $X_2(t)$, respectively. When the system is in equilibrium, it is remarkable that

(i) $X_1(t)$ and $X_2(t)$ are independent at any time t.
(ii) For any given customer, the total time spent in the first queue is independent of the total time spent in the second.
(iii) For any given customer, the total times spent waiting before service in each queue are not independent.

Proof.

(i) In the previous section (4.7.8), we showed that the departure process from the first queue up to time t is a Poisson process of rate λ, and independent of $X_1(t)$. Thus $X_2(t)$, which depends only on this Poisson process up to t, and the independent service times of the second server, is independent of $X_1(t)$, and each is a simple M/M/1 queue with joint distribution (in equilibrium),

$$(2) \quad P(X_1 = j; X_2 = k) = \left(\frac{\lambda}{\mu_1}\right)^j \left(1 - \frac{\lambda}{\mu_1}\right)\left(\frac{\lambda}{\mu_2}\right)^k \left(1 - \frac{\lambda}{\mu_2}\right),$$

$$j, k \geq 0.$$

(ii) We now seek to establish a similar independence between the departure process and waiting times. Consider arrivals at the first queue, in equilibrium.

Suppose a customer arrives at time s, and departs at time t. In the reversed process, this corresponds to an arriving customer at $-t$, and the departure of the same customer at $-s$. (Draw a graph of the size of the queue over a busy period to see this.) But in the original, the time spent by that customer in the queue is independent of arrivals after s, which is to say that in the reversed version the time spent in the queue is independent of departures before $-s$. We have shown (using reversibility) that for a customer C arriving in equilibrium, the waiting time in the queue is independent of all departures prior to the departure of C.

But the time spent in the second queue depends only on departures up to the time of quitting the first queue, and the independence follows.

(iii) If V_1 and V_2 are the respective events that the customer in question finds each queue empty, it is an exercise for you to show that

$$P(V_1 \cap V_2) \neq P(V_1)P(V_2).$$

○

The arguments used in this example clearly work for a similar sequence of any finite number of M/M/1 queues. In equilibrium their lengths at time t are independent, and an arriving customer spends independent total times in each.

Most remarkably, this property may be preserved in more general networks of queues or equivalent processes. Here is one classic example.

(3) **Closed migration process.** There is a set V of K sites (or stations, or queues), and at time t the rth site contains $X_r(t)$ individuals. We assume that $\mathbf{X} = (X_1(t), \ldots, X_K(t))$ is a Markov process; specifically, given $X_i(t) = n_i$,

in the interval $(t, t+h)$ one individual is transferred from the ith site to the jth site with probability

(4) $$\lambda_{ij}\phi_i(n_i)h + o(h).$$

We require $\lambda_{ii} = \phi_i(0) = 0$ and $\phi_i(n) > 0$ for $n > 0$; the freedom to specify the function ϕ makes this model flexible enough to apply to widely disparate situations. For example, if each station is a single-server M/M/1 queue, and customers completing service at the ith queue join the jth with probability r_{ij}, then

(5) $$\lambda_{ij} = \mu r_{ij} \quad \text{and} \quad \phi_i(n) = 1, \quad n > 0.$$

For another example, if individuals move around the network independently of each other, then $\phi_i(n) = n$.

Our assumptions entail that the number of individuals in the system is constant, $\sum_{r=1}^{K} X_r(t) = N$, so this is a *closed migration process*. When $N = 1$, and $\phi_i(1) = 1$ for all i, we have an ordinary Markov process with transition rates $\lambda_{ij}; i \neq j$. We assume this chain with $N = 1$ is irreducible, so that there is a stationary distribution π satisfying

(6) $$\pi_j \sum_k \lambda_{jk} = \sum_k \pi_k \lambda_{kj}.$$

Then we have the following remarkable result.

(7) **Theorem.** The stationary distribution of $\mathbf{X}$ when $X_1(t) + \cdots + X_K(t) = N$ is given by

(8) $$\pi(\mathbf{x}) = c \prod_{i=1}^{K} \left\{ \pi_i^{x_i} \bigg/ \prod_{r=1}^{x_i} \phi_i(r) \right\}, \quad \Sigma x_i = N,$$

where c is a constant chosen to make $\pi(\mathbf{x})$ a probability distribution summing to 1.

Proof. Let $\mathbf{u}_i$ be the vector of length K with zero entries except for a 1 in the ith place. Then, as usual, any stationary distribution $\pi(\mathbf{x})$ must satisfy

(9) $$\sum_{j,k} \pi(\mathbf{x} + \mathbf{u}_k - \mathbf{u}_j)\lambda_{kj}\phi_k(x_k+1) = \pi(\mathbf{x}) \sum_{j,k} \lambda_{jk}\phi_j(x_j).$$

This is rather formidable, but these equations will surely be satisfied if for all j we have

(10) $$\sum_k \pi(\mathbf{x} + \mathbf{u}_k - \mathbf{u}_j)\lambda_{kj}\phi_k(x_k+1) = \pi(\mathbf{x}) \sum_k \lambda_{jk}\phi_j(x_j).$$

So now we only need to check that (8) does satisfy (10). We have from (8) that

$$\pi(\mathbf{x} + \mathbf{u}_k - \mathbf{u}_j) = \frac{\pi(\mathbf{x})\pi_k \phi_j(x_j)}{\pi_j \phi_k(x_k + 1)}.$$

Substitution of this into (10) gives exactly the equation (6), which completes the proof. □

We conclude this section with a famous result that has many applications in queues, and also in other similar systems.

(11) **Little's theorem.** Let $X(t)$ be the length of a queue at time t, where the system in question has the following properties.

I. Returns to the empty queue form a renewal process at the instants $\{0, S_1, S_2, \ldots\}$, where $S_n - S_{n-1} = X_n$, $S_0 = 0$, and $EX_n < \infty$.
II. Customers arrive singly, and W_n is the time spent in the system (including service) by the nth such customer.
III. The number of arrivals in $[S_{n-1}, S_n]$ is A_n, where the A_n are mutually independent with finite mean, and this sequence of random variables gives rise to a renewal process at the instants $B_n = \sum_{r=1}^{n} A_r$.

The number of arrivals by time t will be denoted by $A(t)$, and we assume that $E(A_1 X_1) < \infty$.

Note the important fact that

$$(12) \qquad \int_0^{X_1} X(u)\, du = \sum_{r=1}^{A_1} W_r,$$

because both sides count the total amount of customer time spent in the queue during a busy period; the left-hand side counts it by queue length over time, and the right-hand side counts it by customer.

The entire proof rests on the fact that (S_n) and (B_n) comprise renewal sequences to which we may apply renewal–reward theorems.

Little's theorem deals with these three quantities (where the limits are taken with probability 1).

(i) The long-run average arrival rate

$$\lambda = \lim_{t \to \infty} \frac{A(t)}{t}.$$

In the renewal sequence (S_n), let the reward during $[S_{n-1}, S_n]$ be A_n. Then by the renewal–reward theorem, $\lambda = EA_1/EX_1$.

(ii) The long-run average wait

$$w = \lim_{n\to\infty} \frac{1}{n} \sum_{r=1}^{n} W_r.$$

In the renewal sequence (B_n), let the reward during $[B_{n-1}, B_n]$ be $\sum_{r=B_{n-1}}^{B_n} W_r$. Then, by the renewal–reward theorem, $w = \mathrm{E}\sum_{1}^{A_1} W_r / \mathrm{E}A_1$.

(iii) The long-run average queue length

$$L = \lim_{t\to\infty} \frac{1}{t} \int_0^t X(u)\, du.$$

In the renewal sequence (S_n), let the reward in $[S_{n-1}, S_n]$ be $R_n = \int_{S_{n-1}}^{S_n} X(u)\, du$. By the renewal–reward theorem $L = \mathrm{E}R_1/\mathrm{E}X_1$. Then, by the above and (12),

$$\frac{\lambda w}{L} = \frac{\mathrm{E}\sum_1^{A_1} W_r}{\mathrm{E}A_1} \cdot \frac{\mathrm{E}A_1}{\mathrm{E}X_1} \cdot \frac{\mathrm{E}X_1}{\mathrm{E}R_1} = 1,$$

which is Little's theorem: $L = \lambda w$. □

Exercises
(a)* Prove the assertion 1(iii), that in a tandem queue the times spent by a given customer waiting for service are not independent. [Hint: Consider the probability that both waits are zero in equilibrium.]
(b) Verify directly that (2) is the equilibrium distribution of the tandem queue, by substitution in $\pi Q = 0$.
(c)* Show that the total time spent by a customer in each of a sequence of M/M/1 queues in equilibrium is exponentially distributed.

4.9 Miscellaneous models

Finally, we look at some special cases and methods arising in Markov processes.

In general, we have required our Markov processes to be time-homogeneous, for simplicity. Non-homogeneous processes are generally very intractable, but one classic example can be tackled.

(1) Non-homogeneous birth–death. Let $X(t)$ be a birth–death process with transition rates

$$q_{j,j+1} = \nu(t) + j\lambda(t),$$
$$q_{j,j-1} = j\mu(t), \quad j \geq 0.$$

The forward equations are

(2) $$\frac{d}{dt} p_{j0}(t) = -\nu(t) p_{j0}(t) + \mu(t) p_{j1}(t),$$

$$\frac{d}{dt} p_{jk}(t) = -\{\nu(t) + k(\lambda(t) + \mu(t))\} p_{jk}(t) + \{\nu(t) + (k-1)\lambda(t)\} p_{j,k-1}(t)$$
(3) $$+ (k+1)\mu(t) p_{j,k+1}(t), \quad k > 0.$$

Introducing the usual pgf $G(s,t) = \mathrm{E} s^{X(t)}$, we can, (for any j), condense these, (multiply (3) by s^k and sum over k), into

(4) $$\frac{\partial G}{\partial t} = (\lambda(t)s - \mu(t))(s-1)\frac{\partial G}{\partial s} + \nu(t)(s-1)G.$$

Note that a complete solution needs a boundary condition, such as $X(0) = I$, in which case $G(s, 0) = s^I$.

We examine some solutions of this shortly, but first let us differentiate (4) with respect to s, and set $s = 1$. Because $G_s(1,t) = \mathrm{E}X(t) = m(t)$, say, this gives

(5) $$\frac{d}{dt} m(t) = (\lambda(t) - \mu(t)) m(t) + \nu(t),$$

where we have assumed that for all t

$$\lim_{s \to 1} (s-1) \frac{\partial^2 G}{\partial s^2} = 0.$$

This first-order equation (5) is readily integrated to give

(6) $$m(t) = \mathrm{E}X(t) = X(0) \exp\{-r(t)\} + \int_0^t \nu(u) \exp(r(u) - r(t))\, du,$$

where

$$r(t) = \int_0^t (\mu(u) - \lambda(u))\, du.$$

An interesting special case arises when λ, μ, and ν are periodic; this provides a model for any seasonal effects in this simple population model. We omit the details.

Returning to (4): if $\lambda(t) = \nu(t) = 0$, then $X(t)$ is a pure death process, and it is easy to verify that

(7) $$G(s,t) = \left\{1 + (s-1)\exp\left\{-\int_0^t \mu(u)\,du\right\}\right\}^I,$$

which corresponds to a binomial distribution.

If $\lambda(t) = \mu(t) = X(0) = 0$, the process is a non-homogeneous Poisson process, where

(8) $$G(s,t) = \exp\left\{(s-1)\int_0^t \nu(u)\,du\right\}.$$

If $\mu(t) = \nu(t) = 0$, then $X(t)$ is a pure birth process, and then, when $X(0) = I$,

(9) $$G(s,t) = \left\{\left(1 + (1-s)\exp\left[-\int_0^t \lambda(u)\,du\right]\right)\bigg/ s\right\}^{-I}.$$

When $I = 1$, this shows that $X(t)$ is Geometric with parameter $\exp[-\int_0^t \lambda(u)\,du]$; when $I > 1$, $X(t)$ has a negative binomial distribution.

We could, in principle, find an expression for $G_X(z,t) = \mathrm{E} z^{X(t)}$ in general, but to do so we would need the new technique displayed in the following example. It is often useful when processes involve a Poisson process of recruitment. ○

(10) Poisson arrivals. Suppose that fresh individuals are recruited to a population at the instants of a Poisson process with intensity ν. On arrival, the rth new member M_r founds a Markov birth–death process B_r independently of all the other individuals. At a time s after the arrival of M_r, this process is of size $B_r(s)$ and we denote

$$\mathrm{E} z^{B_r(s)} = G(z,s).$$

The total size $X(t)$ of the population at time t is a Markov process, and one could write down the forward equations. (We assume $X(0) = 0$.)

For a different approach that supplies a very neat solution, let us consider conditioning on the number $N(t)$ of fresh arrivals up to time t, that have founded processes $B_1, \ldots, B_N$, whose sum is $X(t)$.

We first recall from (2.5.15) that conditional on $N(t) = n$, the arrival times of these n recruits in $[0,t]$ are independently and identically distributed uniformly on $[0,t]$.

Let us denote these arrival times by $\{U_1, \ldots, U_n\}$, and observe that for each r

(11) $$\mathrm{E}z^{B_r(t-U_r)} = \mathrm{E}z^{B_r(U_r)} = \int_0^t \frac{1}{t} G(z, u)\, du.$$

Hence, recalling the result (1.8.13) about random sums,

(12) $$\begin{aligned}\mathrm{E}z^{X(t)} &= \mathrm{E}\{\mathrm{E}(z^{X(t)} \mid N(t))\} \\ &= G_N\left(\frac{1}{t}\int_0^t G(z, u)\, du\right) \\ &= \exp\left\{vt\left\{\frac{1}{t}\int_0^t G(z, u)\, du - 1\right\}\right\} \\ &= \exp\left\{v\int_0^t \{G(z, u) - 1\}\, du\right\}.\end{aligned}$$

The same argument works when recruits arrive as a non-homogeneous Poisson process; see Problem (4.8). ○

Here is another application of a similar idea.

(13) **Disasters.** Suppose that new arrivals appear in a population at the instants of a Poisson process of intensity v, and the entire population is annihilated at the instants of an independent Poisson process of intensity δ. Let $X(t)$ be the size of the population at time t, and define $C(t)$ to be the time that has elapsed since the most recent annihilation prior to t. By construction of the Poisson process, $C(t)$ has the distribution

$$F_C(x) = \begin{cases} 1, & x \geq t, \\ 1 - e^{-\delta x}, & 0 \leq x < t. \end{cases}$$

If $N(t)$ is the Poisson process of intensity v, then, by definition of $X(t)$,

$$X(t) = N(C(t)).$$

Hence, by conditional expectation

$$\begin{aligned}\mathrm{E}s^{X(t)} &= \mathrm{E}\{\mathrm{E}(s^{N(C(t))} \mid C(t))\} \\ &= \mathrm{E}\, e^{vC(t)(s-1)} = \int_0^t \delta e^{-\delta x} e^{vx(s-1)}\, dx + e^{-\delta t} e^{vt(s-1)} \\ &= \{\delta + v(1-s)\exp\{(vs - \delta - v)t\}\}\{\delta + v(1-s)\}.\end{aligned}$$

Hence, or directly by a similar argument, it is easy to obtain

$$\mathrm{E}X(t) = \frac{v}{\delta}(1 - e^{-\delta t}).$$

Exercises

(a)* Let $s(t)$ be the second factorial moment of the process $X(t)$ defined in (1); $s(t) = E(X(t)^2 - X(t))$, and $m(t) = EX(t)$. Show that

$$\frac{ds(t)}{dt} = 2(\lambda(t) - \mu(t))s(t) + 2[\nu(t) + \lambda(t)]m(t).$$

(b)* For the process $X(t)$ defined in (1), suppose that λ, μ, and ν are constants and use the result of (10) to show that when $X(0) = 0$, and $\lambda \neq \mu$,

$$Es^{X(t)} = \left(\frac{\mu - \lambda}{\mu - \lambda s + \lambda(s-1)\exp\{(\lambda - \mu)t\}}\right)^{\nu/\lambda}.$$

What happens when $\lambda = \mu$?

(c) For the process $X(t)$ defined in (1), suppose that λ, μ, and ν are constants, and show that if $\lambda < \mu$,

$$\lim_{t \to \infty} G(s,t) = \left\{\frac{\mu - \lambda}{\mu - \lambda s}\right\}^{\nu/\lambda}.$$

(d) Obtain the forward equations for the disaster process defined in (13).

Problems

1.* Verify that the transition probabilities of the Poisson process satisfy the Chapman–Kolmogorov equations (4.1.11).

2. A highly simplified model for DNA substitution at a single site, postulates a Markov chain with Q-matrix

$$Q = \begin{pmatrix} -(\lambda + 2\mu) & \lambda & \mu & \mu \\ \lambda & -(\lambda + 2\mu) & \mu & \mu \\ \mu & \mu & -(\lambda + 2\mu) & \lambda \\ \mu & \mu & \lambda & -(\lambda + 2\mu) \end{pmatrix}.$$

What is the stationary distribution of this chain? Labelling $S = \{1, 2, 3, 4\}$, show that $p_{13}(t) = \frac{1}{4}(1 - e^{-4\mu t}) = p_{14}(t)$, and $p_{11}(t) = \frac{1}{4}(1 + e^{-4\mu t} + 2 e^{-2(\lambda+\mu)t})$. [Hint: Exploit the symmetry, and simply substitute the given solution into the forward equations.]

3.* Find the transition probabilities for the Telegraph process defined in (4.1.23), and verify that they satisfy the Chapman–Kolmogorov equations.

4.* **Balking.** Let $X(t)$ be the length of the queue for a single server at time t. Service times are independent Exponential (μ). Potential customers arrive at the instants of an independent Poisson process of intensity λ, but if there are (at that instant) n customers queueing, they only join the queue with probability b_n; otherwise they leave forever. Write down the forward equations, and find the equilibrium distribution, stating when it exists.

5.* Find the transition probabilities for the pure simple death process with $\mu_n = n\mu$.

6.* Let $X(t)$ be a simple birth–immigration process; that is to say new external recruits arrive at the instants of a Poisson process with intensity ν, and each recruit (independently) generates a simple birth process with parameter λ from the instant of arrival (so that $p_{n,n+1}(h) = \lambda n h + \nu h + o(h)$). Show that, if $X(0) = 0$,

$$EX(t) = \frac{\nu}{\lambda}(e^{\lambda t} - 1).$$

7. Let $X(t)$ be a birth process such that $X(0) = 1$, and

$$p_{n,n+1}(h) = n(n+1)h + o(h), \quad n \geq 1.$$

Show that $X(t)$ is dishonest, and find the expected time until explosion.

8.* Let $N(t)$ be a non-homogeneous Poisson process with intensity $\lambda(t)$, ≥ 0; thus the transition probabilities for small h are

$$P(N(t+h) - N(t) = 1) = \lambda(t)h + o(h),$$
$$P(N(t+h) = N(t)) = 1 - \lambda(t)h + o(h).$$

(a) Write down the forward equations and verify that

$$Es^{N(t)} = \exp\left\{\int_0^t \lambda(u)\, du(s-1)\right\}.$$

(b) Show that conditional on $N(t) = n$, the n arrival times are independent having the common density

$$f(u) = \lambda(u) \Big/ \int_0^t \lambda(u)\,du, \quad 0 \le u \le t.$$

(c) Now generalize the result of (4.9.10) to the case when recruits arrive at the instants of a non-homogeneous Poisson process.

(d) Find the expected population in a Disaster process (4.8.13), when arrivals and disasters appear as non-homogeneous Poisson processes.

9. Let k be a transient state of a Markov process $X(t)$, with $X(0) = k$. Find the distribution of the total time spent in k.

10.* **Yule process.** Each individual in a population behaves independently as follows: every lifetime is independent and Exponential (λ), at the end of which the individual is replaced by two fresh individuals, of age zero. Assume $X(0) = 1$.

By conditioning on the time of the first split, show that the generating function $G = E z^{X(t)}$, of the population $X(t)$ at time t, satisfies

$$G(z,t) = z e^{-\lambda t} + \int_0^t G(z, t-u)^2 \lambda e^{-\lambda u}\, du.$$

Hence show that $X(t)$ is Geometric $(e^{-\lambda t})$. Show further that for $s < t$, the conditional distribution of $X(s)$, given $X(t) = n$, is binomial.

11.* **M/M/1.** Customers arrive at the instants of a Poisson process of rate λ; service times are independent Exponential (μ); there is a single server.

If at a given moment there are $k \ge 1$ customers in the queue (including the one being served), find the distribution of the number of customers completing service before the next arrival.

12. **Simple birth–death.** Let $X(t)$ be the simple birth–death process for which $p_{n,n+1}(h) = \lambda n h + o(h)$ and $p_{n,n-1}(h) = \mu n h + o(h)$, $n \ge 1$. The origin is absorbing and $X(0) = 1$. Let T be the time of extinction when $\lambda < \mu$. Show that

$$ET = \int_0^\infty \frac{\mu - \lambda}{\mu \exp((\mu - \lambda) t) - \lambda}\, dt = \frac{1}{\lambda} \log \frac{\mu}{\mu - \lambda}.$$

What happens when $\lambda = \mu$? [Hint: See Exercise 4.9(b) for useful background.]

13.* **Busy period for M/M/1.** Customers arrive at the instants of a Poisson process of rate λ; service times are independent Exponential (μ), where $\mu > \lambda$. Let B be the duration of the time for which the server is busy, starting at the moment the first customer C arrives, let Z be the number of customers arriving while C is being served, and let the service time of C be S. Argue that

$$B = S + \sum_{j=1}^{Z} B_j,$$

where B_j are independent, having the same distribution as B. Deduce that B has a moment-generating function (mgf) $M(\theta) = \mathrm{E}\,e^{\theta B}$ that satisfies

$$2\lambda M(\theta) = (\lambda + \mu - \theta) - ((\lambda + \mu - \theta)^2 - 4\lambda\mu)^{1/2}.$$

Find the mean and variance of B.

14. **Last in first out.** Customers arrive at the instants of a Poisson process of rate λ; service times are Exponential (μ), where $\mu > \lambda$; but the queue discipline is 'last come, first served'. That is to say, if you arrive to find the server busy, everyone who joins the line behind you will be served before you, unless nobody arrives before the end of the current service period. (This queue is called M/M/1/LIFO.)
    (a) Show that the equilibrium distribution of queue length is the same as in the ordinary M/M/1 queue under 'first in first out'.
    (b) Show that for a customer arriving in equilibrium the expected time spent in the queue before service is the same as the expected duration of a busy period in M/M/1.
    (c) Show that for a customer arriving in equilibrium the total expected time in the queue (including service) is $(\mu - \lambda)^{-1}$, the same as M/M/1.
    (d) Show that the distribution of the time spent in the queue is not the same as M/M/1.
    (e) Calculate the variance of the time until service.

15. **Immigration–birth.** For the process defined in (4.9.1), suppose that $X(0) = \mu = 0$, and λ and ν are constants. By conditioning on the time of the first arrival, show that $G(z, t) = \mathrm{E}z^{X(t)}$ satisfies

$$G(z, t) = \int_0^t G(z, t-s) z \nu\, e^{-\nu s} \{z + (1-z)\, e^{\lambda(t-s)}\}^{-1}\, ds + e^{-\nu t}.$$

Hence show that

$$G(z, t) = (z + (1-z)\, e^{\lambda t})^{-\nu/\lambda}.$$

[Hint: Recall that, for the Yule process $Y(t)$ with $Y(0) = 1$, $\mathrm{E} z^{Y(t)} = z\{z + (1-z)e^{\lambda t}\}^{-1}$.]

16. **Birth process with disasters.** Immigrants arrive according to a Poisson process of rate ν, and each arrival initiates a birth process (Yule process) of rate λ, independently of other arrivals. Disasters annihilate the population at the instants of a Poisson process of rate δ. Initially the population $X(t)$ is zero. Show that

$$\mathrm{E} X(t) = \int_0^t \frac{\delta \nu}{\lambda} e^{-\delta x}(e^{\lambda x} - 1)\, dx + \frac{\nu}{\lambda} e^{-\delta t}(e^{\lambda t} - 1).$$

17.* **M/M/∞.** A service point has an unbounded number of servers, so no queue forms. Customers arrive for service at the instants of a Poisson process of rate λ. Service times (which start immediately on arrival) are independent and Exponential (μ). Let $X(t)$ be the number of busy servers at time t. Find the distribution of $X(t)$ and $m = \lim_{t \to \infty} \mathrm{E} X(t)$.

    Now suppose that the Poisson process of arrivals is non-homogeneous with rate $\lambda(t)$. When $X(0) = 0$, find the distribution of $X(t)$. If, as $t \to \infty$, $(1/t) \int_0^t \lambda(u)\, du \to c$, find m, and the limiting distribution of $X(t)$.

18.* **M/G/∞.** Repeat the preceding question with the one difference that service times are independent having the common distribution $F(x)$.

19. **Renewals.** Show that the age and excess life processes, $A(t)$ and $E(t)$, defined in (2.6.40), both have the Markov property. Deduce that their distributions converge as $t \to \infty$.

5 Diffusions

> Extremely minute particles of solid matter, whether obtained from organic or inorganic substances, when suspended in pure water, or in some other aqueous fluids, exhibit motions for which I am unable to account.
>
> *Robert Brown (1829)*

5.1 Introduction: Brownian motion

All the Markov processes considered in the previous two chapters have been 'jump processes', which is to say that they make transitions between discrete states at times that may be fixed or random. But many stochastic processes in the real world do not have this property. In sharp contrast to jumps, there are random processes that are seen to move continuously between their possible states, which commonly lie in some interval subset of the real line. Examples include meteorological records, physiological data, and the noise that corrupts electronic signals.

Of course it remains true that many of these processes still have the Markov property, and it is therefore natural to seek a general theory of such processes with continuous parameter and continuous state space. As usual, we begin with a simple but important example that illustrates many of the key ideas.

(1) **Brownian motion.** Suppose you were to examine very small particles (such as fine dust, or the very smallest fragments of pollens), suspended in a suitable fluid, such as water. You would observe (as Robert Brown did in 1827), that the particles are in ceaseless erratic motion, and that the smaller the particle, the more vigorous and erratic the motion. This is caused by the ceaseless bombardment of the particle by the molecules of the fluid in question. Indeed, if you could observe them, the molecules themselves would be seen to be in even more vigorous and erratic motion than the pollen or dust.

We model this behaviour mathematically as follows: let $X(t)$ be the x-coordinate (say) of such a particle; we may describe its motion by these rules.

[A] Without loss of generality we can assume for the moment that $X(0) = 0$.

[B] The physical nature of particles in the real world entails that $X(t)$ varies continuously, and does not jump.

[C] The random battering by the molecules naturally leads us to think of $X(t)$ as a type of symmetric random walk which takes a large number of small independent steps in the interval $[0, t], t > 0$.

We can refine this coarse specification into a subtler mathematical model by applying the central limit theorem to this symmetric random walk; recall (2.3.12). That then asserts that $X(t)$ is approximately normally distributed with mean zero and variance proportional to t. This approximate property of the Brownian motion of particles is then idealized to supply us with an appropriate mathematical stochastic process, as follows.

(2) **Wiener process.** A random process $W(t), t \geq 0$, is said to be a *Wiener process* if,

(a) $W(0) = 0$
(b) $W(t)$ is continuous in $t \geq 0$
(c) $W(t)$ has independent increments such that $W(t + s) - W(s)$ has the Normal distribution $N(0, \sigma^2 t)$, for all $s, t \geq 0$, and some $0 < \sigma^2 < \infty$.

△

We make the following remarks about this important definition.

(i) Some writers refer to $W(t)$ as Brownian motion; we do not.
(ii) If $\sigma^2 = 1$, then $W(t)$ is said to be the standard Wiener process; we always make this assumption unless stated otherwise.
(iii) In fact the assumption (b) is not strictly necessary, in that one can construct (by a limiting procedure) a random process $W(t)$ that obeys (a) and (c) and is almost surely continuous. This is Wiener's theorem, but it is beyond our scope to prove it. Note that the word 'construct' is not being used in its everyday sense here. The Wiener process is defined as a limit, so the 'construction' is strictly mathematical. This fact accounts in part for the remarkable properties of the process.
(iv) Because $W(t)$ has independent increments, it follows that it must have the Markov property; the proof is an easy exercise if you recall the same result for the Poisson process; see (2.5.8).

As with all Markov processes, the joint and conditional distributions are natural objects of interest. Of course, the probability mass functions of the discrete case are now replaced by densities.

(3) **Density and diffusion equations.** We define the *transition density function* of the standard Wiener process thus:

(4) $$f(t, y \mid s, x) = \frac{\partial}{\partial y} P(W(t) \leq y \mid W(s) = x), \quad s \leq t.$$

Because of our assumption (c) that $W(t)$ has normally distributed increments, this gives

(5) $$f(t, y \mid s, x) = \frac{1}{\sqrt{2\pi(t-s)}} \exp\left\{-\frac{(y-x)^2}{2(t-s)}\right\}, \quad y \in \mathbb{R}.$$

Hence, after differentiating ad nauseam, we find that f satisfies the following so-called *diffusion equations*.

Forward equation

(6) $$\frac{\partial f}{\partial t} = \frac{1}{2} \frac{\partial^2 f}{\partial y^2}.$$

Backward equation

(7) $$\frac{\partial f}{\partial s} = -\frac{1}{2} \frac{\partial^2 f}{\partial x^2}.$$

The reason for choosing these names will become apparent shortly. By conditioning on the value of $W(u)$ for $s < u < t$, just as in the discrete case, we obtain the

(8) **Chapman–Kolmogorov equations** for the transition density:

(9) $$f(t, y \mid s, x) = \int_{-\infty}^{\infty} f(u, z \mid s, x) f(t, y \mid u, z) \, dz, \quad s < u < t.$$

If you have time you can show that (5) does indeed satisfy (9). Since this is a non-linear functional equation, we rarely make progress by seeking to solve (9) as it stands.

We shall return to examine the properties of the Wiener process in greater detail later on. At this point we take the opportunity to reconsider the Brownian motion from a different point of view. The definition of the Wiener process given above effectively specifies the transition density $f(t, y \mid s, x)$ which may be regarded as the equivalent (in this context) of the transition matrix $\mathbf{P}(t)$ for discrete state space Markov processes.

Given our experience of the Poisson process, and other Markov chains, we may naturally seek to specify the behaviour of the process over arbitrarily small time-intervals $(t, t + h)$, and consider what happens as $h \downarrow 0$. This procedure is natural on physical grounds also, as many real-world processes are most readily specified by short-term behaviour. It is exactly the long-term patterns that we seek to explain by the model.

In the case of discrete state space this yielded the Q-matrix or generator $\mathbf{Q}$. We shall see that the same procedure for diffusion processes yields a slightly more complicated object to play the role of $\mathbf{Q}$.

In view of our discussion above, the following is the obvious description of the short-term properties of Brownian motion $X(t)$ introduced in (1).

(10) **Infinitesimal parameters.** First, we require that

(α) $X(t)$ is a continuous Markov process, $t \geq 0$.

Second, we seek to specify the properties of $dX = X(t+h) - X(t)$, the increment in X over the small time-interval $(t, t+h)$. To be consistent with our earlier remarks about the physical properties of the process, we must require at least that $X(t)$, and hence dX, have no jumps, and that the second moment $E(dX)^2$ is of order h. Formally we express these constraints thus; they are said to specify the infinitesimal parameters of $X(t)$.

(β) $(P|X(t+h) - X(t)| > \epsilon \mid X(t)) = o(h)$, for all $\epsilon > 0$.
(γ) $E(X(t+h) - X(t) \mid X(t)) = 0 + o(h)$.
(δ) $E((X(t+h) - X(t))^2 \mid X(t)) = \sigma^2 h + o(h)$.

It is a remarkable fact that these assumptions are sufficient to determine the transition density function $f(t, y \mid s, x)$ of the process $X(t)$. Following the same procedures as in the discrete case, we consider the Chapman–Kolmogorov equations (9), first with $u = t - h$, and second with $u = s + h$, and allow $h \downarrow 0$ in each case. These yield, respectively, the *forward equation*

$$(11) \qquad \frac{\partial f}{\partial t} = \frac{1}{2}\sigma^2 \frac{\partial^2 f}{\partial y^2}$$

and the *backward equation*

$$(12) \qquad \frac{\partial f}{\partial s} = -\frac{1}{2}\sigma^2 \frac{\partial^2 f}{\partial x^2}.$$

These are, gratifyingly, the same as (6) and (7) when $\sigma^2 = 1$, which explains our earlier choice of names. Furthermore, we now see that (5) supplies the required solution of (11) and (12), without further ado. Less pleasingly, the rigorous derivation of (11) and (12) from (α)–(δ) is lengthy and technically difficult, so we omit it.

We make some remarks here:

(i) By analogy with the discrete case, it is natural to see this as an alternative characterization of the Wiener process, just as the Poisson process may be characterized either by its independent Poisson increments, or by the probability distribution of events in $(t, t+h)$, together with the Markov property.

(ii) Clearly the role of the generator matrix **Q**, in (4.2.10) and (4.2.24), is now being played by the so-called 'differential operator', $\mathbf{Q} = \frac{1}{2}\sigma^2(\partial/\partial x^2)$. It is far beyond our scope to explore the consequences of this, even informally.

However, one important corollary that we stress is this: it follows that we can solve some special types of partial differential equation numerically, by simulating a random walk. This point was first made by Rayleigh in 1899, and is extensively exploited in many modern applications.

(iii) The appearance of these higher derivatives in the forward and backward equations is often a little surprising at first sight. We therefore sketch a very informal outline of a derivation of an equation of this type. For variety, let us consider a Brownian particle suspended in a body (or stream) of liquid moving with constant velocity a along the x-axis. Let its x-coordinate (with respect to some fixed origin) at time t be $D(t)$, which we assume to be a continuous Markov process. We set

(13) $$P(D(t) \leq y \mid D(0) = x) = F(t, y \mid x)$$

and assume that the distribution function F is differentiable as much as we require in all its arguments. The gross motion of the fluid leads to the assumption

(14) $$E(D(t+h) - D(t) \mid D(t)) = ah + o(h),$$

where this is called the instantaneous mean of $D(t)$, and the properties of Brownian motion discussed above lead to the assumption

(15) $$E(\{D(t+h) - D(t)\}^2 \mid D(t)) = bh + o(h),$$

where this is called the instantaneous variance of $D(t)$. In this case all higher instantaneous moments are $o(h)$; we do not actually need to assume this, nor is it necessarily true for all diffusions, but it simplifies the analysis here.

Now if we set $D(t+h) = D(t) + \Delta$, where $\Delta = D(t+h) - D(t)$ is the increment in $D(t)$ over $(t, t+h)$, then we can seek to derive a forward equation in the usual way; we condition on $D(t)$, use the Markov property, and ultimately let $h \downarrow 0$.

Let the conditional density of Δ given that $D(t) = u$ be $f_\Delta(u, z)$, assume that $D(0) = 0$, and denote the density of $D(t)$ by $f(t, y)$. Then the above plan yields (non-rigorously):

(16) $F(t+h, y) - F(t, y)$
$$= P(D(t) \leq y - \Delta) - P(D(t) \leq y)$$
$$= \int_{-\infty}^{\infty} \int_{y}^{y-z} f(t, u) f_\Delta(u, z) \, du \, dz$$
$$\simeq \int_{-\infty}^{\infty} -z f(t, y) f_\Delta(y, z) + \frac{1}{2} z^2 \frac{\partial}{\partial y} \{f(t, y) f_\Delta(y, z)\} \, dz,$$

assuming Taylor's theorem, and neglecting higher-order terms in z,

$$= -f(t, y) E(\Delta \mid D(t) = y)$$
$$+ \frac{1}{2} \frac{\partial}{\partial y} \{f(t, y) E(\Delta^2 \mid D(t) = y)\}$$
$$= -ah \, f(t, y) + bh \frac{1}{2} \frac{\partial f}{\partial y}(t, y).$$

Now dividing by h, allowing $h \downarrow 0$, and differentiating with respect to y yields (17), the forward equation for the density of $D(t)$:

(17) $$\frac{\partial f}{\partial t} = -a \frac{\partial f}{\partial y} + \frac{1}{2} b \frac{\partial^2 f}{\partial y^2}.$$

It is straightforward to verify that one possible solution of (17) is given by the normal density with mean at and variance bt. This is the same as the density of $at + \sqrt{b} W(t)$, where $W(t)$ is the standard Wiener process, which is of course intuitively to be expected from the original definition of $D(t)$. For obvious reasons, $D(t)$ may often be referred to as the *drifting Wiener process*, with drift a.

Clearly this way of discovering that $D(t) = at + \sqrt{b} W(t)$ is a somewhat round-about route. And as we have noted, it is a nontrivial task to make the argument rigorous. That being the case, you may well wonder why anyone would ever do it this way. The answer is clear when we come to consider general diffusions that are not (as the Wiener process is) homogeneous in time and space with independent increments. In view of our discussion above, the following is a natural definition.

(18) **Diffusion processes.** The process $X(t), t \geq 0$, is a *diffusion process* if it is a continuous (with probability 1) Markov process

satisfying

(19) $\quad P(|X(t+h) - X(t)| > \epsilon \mid X(t)) = o(h), \quad \text{for } \epsilon > 0.$

(20) $\quad E(X(t+h) - X(t) \mid X(t)) = a(t, X(t))h + o(h).$

(21) $\quad E\{(X(t+h) - X(t))^2 \mid X(t)\} = b(t, X(t))h + o(h).$

We may refer to the functions $a(t,x)$ and $b(t,x)$ as the instantaneous parameters of the process $X(t)$; similarly, ah and bh are often called the instantaneous mean and variance, respectively.

A convenient shorthand is to denote the small increment in $X(t)$ (over any small interval dt), by $dX(t)$. Then (after a little thought, and a glance at the drifting Wiener process above) it will seem natural to encapsulate these properties (19)–(21) in the usual notation

(22) $\quad dX = a(t, X)\, dt + (b(t, X))^{1/2}\, dW.$

Here dW is the increment of the standard Wiener process over a small time-interval $(t, t+dt)$ having the property that

$$E\{(dW)^2 \mid X(t)\} = E(dW)^2 = dt + o(dt).$$

Substituting (22) in (20) and (21) verifies the formal validity of this representation, to which we return in later sections. Note that we assume the process $X(t)$ defined in (18) does actually exist; note also that the conditions (19)–(21) can be weakened a little, but to explore all this would take us too far afield. △

Finally, we stress that at this point (22) is to be seen as strictly a notational device; although later we shall give it a more powerful interpretation in the framework of a stochastic calculus.

For the moment we record that (subject to further minor conditions) the transition density function $f(t, y \mid s, x)$ of the diffusion denoted in (22) and defined in (19)–(21), can be shown to satisfy these two equations:

Forward equation

(23) $\quad \dfrac{\partial f}{\partial t} = -\dfrac{\partial}{\partial y}(a(t, y)f) + \dfrac{1}{2}\dfrac{\partial^2}{\partial y^2}(b(t, y)f).$

Backward equation

(24) $\quad \dfrac{\partial f}{\partial s} = -a(s, x)\dfrac{\partial f}{\partial x} - \dfrac{1}{2}b(s, x)\dfrac{\partial f}{\partial x^2}.$

(See equation (4) if you need to be reminded what a transition density is.) In fact most processes of interest here are time-homogeneous, so that $a(t, y) = a(y)$ and $b(t, y) = b(y)$; (but not all: see the Brownian Bridge in (5.4.1)). And sometimes equations (23) and (24) can then be solved, but to specify the solution we do need suitable boundary conditions.

These come in several kinds: first we often have an initial condition; for example, one may specify the density of $X(0)$. If the diffusion starts at a fixed point, such as $X(0) = 0$, then we must have

(25) $$f(0, y \mid 0, x) = \delta(x - y),$$

where this is the delta-function corresponding to a unit probability mass at $y = x$. Next, we consider barriers. As the diffusion $X(t)$ moves randomly in $\mathbb{R}$, it may encounter a wide variety of different types of boundary. They come in two broad classes:

(a) *Intrinsic* boundaries are determined completely by the functions $a(y)$ and $b(y)$ in (20) and (21). We shall not be much concerned with this type.
(b) *Extrinsic* or *regular* boundary points are those at which it is possible to choose conditions to model external constraints of practical interest.

Once again, there are several types, but the two simplest and commonest barriers are called *absorbing* and *reflecting*. These names convey exactly the physical circumstances in which they are relevant:

(26) **Absorbing barrier.** If $x = d$ is an absorbing barrier for $X(t)$, then the process stops at the instant it reaches d, and takes the value d thereafter. This property is expressed as a condition on $f(t, y \mid s, x)$ by

(27) $$f(t, d \mid s, x) = 0, \quad t > s; \quad x \neq d.$$

We omit the proof of this intuitively natural condition.

(28) **Reflecting barrier.** If $x = d$ is a reflecting barrier for $X(t)$, then the process is not permitted to pass the barrier, but there is no other interaction between them. (For a physical analogy, think of heat or electricity in the presence of a perfect insulator.) The appropriate mathematical condition here is not quite so simple or natural, but it can be proved to take the form

(29) $$\lim_{y \to d} \left(af - \frac{1}{2} \frac{\partial}{\partial y}(bf) \right) = 0, \quad \text{for all } t > s.$$

(If you have any familiarity with related diffusion problems in mathematical physics, this looks more obvious; it asserts that, e.g. the flux of heat (or electricity) across $x = d$ is zero. So, if it helps you to do so, you can think of (29) as expressing the idea that the flux of probability 'mass' across the reflecting barrier $x = d$ is zero.)

For the Wiener process, with all its extra symmetry, the condition is even more natural; effectively the paths of the process are simply reflected at d as if by a mirror.

(30) **Reflected Wiener process.** Let us seek to consider a Wiener process with a reflecting barrier at the origin. The density $f(t, y)$ of this reflected process $R(t)$ must satisfy the forward equation

$$\frac{\partial f}{\partial t} = \frac{1}{2}\frac{\partial^2 f}{\partial y^2}, \quad y \neq 0.$$

Let us start the process at $x > 0$, so that the initial condition $R(0) = x$ corresponds to the condition $f(0, y) = \delta(x - y)$, as in (25). The boundary condition (29) applied at $y = 0$ now tells us that $(\partial/\partial y) f(t, y) = 0$, at $y = 0$.

If we denote the density of a normal $N(0, t)$ random variable Z by $\phi_t(z)$, it is easy to check that the required density $f(t, y)$, satisfying all the above constraints, is

$$f(t, y) = \phi_t(y - x) + \phi_t(y + x).$$

If we let $x \to 0$, so that $W(0) = 0 = R(0)$, we obtain the important special case of the standard Wiener process reflected at the origin. This gives the required density as $2\phi_t(y)$; or you can verify directly that

(31) $$f(t, y) = \sqrt{\frac{2}{\pi t}} \exp\left(-\frac{y^2}{2t}\right), \quad y \geq 0$$

satisfies all these constraints. It is also easy to see (by symmetry) that the process

$$Y(t) = |W(t)|$$

has exactly the same density (31) as $R(t)$. This explains our earlier remark about the mirror image of the Wiener process; we can equally well take $|W(t)|$ as a definition of the Wiener process reflected at the origin. Note, however, that it is not immediately obvious that $|W(t)|$ should have the Markov property; happily, it does, and this is an exercise for you to prove. ○

Just as in the discrete case, the introduction of reflecting barriers may change the behaviour of the process quite dramatically. For example, consider the Wiener process with drift $a < 0$, which is (when unrestrained) naturally transient. However, suppose we introduce a reflecting barrier at the origin, and consider the process for $y \geq 0$. We may seek a stationary distribution by setting $\partial f/\partial t = 0$ in the forward equation (17), yielding

$$(32) \qquad \frac{1}{2}\frac{\partial^2 f}{\partial y^2} - \frac{a}{b}\frac{\partial f}{\partial y} = 0, \quad y \geq 0,$$

with the boundary condition $af(0) = \frac{1}{2}f_y(0)$. It is easy to check that the density

$$(33) \qquad f(y) = 2|a/b|e^{-2|a/b|y}, \quad y \geq 0$$

is the solution yielding a stationary distribution; this is Exponential $(-2a/b)$. It can further be shown that the density of such a drifting Wiener process started with a non-negative value converges to (33), when $a < 0$, but it is beyond our scope to give the details.

Exercises
(a) Show that the Wiener process, (as defined in (2)), and the reflected Wiener process $|W(t)|$ have the Markov property.
(b)* Show that if $W(t)$ is a Wiener process then so are
  (i) $W(t)$;
  (ii) $W(a+t) - W(a)$, $a > 0$.
(c)* Let $W(t)$ be a Wiener process with an absorbing barrier at $a > 0$. Show that the density of $W(t), t > 0$ is $f(t, y) - f(t, y - 2a)$, $y < a$, where $f(t, y)$ is the density of $W(t)$ in the absence of any barrier.
(d) **Wright's formula.** Let $X(t)$ be a diffusion with instantaneous parameters $a(x)$ and $b(x)$, respectively. There are reflecting barriers at b_1 and b_2, $b_1 < X(0) < b_2$, where $a(x)$ and $b(x)$ are both bounded with no zeros in $[b_1, b_2]$. Show that the stationary density of $X(t)$ is given by

$$f(x) = \frac{A}{b(x)} \exp\left\{\int^x 2\frac{a(y)}{b(y)}\,dy\right\}$$

and determine the constant A.

5.2 The Wiener process

Of all diffusion processes, the Wiener process $W(t)$ is the most important, and we devote this section to a minimal exploration of some of its properties.

Roughly speaking, they fall into two types: namely sample-path properties, and distributional properties. Here are some of the first kind: with probability 1

(a) $W(t)$ is continuous;
(b) $W(t)$ is not differentiable anywhere;
(c) $W(t)$ changes sign infinitely often in any interval $[0, \epsilon]$, $\epsilon > 0$;
(d) $W(t)$ eventually visits every point in $\mathbb{R}$;
(e) $W(t)/t \to 0$, as $t \to \infty$.

Broadly interpreted, this says that $W(t)$ is incredibly spiky, but stays reasonably near its starting point over moderate time-spans.

Such properties are not easy to prove, and this section confines itself to the second kind. We begin with an easy but important observation about the joint moments of $W(s)$ and $W(t)$.

(1) **Covariance.** Because $W(t)$ has independent Normal increments, we can calculate, for $s < t$,

$$\begin{aligned}
\mathrm{E}(W(s)W(t)) &= \mathrm{E}(\{W(t) - W(s)\}W(s) + W(s)^2) \\
&= \mathrm{E}\{W(s)^2\}, \quad \text{since } \mathrm{E}\{W(t) - W(s) \mid W(s)\} = 0 \\
&= s, \quad \text{since } W(s) \text{ is } N(0, s).
\end{aligned}$$
(2)

Hence, $\mathrm{cov}(W(s), W(t)) = \min\{s, t\} = s \wedge t$ and the correlation is

(3) $$\rho(W(s), W(t)) = \sqrt{\frac{s}{t}}, \quad \text{for } s < t.$$

○

Joint distributions. In considering the joint distributions of $W(t)$, it is convenient to denote the $N(0, t)$ density function by $\phi_t(x)$, so that

(4) $$\phi_t(x) = \frac{1}{\sqrt{2\pi t}} \exp\left(-\frac{1}{2}\frac{x^2}{t}\right)$$

with corresponding distribution function $\Phi_t(x)$. Thus, using independence of increments, the joint density of $W(s)$ and $W(t)$ is seen to be, for $0 < s < t$,

$$f(x, y) = \phi_s(x)\phi_{t-s}(y - x)$$

$$= \frac{1}{2\pi}\frac{1}{\sqrt{s(t-s)}} \exp\left\{-\frac{1}{2}\frac{x^2}{s} - \frac{1}{2}\frac{(y-x)^2}{t-s}\right\}$$

(5) $$= \frac{1}{2\pi\sqrt{st(1 - s/t)}} \exp\left\{-\frac{1}{2(1 - s/t)}\left\{\frac{x^2}{s} - \frac{2}{\sqrt{st}}\sqrt{\frac{s}{t}}xy + \frac{y^2}{t}\right\}\right\},$$

which we now recognize as the bivariate normal density with parameters $\sqrt{s/t}$, s, and t. This joint normal distribution of values of the Wiener process greatly simplifies many routine calculations. For example, using the results established in (1.9), especially Exercise (1.9.6), we obtain immediately that for $0 < s < t$,

(6) $$E(W(s) \mid W(t)) = \frac{s}{t}W(t),$$

(7) $$\text{var}(W(s) \mid W(t)) = s\left(1 - \frac{s}{t}\right).$$

Of course, for $0 < t < s$,

$$E(W(s) \mid W(t)) = W(t),$$
$$\text{var}(W(s) \mid W(t)) = s - t$$

and so for any s and t,

(8) $$E(W(s) \mid W(t)) = \frac{s \wedge t}{t}W(t),$$

(9) $$\text{var}(W(s) \mid W(t)) = \frac{s \wedge t}{t}|s - t|,$$

where $s \wedge t$ is the smaller of s and t. It is interesting that the conditional variance does not depend on $W(t)$.

Once again using the independence of the increments of $W(t)$, we have that the joint density of $W(t_1), W(t_2), \ldots, W(t_n)$, (where $t_1 < t_2 < \cdots < t_n$), is

(10) $$f(w_1, \ldots, w_n) = \phi_{t_1}(w_1)\phi_{t_2-t_1}(w_2 - w_1) \ldots \phi_{t_n-t_{n-1}}(w_n - w_{n-1}).$$

It follows from this, and equally from Definition (1.9.11), that $W(t_1), \ldots, W(t_n)$ have a multivariate normal distribution. We remarked in (1.9) that such distributions are completely determined by their first and second moments. This fact often simplifies the exploration of properties of the Wiener process and processes derived from it. For example, to show that a given process $X(t)$ is a Wiener process, it is sufficient to show that it is continuous, with multivariate normal joint distributions having zero mean and independent increments such that $\text{var}\,(X(t) - X(s)) = |t - s|$, or equivalently that

$$\text{cov}(X(s), X(t)) = s \wedge t.$$

We illustrate this idea in looking at some simple but interesting functions of the Wiener process.

(11) **Scaled Wiener process.** For any constant $a > 0$, the process $W_a(t) = \sqrt{a}\,W(t/a)$ is also a Wiener process.

To see this, note that W_a is continuous, has independent zero-mean normal increments, and starts at $W(0) = 0$, because $W(t)$ has those properties. Therefore, the only thing remaining to be proved is that for $s < t$

$$\text{var}\,\{W_a(t) - W_a(s)\} = a\,\text{var}\,\left\{W\left(\frac{t}{a}\right) - W\left(\frac{s}{a}\right)\right\}$$
$$= t - s$$

as required. ○

(12) **Time-inverted Wiener process.** If we define

$$W_i(t) = \begin{cases} tW(1/t), & t > 0, \\ 0, & t = 0, \end{cases}$$

then $W_i(t)$ is a Wiener process.

As above, it is easy to check that $W_i(t)$ has independent Normal increments with the correct mean and variance, and $W_i(0) = 0$ by definition. The only thing remaining to be proved is that $W_i(t)$ is continuous at $t = 0$; this follows from the claim at the beginning of this section that

$$W(t)/t \to 0, \quad \text{as } t \to \infty.$$

Setting $t' = 1/t$, the required result follows. ○

We may consider other interesting functions of the Wiener process.

(13) **Ornstein–Uhlenbeck process.** This can be obtained from $W(t)$ by yet another form of scaling and time change,

(14) $$U(t) = e^{-t} W(e^{2t}).$$

It is clearly continuous, with multivariate normal joint distributions. Since $EU(t) = 0$, and

$$\operatorname{cov}(U(t), U(t+s)) = e^{-2t-s}E(W(e^{2t})W(e^{2(t+s)}))$$
$$= e^{-s}$$

it follows that $U(t)$ is also a stationary process. It does not have independent increments, but it is a Markov process; $U(t)$ arises naturally as a model for the velocity of a particle in a fluid under the influence of Brownian motion and drag. To see this most directly, we simply calculate the instantaneous mean and variance of $U(t)$ as

$$E(U(t+dt) - U(t) \mid U(t)) = -U(t)\,dt = a(t, U)\,dt$$

and

$$E((U(t+dt) - U(t))^2 \mid U(t)) = 2dt = b(t, U)\,dt.$$

The details of this calculation are an exercise for you, (Exercise 5.2(d)); clearly the term $-U(t)\,dt$ models the frictional, or drag, force, and in this case the defining equation (5.1.22) takes the form

(15) $$dU = -U(t)\,dt + \sqrt{2}\,dW(t).$$

Notice that the process defined in (14) is stationary because, by definition, $U(0) = W(1)$, so that $U(t)$ starts with its stationary distribution. If we seek the motion of a particle starting at rest, so $U(0) = 0$, then we may set

(16) $$U_0(t) = e^{-t}W(e^{2t} - 1).$$

It is an easy exercise for you to check that $U_0(t)$ has the same instantaneous moments as $U(t)$. However, $U_0(t)$ is not stationary, and the covariance function is more complicated; for $s, t > 0$ we have

$$\operatorname{cov}(U_0(t), U_0(t+s)) = e^{-t}e^{-(t+s)}\{e^{2t} - 1\}$$
$$= e^{-s} - e^{-(2t+s)}$$
$$\to e^{-s}, \quad \text{as } t \to \infty.$$

○

More generally, the same argument as that leading from (14) to (15) shows that the equation

$$dU(t) = -\beta U(t)\,dt + \alpha\,dW(t),$$

corresponds to the process $U(t)$ obtained from $W(t)$ by

$$U(t) = e^{-\beta t} W\left(\frac{\alpha^2}{2\beta} e^{2\beta t}\right).$$

Note that, in the particular case when $\alpha = \beta = 1$, so that $U(t) = e^{-t} W(\frac{1}{2} e^{2t})$, we obtain what is occasionally called the standard Ornstein–Uhlenbeck process, such that

$$dU(t) = -U(t)\,dt + dW(t).$$

(17) Exponential (or geometric) Wiener process. This is obtained from $W(t)$ by writing
$$Y(t) = \exp(\mu t + \sigma W(t)).$$

This is easily seen to be a Markov process; in this case it is not $Y(t)$ itself but $\log Y(t)$ that has multivariate normal joint distributions. Since $Y(t)$ cannot be negative, and $\log Y(t)$ has independent increments, the exponential Wiener process is popular as a model for stock prices. ○

Another important method of deriving new and interesting stochastic processes from $W(t)$ is to use conditioning.

Our knowledge of all the joint distributions of $W(t)$ enables us to consider the process conditional on a number of possible events of interest. For example, an important case arises when we consider $W(t)$ conditional on its value at some future time $s > t$. This so-called *Brownian Bridge* has been useful in the statistical analysis of empirical distribution functions; more recently it has found applications in financial mathematics. For example, one could take it as a first (simple) model for the price of a gilt (government bond) that has a fixed redemption price in the future.

(18) Brownian Bridge. This is the process $B(t)$, $0 \leq t \leq 1$, obtained by taking $W(t)$ conditional on the event that $W(1) = 0$. (It may also be called the '*tied-down Wiener process*', for obvious reasons.) Since $W(t)$ has multivariate normal joint distributions and is a Markov process, it follows that $B(t)$ has both these properties also. But, clearly, $B(t)$ does not have independent increments.

Recalling that any multivariate normal density is determined by its first and second moments, our natural first task is therefore to find $EB(t)$ and $\text{cov}(B(s), B(t))$. Using the definition of $B(t)$, and the results displayed in (6) and (7), we calculate these as follows. For $t < 1$,

$$EB(t) = E(W(t) \mid W(1) = 0) = tW(1) = 0.$$

For the covariance, suppose $0 < s < t < 1$, and use conditional expectation, thus:

$$\begin{aligned}
E(B(s)B(t)) &= E(W(s)W(t) \mid W(1) = 0) \\
&= E(E(W(s)W(t) \mid W(t), W(1) = 0) \mid W(1) = 0) \\
&= E(W(t)E(W(s) \mid W(t)) \mid W(1) = 0),
\end{aligned}$$

by the pull-through property (1.7.20)

$$\begin{aligned}
&= E\left(\frac{s}{t}W(t)^2 \mid W(1) = 0\right), \quad \text{by (5.2.6)} \\
&= \frac{s}{t}\text{var}(W(t) \mid W(1) = 0) \\
&= \frac{s}{t}t(1-t), \quad \text{by the above.}
\end{aligned}$$

Hence,

(19) $$\text{cov}(B(s), B(t)) = s \wedge t - st$$

and for $s < t$,

(20) $$\rho(B(s), B(t)) = \left(\frac{s}{t}\frac{(1-t)}{(1-s)}\right)^{1/2}.$$

One possible source of unease about the above results (especially when we consider the task of simulating $B(t)$) arises when we observe that, in the definition of $B(t)$, the conditioning event $W(1) = 0$ has probability zero. Fortunately, the following result immediately resolves the question.

(21) Lemma. Let $B^*(t) = W(t) - tW(1)$, $0 \leq t \leq 1$, where $W(t)$ is the standard Wiener process. Then $B^*(t)$ is the Brownian Bridge.

Proof. Because $W(t)$ has multinormal joint distributions, so does $B^*(t)$. By the remark above, it is therefore sufficient to check that the first and second moments of $B^*(t)$ are the same as those of $B(t)$. First, trivially, $EB^*(t) = 0$. Second, for $0 < s < t < 1$,

$$\begin{aligned}
\text{cov}(B^*(s), B^*(t)) &= E(W(s)W(t) - sW(t)W(1) - tW(s)W(1) + stW(1)^2) \\
&= s - st - st + st = s - st \\
&= \text{cov}(B(s), B(t)).
\end{aligned}$$

$\square$

There are several more ways of representing the Brownian Bridge; we shall see some later.

Exercises

(a)* If $W(t)$ is a standard Wiener process, show that $\hat{B}(t) = (1-t)W(t/(1-t))$ defines a Brownian Bridge, if we set $\hat{B}(1) = 0$.

(b) Show explicitly that the Brownian Bridge has the Markov property.

(c) Let $V(t) = \int_0^t W(u)\, du$, where $W(t)$ is the standard Wiener process. Show that

  (i) $V(t)$ has multinormal joint distributions;

  (ii) $V(t)$ does not have the Markov property;

  (iii) $\operatorname{cov}(V(s), V(t)) = \frac{1}{2}s^2(t - \frac{1}{3}s)$, $0 \le s \le t$.

(d)* Verify that the instantaneous parameters of the Ornstein–Uhlenbeck processes defined in (13) are given by $a(t, u) = -u$ and $b(t, u) = 2$.

5.3 Reflection principle; first-passage times

It is a trite observation that the Normal density is symmetric about its mean. Nevertheless, this simple symmetry turns out to be very useful in the analysis of the Wiener process. Here is a classic example.

(1) **The reflection principle.** Let $m > 0$, fix $t > 0$, and consider the event that $W(t) > m$. Since $W(s)$ is continuous $0 \le s \le t$, and $W(0) = 0$, it follows that in this event we have $W(s) = m$ for at least one $s \in [0, t]$. It is natural to be interested in

(2) $$T_m = \inf\{s > 0 : W(s) = m\}$$

and this is called the *first-passage time* to m.

Now (a little less naturally) let us derive another possible path from that considered above. We set

(3) $$R(s) = \begin{cases} W(s), & \text{for } s < T_m, \\ 2m - W(s), & \text{for } s \ge T_m. \end{cases}$$

This may be envisaged as reflecting that portion of the path after T_m in the horizontal line $y = m$. The reflected path is dotted, and of course $R(t) \le m$ when $W(t) \ge m$; (see Figure 5.1 on the following page).

The argument of the reflection principle first asserts that the original and reflected paths are equally likely, because of the symmetry of the normal

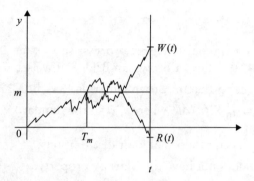

Figure 5.1 Sketch illustrating the reflexion principle.

density. Second, it observes that reflection is a one–one transformation, and finally it claims that we therefore have

(4) $$P(T_m \leq t, W(t) > m) = P(T_m \leq t, W(t) < m).$$

This observation yields a remarkably neat way to find the distribution of T_m. Simply note that the event $\{W(t) > m\}$ is included in the event $\{T_m \leq t\}$. Hence,

(5) $$P(T_m \leq t, W(t) > m) = P(W(t) > m).$$

Now we can write

$$P(T_m \leq t) = P(T_m \leq t; W(t) > m) + P(T_m \leq t, W(t) < m)$$
$$= 2P(T_m \leq t, W(t) > m), \quad \text{by the reflection principle (4)}$$
$$= 2P(W(t) > m), \quad \text{by (5)}$$
$$= P(|W(t)| > m), \quad \text{by symmetry}$$
$$= \sqrt{\frac{2}{\pi t}} \int_m^\infty \exp\left(-\frac{x^2}{2t}\right) dx, \quad \text{since } W(t) \text{ is } N(0,t)$$

(6) $$= \int_0^t \frac{|m|}{(2\pi y^3)^{1/2}} \exp\left(-\frac{m^2}{2y}\right) dy, \quad \text{setting } x^2 = tm^2/y.$$

Hence T_m has density

(7) $$f(t) = \frac{|m|}{(2\pi t^3)^{1/2}} \exp\left(-\frac{m^2}{2t}\right), \quad t \geq 0.$$

If you like integrating, you can show that $P(T_m < \infty) = 1$, by evaluating $\int_0^\infty f(t)\,dt = 1$. We shall discover a neater method in Section (5.5). ○

This is a clever and ingenious argument, but this formulation of the reflection principle leaves the argument incomplete. The statement (4) in (1) has surreptitiously included the assumption that the development of $W(t)$ after T_m is independent of its track record up to T_m. From our previous discussion in (4.1.27) et seq. we recognize this as the Strong Markov property, as discussed in (4.1.27)–(4.1.29) in the context of Markov chains; the key result is this, which we state without proof.

(8) **Strong Markov property.** The first-passage time T_m is a stopping time for $W(t)$, and the process

$$W^*(t) = W(T_m + t) - W(T_m)$$

is a standard Wiener process independent of $W(s)$, $s \leq T_m$.

Defining a stopping time rigorously for $W(t)$, and then proving this theorem would take us too far afield, but we use the result freely to obtain several attractive corollaries. We begin with a simple consequence of (7).

(9) **Maximum of the Wiener process.** Define

$$M(t) = \max\{W(s); 0 \leq s \leq t\}.$$

For $m > 0$ it is obvious that

(10) $\qquad T_m \leq t \quad \text{if and only if} \quad M(t) \geq m.$

Hence, from result (7) about the distribution of T_m,

$$P(M(t) \geq m) = P(T_m \leq t)$$

$$= \int_m^\infty \sqrt{\frac{2}{\pi t}} \exp\left(-\frac{x^2}{2t}\right) dx$$

and differentiating gives the density of $M(t)$ immediately as

(11) $\qquad f(m) = \sqrt{\dfrac{2}{\pi t}} \exp\left(-\dfrac{m^2}{2t}\right), \quad m \geq 0.$

○

An interesting result is that which arises when we look at the zeros of the Wiener process.

(12) **Arcsine law for zeros of $W(t)$.** Let Z be the event that the Wiener process $W(t)$ has at least one zero in (s, t); $0 \le s < t$. We shall prove the remarkable fact that

(13) $$P(Z^c) = 1 - P(Z) = \frac{2}{\pi} \sin^{-1} \sqrt{\frac{s}{t}} = \frac{2}{\pi} \arcsin \sqrt{\frac{s}{t}}.$$

Proof. Suppose that $W(s) = w$. Then, using the strong Markov property (or the reflection principle),

$$P(Z \mid W(s) = w) = P(T_{-w} \le t - s), \quad \text{by definition of } T_{-w}$$
$$= P(T_w \le t - s), \quad \text{by symmetry.}$$

Without loss of generality we can take $w > 0$, and use the fact that the density $\phi_s(w)$ of $W(s)$ is symmetric. Then, using the Markov property at s, to remove the condition on $W(s) = w$, we obtain

$$P(Z) = 2 \int_{w=0}^{\infty} \phi_s(w) \int_{u=0}^{t-s} f_{T_w}(u) \, du \, dw$$

$$= \frac{1}{\pi\sqrt{s}} \int_{u=0}^{t-s} u^{-3/2} \int_{w=0}^{\infty} w \exp\left\{-\frac{1}{2}\frac{u+s}{us} w^2\right\} dw \, du$$

(14) $$= \frac{\sqrt{s}}{\pi} \int_0^{t-s} \frac{du}{(u+s)\sqrt{u}}, \quad \text{on evaluating the inner integral}$$

$$= \frac{2}{\pi} \tan^{-1}\left(\frac{t}{s} - 1\right)^{1/2}, \quad \text{using the substitution } u = sv^2$$

$$= \frac{2}{\pi} \cos^{-1} \sqrt{\frac{s}{t}} = 1 - \frac{2}{\pi} \sin^{-1} \sqrt{\frac{s}{t}},$$

where we used the right-angled triangle with sides 1, $\sqrt{s/t}$, and $\sqrt{1 - s/t}$ for the last two equalities.

This gives (13), and we can immediately obtain a surprising corollary. Let V be the time of the last zero of $W(t)$ in $[0, t]$;

$$V = \sup\{s < t : W(s) = 0\}.$$

The distribution of V is simply $P(Z^c)$, so differentiating (14) yields the density

(15) $$f_V(s) = \frac{1}{\pi} \frac{1}{\{s(t-s)\}^{1/2}}, \quad 0 < s < t.$$

The most striking and curious features of this density are that

(a) it is unbounded as $s \to 0$, and as $s \to t$,
(b) it is symmetric about the halfway time $s = \frac{1}{2}t$,
(c) $P(V > 0) = 1$.

The last property may not seem remarkable at first sight, but it follows that
$$T(0) = \inf\{t > 0: W(t) = 0\} = 0.$$

Hence, with more work (which we omit), it can be shown that for any $t > 0$, $W(t)$ has infinitely many zeros in $[0, t]$.

A process linked with the last zero of the Wiener process before time t is this.

(16) **Brownian meander.** Consider a Wiener process starting at $W(0) = \epsilon > 0$; say $W_\epsilon(t)$. The so-called *meander* process $W_\epsilon^+(t)$ is this Wiener process conditional on the event of its never becoming zero in $0 < u < t$, an event denoted by Z_t^c. That is to say, we seek

(17) $$P(W_\epsilon^+(t) > y) = P(W_\epsilon(t) > y \mid Z_t^c)$$
$$= P(\{W_\epsilon(t) > y\} \cap Z_t^c)/P(Z_t^c).$$

Now, by the reflection principle,

$$P(W_\epsilon(t) > y) = P(\{W_\epsilon(t) > y\} \cap Z_t^c) + P(\{W_\epsilon(t) > y\} \cap Z_t)$$
$$= P(\{W_\epsilon(t) > y\} \cap Z_t^c) + P(W_\epsilon(t) < -y).$$

Therefore,

$$P(\{W_\epsilon(t) > y\} \cap Z_t^c) = P(W(t) > y - \epsilon) - P(W(t) < -y - \epsilon)$$
$$= \Phi_t(y + \epsilon) - \Phi_t(y - \epsilon)$$

using the symmetry of the Normal distribution. Hence, substituting into (17),

$$P(W_\epsilon^+(t) > y) = \frac{\Phi_t(y + \epsilon) - \Phi_t(y - \epsilon)}{\Phi_t(\epsilon) - \Phi_t(-\epsilon)}.$$

Allowing $\epsilon \to 0$ gives the transition distribution for the meander starting at the origin:

$$P(W_0^+(t) > y) = \phi_t(y)/\phi_t(0)$$
(18)
$$= \exp(-\tfrac{1}{2}y^2/t).$$

This is often called a Rayleigh distribution.

The connexion between the Brownian meander $W_0^+(t)$ and the last zero V of the Wiener process in $[0,t]$ is clear. After V, $W(t)$ follows the meander process up to time t because it is conditioned not to revisit the origin. (Or it may follow the negative meander, which is essentially the same process, by symmetry.) This is a convenient property, because the meander as originally introduced above is defined on an event of zero probability. (Recall that we made the same observation about the Brownian Bridge.)

Finally, in this section we note that one can often obtain the joint distributions of these random variables that we have considered separately; here is just one example.

(19) **Maximum and first-passage time.** As above, we let T_m be the first-passage time to $m > 0$, and let M_t be the maximum of $W(t)$ in $[0,t]$

$$M_t = \sup\{W(s) : s \leq t\}.$$

For $0 \leq m \leq y$, we can use the Markov property at $T_m = s$ to give

$$P(M_t \leq y \mid T_m = s) = P(m + M_{t-s} \leq y).$$

Hence, by conditional probability,

$$P(M_t \leq y, T_m \leq u) = \int_0^u P(M_t \leq y \mid T_m = s) f_{T_m}(s) \, ds$$

$$= \int_0^u P(M_{t-s} \leq y - m) f_{T_m}(s) \, ds$$

$$= \frac{m}{\pi} \int_0^u \int_0^{y-m} \exp\left\{-\frac{1}{2}\left[\frac{v^2}{t-s} + \frac{m^2}{s}\right]\right\} \frac{dv \, ds}{s\sqrt{s(t-s)}}$$

on substituting the known distributions of f_{T_m} and M_{t-s} from (7) and (11), respectively. Finally, differentiating for y and u, the joint density of M_t and T_m is

(20) $$f(y,u) = \frac{m}{\pi u \sqrt{u(t-u)}} \exp\left\{-\frac{1}{2}\frac{(y-m)^2}{t-u} - \frac{1}{2}\frac{m^2}{u}\right\},$$

$$0 < u < t, \quad 0 \leq m \leq y.$$

This also has a curious corollary:

(21) **Arcsine law for attaining the maximum.** Let U be the time at which $W(t)$ first takes the value M_t:

(22) $$U = \inf\{s \leq t : W(s) = M_t\}.$$

Now, on the event $M_t = y$, we have

$$U = T_y.$$

Therefore, we can set $m = y$, and hence $T_y = T_m = U$ in (20), to obtain the joint density of M_t and U as

(23) $$f_{M_t,U}(y,u) = \frac{y}{\pi u \sqrt{u(t-u)}} \exp\left\{-\frac{y^2}{2u}\right\}, \quad 0 < u < t; \; y > 0.$$

Integrating with respect to y, we find

$$f_U(u) = \frac{1}{\pi \sqrt{u(t-u)}} \quad 0 < u < t$$

so that the time U at which W attains its maximum M_t in $(0,t)$ has the same arcsine distribution as V, the time of the last zero. ○

Exercises

(a)* Show that the first-passage density in (7) is the same as the density of $T = X^{-2}$, where X is normal $N(0, m^{-2})$.

(b) (i) Let T_m be the first-passage time of the Wiener process $W(t)$ to $m > 0$. Use the reflection principle to show that for $\epsilon > 0$,

$$P(\{T_m < 1\} \cap \{0 \leq W(1) < \epsilon\}) = P(2m - \epsilon < W(1) \leq 2m)$$
$$= \frac{\epsilon}{\sqrt{2\pi}} e^{-2m^2} + o(\epsilon).$$

(ii) Let $B(t)$ be the Brownian Bridge on $0 \leq t \leq 1$. Use the above result to show that

$$P(\max_{0 \leq t \leq 1} B(t) > m) = \lim_{\epsilon \to 0} P(T_m < 1 \mid 0 \leq W(1) \leq \epsilon)$$
$$= e^{-2m^2}.$$

(c)* If $0 \leq s \leq t \leq u$, show that the probability that $W(t)$ has no zero in (s, u), given that it has no zero in (s, t) is $\sin^{-1} \sqrt{s/u} / \sin^{-1} \sqrt{s/t}$; show also that the probability that $W(t)$ has no zero in $(0, t)$, given that it has no zero in $(0, s)$, is $\sqrt{s/t}$.

(d)* Show that $P(T_m < \infty) = 1$.

5.4 Functions of diffusions

We remarked in Section (5.1) that any diffusion (continuous Markov process) is characterized by its instantaneous parameters, denoted by $a(t, x)$ and $b(t, x)$. Let us revisit some of the examples seen since then, bearing this in mind.

(1) Brownian Bridge. This is the Wiener process conditional on the event that $W(1) = 0$. By this definition, we may calculate the instantaneous mean

(2)
$$\begin{aligned} a(t, x) \, dt &= E(B(t + dt) - B(t) \mid B(t) = x) \\ &= E(W(t + dt) - W(t) \mid W(t) = x, W(1) = 0) \\ &= E(W(dt) \mid W(1 - t) = -x), \quad \text{on changing the origin} \\ &= -x \frac{dt}{1 - t}, \quad \text{recalling (5.2.6)}. \end{aligned}$$

Likewise using (5.2.7) we obtain $b(t, x) \, dt = dt + o(dt)$, so that the Brownian Bridge has instantaneous parameters $-x/(1 - t)$ and 1, respectively, leading to the representation introduced in (5.1.22):

(3)
$$dB = \frac{-B}{1 - t} \, dt + dW(t).$$

○

(4) Ornstein–Uhlenbeck process. We defined this as

(5)
$$U(t) = e^{-t} W(e^{+2t}), \quad t \geq 0.$$

Using the fact that $W(e^{2t})$ has independent increments with zero mean, it is easy, (see Exercise 5.2(d)), to show that

(6)
$$E(U(t + dt) - U(t) \mid U(t) = x) = -x \, dt$$

so that $a(t,x) = -x$. Likewise, one readily obtains $b(t,x) = 2$, yielding the representation

(7) $$dU = -U\,dt + \sqrt{2}\,dW(t).$$

Note that this process is homogeneous. ○

(8) **Squared Wiener process.** We remarked above that the reflected Wiener process $|W(t)|$ is a Markov process, and therefore a diffusion. It follows immediately that $W(t)^2$ is a Markov process, and we can easily calculate

$$E\{W(t+dt)^2 - W(t)^2 \mid W(t)^2\}$$
$$= E\{2W(t)(W(t+dt) - W(t)) + (W(t+dt) - W(t))^2 \mid W(t)^2\} = dt$$

and likewise, after a similar calculation,

$$E\{(W(t+dt)^2 - W(t)^2)^2 \mid W(t)^2\} = 4W(t)^2\,dt.$$

Hence the process $S(t) = W(t)^2$ has instantaneous parameters $a(t,x) = 1$, and $b(t,x) = 4x$. We may write

(9) $$dS(t) = dt + 2S(t)^{1/2}\,dW(t)$$

and observe that the process is time-homogeneous. ○

We could continue in this vein, examining individual functions of interest one by one, but it is natural to wonder if there is a general approach. Indeed there is, and we give an introductory sketch here, returning with more details in Sections (5.7) and (5.8).

(10) **Transformed diffusion.** Let the diffusion process $X(t)$ have instantaneous parameters $a(t,x)$ and $b(t,x)$, respectively. We define $Y(t) = g(X(t),t)$, where g is an invertible function of $X(t)$, that is differentiable in both arguments as often as we may require. Because g is invertible, $Y(t)$ is also a Markov process, and continuous because g is. The assumed differentiability of $g(X(t),t)$ means that we can expand $g(X(t+h),t+h)$ as a Taylor series, so we introduce the more compact notation $\delta = X(t+h) - X(t)$, for small h. Thus,

$$E(\delta \mid X(t)) = a(t,X)h + o(h),$$

and

$$E(\delta^2 \mid X(t)) = b(t,X)h + o(h).$$

Using suffices to denote partial derivatives in the natural way, we can then write formally, using Taylor's theorem for functions of two independent variables,

$$Y(t+h) - Y(t) = g(X+\delta, t+h) - g(X,t)$$
(11)
$$= g_t(X,t)h + g_x(X,t)\delta + \tfrac{1}{2}g_{tt}(X,t)h^2$$
$$+ g_{tx}(X,t)h\delta + \tfrac{1}{2}g_{xx}(X,t)\delta^2 + R,$$

where R is the remainder after the first- and second-order terms given above. As usual, we seek to consider the effects as $h \to 0$, but since we are then dealing with sequences of random variables we must specify the type of convergence involved. It turns out that mean-square convergence is an appropriate mode; and we note that it can be shown that $ER = o(h)$. Taking the expected value of (11) conditional on $X(t)$, and rejecting all those terms that are $o(h)$, we find that

$$E(Y(t+h) - Y(t) \mid X(t)) = (g_t + a(t,X)g_x + \tfrac{1}{2}b(t,X)g_{xx})h$$

so that $Y(t)$ has instantaneous parameters given by α and β, where

(12)
$$\alpha = g_t + ag_x + \tfrac{1}{2}bg_{xx}$$

and also, likewise,

$$E\{(Y(t+h) - Y(t))^2 \mid X(t)\} = b(t,X)g_x^2 h$$

so that

(13)
$$\beta = bg_x^2.$$

Obviously, we have omitted an enormous amount of technical detail here, but a rigorous analysis is far beyond our scope. A particularly important case arises when none of a, b, α, or β depends on t, so that we are dealing with homogeneous diffusions. Here are some useful examples.

(14) **Geometric Wiener process.** Let $X(t)$ be the standard Wiener process, and define

(15)
$$Y(t) = g(X(t),t) = \exp\{\mu t + \sigma X(t)\}.$$

We have, given $Y(t)$,

(16)
$$g_t = \mu e^{\mu t + \sigma X} = \mu Y,$$

(17)
$$g_x = \sigma e^{\mu t + \sigma X} = \sigma Y,$$

(18)
$$g_{xx} = \sigma^2 Y.$$

Hence Y has instantaneous parameters

(19) $$\alpha(t, y) = \mu y + \tfrac{1}{2}\sigma^2 y,$$

(20) $$\beta(t, y) = \sigma^2 y^2$$

and therefore it is a homogeneous diffusion.

The geometric Wiener price is a classic and popular model for the market prices of stocks and shares, especially in the famous Black–Scholes option-pricing theorem. We note that if $\mu = -\tfrac{1}{2}\sigma^2$, then the instantaneous mean is zero; this fact turns out to be important in that context. ○

(21) **Bessel process.** Let $W_1(t), \ldots, W_n(t)$ be independent Wiener processes, and define

(22) $$S(t) = \sum_{r=1}^{n} W_r^2(t).$$

It can be shown that $S(t)$ is a Markov process; you are asked to verify the fact for $n = 2$ and $n = 3$ in Problem (5.1) below. (It may be intuitively 'clear' that this is so, but proof is required!) The instantaneous parameters of $S(t)$ are $a(t, s) = n$ and $b(t, s) = 4s$; this follows easily from (8) above.

Now define the radial distance $R(t) = \sqrt{S(t)}$, called the Bessel process. It is clearly Markov because $S(t)$ is, and setting $g(s, t) = \sqrt{s}$ we have, given $R = r$,

(23) $$g_s = \frac{1}{2}\frac{1}{\sqrt{s}} = \frac{1}{2r}, \quad g_{ss} = \frac{-1}{4s^{3/2}} = -\frac{1}{4r^3}.$$

Hence $R(t)$ is a diffusion with instantaneous parameters, from (10) above,

(24) $$\alpha(t, r) = n\frac{1}{2r} - \frac{1}{2}4r^2\frac{1}{4r^3} = \frac{n-1}{2r},$$

(25) $$\beta(t, r) = \frac{4r^2}{4r^2} = 1.$$

This diffusion is also homogeneous. ○

As we have noted many times in preceding work, we are often interested in the first-passage time T of a process $X(t)$ to a barrier. More generally (and especially in two or more dimensions), we may be interested in just where on the boundary the process arrives first. In one dimension, this problem may be seen as analogous to the gambler's ruin problem. The process starts at $X(0) = x$, the first-passage time to $a > x$ is T_a, the first-passage time to $b < x$ is T_b, the first time to hit either a or b is $T_a \wedge T_b = T$. We may seek the distribution and moments of T, or, for example, the probability of hitting a before b:

(26) $$p(x) = P(T_a < T_b \mid X(0) = x).$$

We assume that $X(t)$ is a time-homogeneous diffusion, with instantaneous parameters $a(x)$ and $b(x)$, respectively. We also assume that $a(x)$ and $b(x)$ are well-behaved, in that they allow a and b to be regular (absorbing or reflecting) boundaries from 0.

These quantities, regarded as functions of $X(0) = x$, each turn out to satisfy an appropriate differential equation. We give an informal heuristic demonstration of some simple but important examples. The essential idea is the same as that used in deriving backward equations, and in producing the results of (10) above about transformations. We consider the required expectation (or probability) after the process has run for a short time-interval $[0, h]$, arriving at $X(h)$, and then let $h \to 0$. As usual, we assume that all functions in question have Taylor expansions of any required order, and that we can neglect all those terms that are $o(h)$ in this expansion. We begin with a topic of natural interest for any homogeneous diffusion process $X(t)$; the name arises from the analogy with Gambler's Ruin.

(27) **Ruin probability.** We seek, as discussed before (26),

$$p(x) = P(T_a < T_b \mid X(0) = x), \quad b \leq x \leq a.$$

First, and obviously, note that $p(a) = 1$ and $p(b) = 0$. Then, second, for $b < x < a$ we have by conditioning on the state of the process at time h, and setting $\delta = X(h) - x$,

$$p(x) = E\{p(X(h)) \mid X(0) = x\}, \quad \text{using the Markov property at } h$$
(28) $$ = E\{p(x + \delta)\}$$
$$ = E\{p(x) + p'(x)\delta + \tfrac{1}{2} p''(x)\delta^2 + R\}, \quad \text{by Taylor's theorem,}$$

where primes denote differentiation with respect to x.

As usual, we have the instantaneous moments $E\delta = a(x)h + o(h)$, $E\delta^2 = b(x)h + o(h)$, and the remainder R is such that $ER = o(h)$. Hence, rearranging and letting $h \to 0$, we find that $p(x)$ satisfies the second-order ordinary differential equation

(29) $$\tfrac{1}{2} b(x) p''(x) + a(x) p'(x) = 0, \quad b < x < a.$$

○

Similarly, we may derive a differential equation for the expected value of T, the first-passage time to either of a or b.

(30) **Expected duration.** Let

$$m(x) = E(T \mid X(0) = x)$$
$$= E(T_a \wedge T_b \mid X(0) = x).$$

Then, trivially, $m(a) = m(b) = 0$. For $b < x < a$ we condition on $X(h) = x + \delta$, and write, using the Markov property at h,

(31)
$$\begin{aligned} m(x) &= E\{m(X(h)) + h \mid X(0) = x\} \\ &= E\{m(x+\delta)\} + h \\ &= h + m(x) + m'(x)E\delta + \tfrac{1}{2}m''(x)E\delta^2 + E(R) \\ &= h + m(x) + a(x)m'(x)h + \tfrac{1}{2}m''(x)b(x)h + o(h). \end{aligned}$$

Therefore, as usual, we find that $m(x)$ satisfies

(32) $$\tfrac{1}{2}b(x)m''(x) + a(x)m'(x) = -1.$$

○

Various other functions and functionals can be treated in essentially the same way. For example,

(33) **Functions of T.** Let us define

$$f(x) = E(g(T) \mid X(0) = x)$$
$$= E(g(T(x))), \quad \text{say,}$$

where $g(\cdot)$ has a suitable Taylor expansion. Then for $b < x < a$, conditioning on $X(h)$, and using the Markov property at h,

(34)
$$\begin{aligned} f(x) &= E(g(T(X(h)) + h) \mid X(0) = x) \\ &= E\{g(T(X(h))) + hg'\{T(X(h))\} \mid X(0) = x\} + o(h) \\ &= E\{f(X(h)) \mid X(0) = x\} + hE(g'(T(x)) \mid X(0) = x) + o(h) \\ &= f(x) + a(x)f'(x)h + \tfrac{1}{2}b(x)f''(x)h + hE(g'(T(x))) + o(h), \end{aligned}$$

where the last line follows by the same argument used in (27) and (30). Rearranging and allowing $h \to 0$ now gives

(35) $$\tfrac{1}{2}b(x)f''(x) + a(x)f'(x) = -Eg'\{T(x)\}.$$

Note that if we set $g(T) = T$, this immediately gives (32). More generally, if we set $g(T) = e^{-sT}$, this yields an equation for the moment-generating function (mgf) of T. The details are an exercise for you. ○

Second-order differential equations such as (29), (32), and (35) are well-understood, but it is beyond our scope to pursue all the general results. However, it is amusing to consider special cases.

(36) Bessel process. Let $R(t)$ be a Bessel process (defined in (21)), started at $R(0) = x$, where $0 \leq b < x < a$. Then if T is the first-passage time to either a or b, and $m(x) = ET$, we have, from (32),

$$(37) \qquad \frac{1}{2}m''(x) + \frac{n-1}{2x}m'(x) = -1$$

with $m(b) = m(a) = 0$. Integrating once gives

$$m'(x) = \frac{-2x}{n} + \frac{c}{x^{n-1}}.$$

One further integration, and use of the boundary conditions, easily gives the (rather complicated) answer. Note that the case $n = 2$ requires a slightly different treatment from $n \geq 3$, because of the appearance of the logarithm. When $n = 1$, we have the reflected Wiener process.

In the special case when $b = 0$, we find that

$$m'(x) = \frac{-2x}{n}, \quad n \geq 1$$

and so

$$(38) \qquad m(x) = \frac{a^2 - x^2}{n}.$$

○

See the problems for some additional examples.

Finally, in this section we look at one more process obtained by conditioning the Wiener process. In particular, we consider $W(t)$, started at $W(0) = x$, $b \leq x \leq a$, conditional on $T_a < T_b$. This is interpreted as the Wiener process with absorbing barriers at b and a, conditional on absorption at a. Let us denote it by $C(t)$.

The first point is that $C(t)$ is a Markov process, and also continuous, because $W(t)$ has these properties. Thus $C(t)$ is a diffusion process, and our natural first task is to find its instantaneous mean and variance. It will be helpful to recall that

$$(39) \qquad P(T_a < T_b \mid W(0) = x) = \frac{x-b}{a-b}, \quad b \leq x \leq a.$$

5.4 Functions of diffusions

Let $p(t, y \mid x)$ be the transition density of $C(t)$ started at $C(0) = x$. By conditional probability and the definition of $C(t)$, we have, for $b < y < a$,

$$p(t, y \mid x)\, dy$$
$$= P(C(t) \in (y, y + dy) \mid C(0) = x)$$
$$= P(W(t) \in (y, y + dy) \mid W(0) = x \cap T_a < T_b)$$
$$= \frac{P(W(t) \in (y, y + dy) \mid W(0) = x) P(T_a < T_b \mid W(t) = y \cap W(0) = x)}{P(T_a < T_b \mid W(0) = x)}$$
$$= dy f(t, y \mid x) \frac{y - b}{x - b}, \quad \text{using the Markov property and (39),}$$

where $f(t, y \mid x)$ is the transition density of the Wiener process started at $W(0) = x$.

Now we can use this to calculate the instantaneous moments of $C(t)$; thus

$$E(C(t + dt) - C(t) \mid C(t) = c) = \int (y - c) p(dt, y \mid c)\, dy$$
$$= \int \frac{(y - c)(y - b)}{c - b} f(dt, y \mid c)\, dy$$
$$= \frac{dt}{c - b} + o(dt),$$

when we recall that $f(dt, y \mid c)$ is the normal $N(c, dt)$ density. Likewise it is easy to see that

$$E((C(t + dt) - C(t))^2 \mid C(t) = c) = dt + o(dt).$$

Therefore $C(t)$, the Wiener process conditioned to be absorbed at a before b, has instantaneous parameters $a(t, x) = 1/(x - b)$, and $b(t, x) = 1$. Hence, for example, if we set

$$m(x) = E(T_a \mid T_a < T_b \cap W(0) = x)$$

this expected duration satisfies

(40) $$\frac{1}{2} m''(x) + \frac{1}{x - b} m'(x) = -1$$

with the boundary condition $m(a) = 0$. In particular, we observe that if $b = 0$ this is the same as the expected time for the Bessel process in

$\mathbb{R}^2$ to reach a from x. It is an easy exercise for you to show that the solution of (40) is

(41) $\qquad m(x) = \frac{1}{3}\{(a-b)^2 - (x-b)^2\}, \quad b < x < a.$

Exercises

(a)* Let the diffusion process $X(t)$ have instantaneous parameters $a(x) = 0$ and $b(x)$; suppose that $\beta < x = X(0) < \alpha$. Show that, in general, the probability that $X(t)$ hits α before β is

$$p(x) = \frac{x-\beta}{\alpha-\beta}.$$

When may your argument fail?

(b) **Neutral Fisher–Wright model.** In this model for gene frequency, $X(t)$ is a diffusion in $[0, 1]$ with instantaneous parameters $a(X) = 0$ and $b(X) = \nu X(1-X)$. If $m(x)$ is the expected time to hit either of 0 or 1 when $X(0) = x$, show that

$$\nu m(x) = -x \log x - (1-x) \log(1-x).$$

(c) Verify that the squared Bessel process $S(t)$ defined in (21) does indeed have instantaneous parameters

$$a(t,s) = n \quad \text{and} \quad b(t,s) = 4s.$$

(d) Integrate (40) to obtain (41).

5.5 Martingale methods

We have motivated our interest in diffusions by defining them to be continuous Markov processes in general, and in particular seeing the Wiener process as a continuous limit of a random walk. Now recall that martingales proved useful in analysing both random walks and Markov chains. It seems inevitable that martingales will be useful for tackling the Wiener process and diffusions; and so they are. Of course, we will need to upgrade our original Definition (2.4.1).

(1) **Martingale.** The process $(X(t); t \geq 0)$ is a *martingale* if $\mathrm{E}|X(t)| < \infty$, for all t, and

(2) $\qquad \mathrm{E}(X(t) \mid X(t_1), \ldots, X(t_n)) = X(t_n),$

for any $0 \leq t_1 < t_2 < \cdots < t_n < t$.

We often write condition (2) as

(3) $$E(X(t) \mid X(u), u \leq s) = X(s), \quad s < t$$

but this is a slight abuse of notation; we cannot condition on an uncountable number of random variables. But the convenience outweighs this objection, just as it did for the Markov property in continuous time. △

Here are some simple examples of useful martingales.

(4) **Wiener process.** Let $W(t)$ be the standard Wiener process. Then these are martingales:

(5)
$$\text{(i)} \quad M_1(t) = W(t)$$
$$\text{(ii)} \quad M_2(t) = W(t)^2 - t$$
$$\text{(iii)} \quad Z(t) = \exp\left\{aW(t) - \frac{a^2 t}{2}\right\}, \quad a \in \mathbb{R}.$$

Proof. The key point is that, for $s < t$, the increment $W(t) - W(s)$ is independent of $W(s)$, and indeed of $W(u)$, $0 \leq u \leq s$. Thus, we write

(i)
$$E(W(t) \mid W(u), u \leq s) = E(W(t) - W(s) + W(s) \mid W(u), u \leq s)$$
$$= 0 + E(W(s) \mid W(u), u \leq s)$$
$$= W(s).$$

It is trivial to see that $E|W(t)| < \infty$, so $W(t)$ is a martingale.

(ii) Likewise it is easy to see that $EW(t)^2 < \infty$. Then, using the facts that $W(t) - W(s)$ is normal $N(0, t-s)$ and independent of $W(u)$, $u \leq s$, we have

$$E(W(t)^2 - t \mid W(u), u \leq s)$$
$$= E\{(W(t) - W(s) + W(s))^2 - t \mid W(u), u \leq s\}$$
$$= t - s - t + 2W(s)E(W(t) - W(s) \mid W(u), u \leq s) + W(s)^2$$
$$= W(s)^2 - s.$$

(iii) Using all the same ideas,

$$E(e^{aW(t)-a^2t/2} \mid W(u), u \le s)$$

(6)
$$= E(\exp\{aW(t) - aW(s) + aW(s)\} \mid W(u), u \le s)\, e^{-a^2t/2}$$
$$= e^{aW(s)}\, e^{a^2(t-s)/2 - a^2t/2} = e^{aW(s) - a^2s/2}$$

as required. ○

This final martingale is noteworthy as a source of many more.

(7) **Many martingales.** The identity (6) holds for any real a. If we differentiate the identity with respect to a (which is a procedure that can be justified in this case), and set $a = 0$, we obtain the defining equation for the first martingale in (4):

$$E(W(t) \mid W(u), u \le s) = W(s).$$

Continuing in this manner, if we differentiate (6) twice with respect to a, and set $a = 0$, this yields the second martingale $M_2(t)$ in (4). Repeated differentiation then yields a sequence of higher-order martingales that we have not yet seen; here are the next four:

(8)
$$M_3(t) = W(t)^3 - 3tW(t),$$
$$M_4(t) = W(t)^4 - 6tW(t)^2 + 3t^2,$$
$$M_5(t) = W(t)^5 - 10W(t)^3 + 15t^2W(t),$$
$$M_6(t) = W(t)^6 - 15tW(t)^4 + 45t^2W(t)^2 - 15t^3.$$

It is an exercise for you to write down the details, and you may then check by direct computations that these are indeed martingales as claimed.

As we have seen before, many martingales are most useful when working in conjunction with a stopping time T. It is therefore appropriate for us to insert a reworded version of (4.1.27) appropriate for these new circumstances.

(12) **Stopping time.** The non-negative random variable T is a stopping time for the martingale $X(t)$ if the indicator of the event $\{T \le t\}$ depends only on $X(s)$ for $s \le t$, and does not depend on $X(t+u)$ for all $u > 0$. That is to say, if we know $X(s)$ for all $0 \le s \le t$, then we also know whether or not $T \le t$.

Just as for discrete Markov chains, it is natural to expect that the first-passage time of $X(t)$ to a set A,

$$T_A = \inf\{t \ge 0 : X(t) \in A\}$$

is a stopping time. Fortunately this is in practice indeed the case, provided that we make suitable conditions on A and $X(t)$. It is beyond our scope to explore the details here, but no problems arise when the paths of $X(t)$ are continuous and A is a closed set; we refrain from looking at any other circumstances. △

Just as in the discrete case, the two key results are these; we omit the proofs, which closely mimic those in discrete time.

(13) **Stopping a martingale.** Let $X(t)$ be a martingale, and assume T is a stopping time for $X(t)$, then $X(t \wedge T)$ is a martingale.

As usual, supplementary conditions are required to yield the optional stopping theorem. A useful and straightforward case arises when $P(T < \infty) = 1$, and $|X(t \wedge T)| \leq K < \infty$ for all t. Under these conditions

(14) $$EX(T) = EX(0).$$

There are stronger versions of this result (that is to say, (14) holds under weaker conditions), but we shall not need them here. Just as in the discrete case, we can often bypass formal use of the optional stopping theorem by simply applying Dominated or Monotone convergence directly to $EX(t \wedge T) = EX(0)$, whenever $X(t \wedge T) \to X(T)$ with probability 1, as $t \to \infty$.

Here is a classic example of stopping.

(15) **Wiener process in a strip.** Let $W(t)$ be the standard Wiener process, and define the stopping time T to be the instant at which $W(t)$ hits either a or b, $b < 0 < a$. We call T the *first exit time*; it is a straightforward exercise for you to show directly that $P(T < \infty) = 1$, and indeed $ET < \infty$.

We can use the martingales in (5) to tell us much more about T and $W(T)$. Of course the key observation is that the stopped martingale $W(t \wedge T)$ is uniformly bounded in the finite strip $b \leq W(t \wedge T) \leq a$.

(16) **Hitting probabilities.** Since $P(T < \infty) = 1$ and $|W(t \wedge T)| < a - b = |a| + |b|$, we can apply the optional stopping theorem (14) immediately to the martingale $W(t)$, to find

$$0 = EW(0) = EW(T)$$
$$= b(1 - P(W(T) = a)) + aP(W(T) = a).$$

Hence,

(17) $$P(W(T) = a) = \frac{b}{b-a} \quad \text{and} \quad P(W(T) = b) = \frac{a}{a-b}.$$

As a corollary to this, recall Definition (5.3.2) of the first-passage time T_a. From (17) we have
$$P(T_a < T_b) = \frac{b}{b-a}.$$
Now allowing $b \to -\infty$, and noting that, for any b,
$$P(T_a < \infty) \geq P(T_a < T_b)$$
we obtain $P(T_a < \infty) = 1$.

Expected exit time. Next we turn to the martingale $M_2(t) = W(t)^2 - t$. Because T is a stopping time for $M_2(t)$, we know that $M_2(T \wedge t)$ is a martingale and

(18)
$$\begin{aligned} 0 = EM_2(0) &= EM_2(T \wedge t) \\ &= EW(T \wedge t)^2 - E(T \wedge t). \end{aligned}$$

Now $|W(T \wedge t)^2 - T \wedge t| \leq a^2 + b^2 + T$, and $E(a^2 + b^2 + T) < \infty$, since $ET < \infty$. Therefore, by the Dominated Convergence theorem, we may let $t \to \infty$ to find that

(19)
$$\begin{aligned} ET &= EW(t)^2 \\ &= a^2 P(W(t) = a) + b^2 P(W(T) = b) \\ &= \frac{a^2 b}{b-a} + \frac{b^2 a}{a-b} \\ &= -ab. \end{aligned}$$

Similar arguments applied to $M_3(t)$ and $M_4(t)$ show further that $\mathrm{var}\, T = -\frac{1}{3}ab\,(a^2 + b^2)$. This is an exercise for you; see Problem (5.7). ○

Exponential martingales are equally useful.

(20) **First-passage transform.** Let T be the first-passage time of the Wiener process $W(t)$ to $a > 0$, with probability density $f_T(t)$. We showed in (5) that
$$Z(t) = \exp\left\{\lambda W(t) - \tfrac{1}{2}\lambda^2 t\right\}, \quad \lambda > 0,$$
is a martingale. Because $|Z(t \wedge T)| \leq e^{\lambda a} < \infty$ and $P(T < \infty) = 1$, we deduce immediately from (14) that

(21)
$$1 = EZ(0) = EZ(T) = e^{\lambda a} E e^{-(1/2)\lambda^2 T}.$$

Hence, setting $\theta = \tfrac{1}{2}\lambda^2$, we obtain the mgf of T,

(22)
$$E e^{-\theta T} = e^{-\sqrt{2\theta}\, a}.$$

(Note that this is formally the Laplace transform of $f_T(t)$.) Actually, it is interesting to note that we do not need this prior information that $P(T < \infty) = 1$. All we need is that T is a stopping time, so that $Z(t \wedge T)$ is a martingale such that as $t \to \infty$, with probability 1,

$$Z(t \wedge T) \to \begin{cases} e^{\lambda a - (1/2)\lambda^2 T}, & \text{if } T < \infty, \\ 0, & \text{if } T = \infty. \end{cases}$$

Now $|Z(t \wedge T)| \leq e^{\lambda a} < \infty$. Hence, by the Dominated Convergence theorem we still have (21) and (22). It follows by setting $\theta = 0$ in (22) that $P(T < \infty) = \int_0^\infty f_T(t)\, dt = 1$.

Of course, we have already discovered $f_T(t)$ in (5.3.7) by other methods, so (22) further demonstrates that

$$\int_0^\infty e^{-\theta t} t^{-3/2} e^{-(a^2/2t)}\, dt = \sqrt{2\pi}\, a^{-1} e^{-\sqrt{2\theta} a}$$

a result that you may care to verify by direct integration, if you have time. ○

This idea can be made to work with two barriers, with a little extra effort.

(23) Wiener process in a strip. We use all the notation defined above: $W(t)$ is the Wiener process in the strip $b \leq W(t) \leq a$; T_a and T_b are the first-passage times to a and b, respectively, and $T = T_a \wedge T_b$ is the first exit from (b, a); I_a and I_b are the respective indicators of $\{T_a < T_b\}$ and $\{T_b < T_a\}$. Then, from (22), and the properties of indicators

(24) $$e^{-\sqrt{2\theta} a} = E e^{-\theta T_a} = E\{e^{-\theta T_a}(I_a + I_b)\}.$$

Now we make the crucial observation that on the event $T_b < T_a$, the first-passage time to a then requires a first passage from b to a that is independent of the process up to T_b, by the strong Markov property. We denote this further time by T^*_{a-b}, and note that in particular, by the remarks above,

(25) $$\begin{aligned} E\{I_b\, e^{-\theta T_a}\} &= E\{I_b\, e^{-\theta T_b - \theta T^*_{a-b}}\} \\ &= E\{I_b\, e^{-\theta T_b}\} E\{e^{-\theta T^*_{a-b}}\} \\ &= E\{I_b\, e^{-\theta T_b}\} e^{-(a-b)\sqrt{2\theta}}. \end{aligned}$$

Hence, substituting (25) in (24),

$$e^{-a\sqrt{2\theta}} = E(I_a\, e^{-\theta T_a}) + E(I_b\, e^{-\theta T_b})\, e^{-(a-b)\sqrt{2\theta}}.$$

Interchanging the roles of a and b yields

$$e^{+b\sqrt{2\theta}} = E(I_b\, e^{-\theta T_b}) + E(I_a\, e^{-\theta T_a})\, e^{-(a-b)\sqrt{2\theta}}.$$

These two simultaneous equations now enable us to calculate

$$\begin{aligned} E\, e^{-\theta T} &= E(e^{-\theta T_a} I_a) + E(e^{-\theta T_b} I_b) \\ &= \frac{\sinh(a\sqrt{2\theta}) - \sinh(b\sqrt{2\theta})}{\sinh\{(a-b)\sqrt{2\theta}\}}. \end{aligned}$$

The drifting Wiener process can also be tackled using an appropriate martingale; see the problems.

Exercises
(a)* Show that the first-exit time $T = \inf\{t : W(t) \notin (b, a)\}$ satisfies $P(T < \infty) = 1$ and $ET < \infty$. [Hint: Construct a geometric random variable X such that $T < X$.]

(b)* Let $(V(t), W(t))$ be a pair of independent standard Wiener processes. Let T be the first-passage time of $W(t)$ to x. Show that $V(T)$ has density

$$f_V(v) = \frac{|x|}{\pi(x^2 + v^2)}, \quad -\infty < v < \infty.$$

Interpret this geometrically if the process $(V(t), W(t))$, $t \geq 0$, is regarded as a random path in the (v, w)-plane.

(c)* With the notation of Exercise (b), show that

$$E(e^{i\theta V(T)} \mid T) = e^{-(1/2)\theta^2 T}$$

and deduce that $E\, e^{i\theta V(T)} = e^{-|\theta|a}$, where $a = |x|$.

5.6 Stochastic calculus: introduction

Up to now we have looked at diffusion processes in general, and the Wiener process in particular, through their joint distributions. The fact that they have the Markov property is especially useful, and the links with various martingales have been exploited to good effect. However, it is natural to wonder if we can find some way to deal with these processes directly.

This section gives a very informal and heuristic introduction to the ideas and tasks required to do so.

The formal representation that we introduced in (5.1.22)

$$\text{(1)} \qquad dX(t) = a(t, X)\, dt + \sqrt{b(t, X)}\, dW(t)$$

seems to be the obvious starting point for a direct analysis of the process $X(t)$. This supposition is indeed correct, but some care is required. To prepare ourselves for dealing with the random process $X(t)$, it is convenient to recall briefly the arguments and methods that we use in the non-random case.

(2) **Classical modelling.** In dealing with deterministic systems evolving in continuous time, a long-standing and brilliantly successful approach is to begin by describing its development over a small time-interval $(t, t+dt)$. For example, consider the velocity $v(t)$ of a particle of unit mass falling vertically under the Earth's gravity g. The classical laws of motion are expressed in the statement

$$\text{(3)} \qquad dv(t) = v(t + dt) - v(t) = g\, dt,$$

which gives the change in the velocity, over the small interval $(t, t + dt)$, due to gravity.

From this modelling assumption there are two routine mathematical paths, both of which entail taking the limit as $dt \downarrow 0$.

The first is to divide (3) throughout by dt, and let $dt \to 0$. If this procedure is valid (which it is in this trivial case), we obtain the differential equation

$$\frac{dv}{dt} = g.$$

The second approach is to sum (3) over all such small disjoint time-intervals in $[a, b]$ (say), and let the size of the largest such interval tend to zero. If this procedure is valid (which of course it is in this trivial case) we obtain the integral

$$v(b) - v(a) = \int_a^b g\, dt.$$

Of course, a wealth of classical theorems links these two approaches and tells us ultimately that both routes yield

$$v(t) - v(0) = gt.$$

The key question is whether we can transfer these ideas to stochastic models.

We have already introduced an elementary model of similar type in the Langevin equation (see (5.2.15) and (5.4.7)), so let us start with that.

(4) Ornstein–Uhlenbeck model. If $U(t)$ is the velocity of a particle of unit mass moving under the influence of the Brownian effect, together with a drag proportional to $U(t)$, then as discussed above, we may write

$$(5) \qquad dU(t) = U(t+dt) - U(t) = -\beta U(t)\, dt + dW(t).$$

(Note that, for simplicity of notation, we have assumed that the instantaneous variance is dt.) There is no question here of dividing each side by dt and allowing $dt \to 0$, because as we remarked above, $W(t)$ is not differentiable. However, $W(t)$ is continuous, so the alternative procedure of forming an integral offers some hope of progress. Formally, the integrated form of (5) is

$$(6) \qquad U(t) - U(0) = \int_0^t dU(s) = -\int_0^t \beta U(s)\, ds + \int_0^t dW(s)$$

but we have yet to give any meaning to the final integral on the right, nor does this give an explicit representation of $U(t)$ as a function of W.

We may hope to bypass these difficulties if we seek to integrate (5) by using elementary methods. Simply multiply (5) throughout by $e^{\beta t}$ and obtain

$$e^{\beta t}\, dU(t) + \beta e^{\beta t} U(t) = e^{\beta t}\, dW(t).$$

Now integrating (assuming that the usual integration by parts is valid here) yields

$$(7) \qquad U(t) - U(0) = W(t) - \beta \int_0^t e^{-\beta(t-s)} W(s)\, ds.$$

The integral in (7) certainly exists, because $W(s)$ is continuous, and it is easy for you to check that $U(t)$ given by (7) has multivariate normal joint distributions, and instantaneous parameters $a(t, U) = -\beta U(t)$, and $b(t, U) = 1$, respectively. It seems that we have successfully integrated (5) to give the Ornstein–Uhlenbeck process in (7). In fact this naive line of argument can be made rigorous here (for this special example), but unfortunately we cannot extend it to more general cases of interest.

Here is an important example to show this failure of naive methods.

(8) Stock price model. Let $S(t)$ be the price of a unit of stock or a share, at time t. As a first modelling assumption it should seem reasonable to suppose that changes in $S(t)$ during $(t, t+dt)$ are proportional to $S(t)$, and are otherwise as random as changes in $W(t)$. This leads to the formal representation, or model,

(9) $$dS(t) = \beta S(t)\,dW(t)$$

for a suitable constant β. We would like to write this in integrated form, (if we can define the integral), as

(10) $$S(t) - S(0) = \int_0^t \beta S(u)\,dW(u).$$

More familiarity with financial matters may well lead you to model the effects of inflation likewise; if we assume that the rate of inflation is a constant ρ, the natural model for the stock price $S(t)$ is revised to take the form:

(11) $$dS(t) = \rho S(t)\,dt + \beta S(t)\,dW(t).$$

Again, we would like to be able to write this as

(12) $$S(t) - S(0) = \int_0^t \rho S(u)\,du + \beta \int_0^t S(u)\,dW(u).$$

However, we stress once again that we have not yet defined the final integral in (10), (11), or (12). And, furthermore, these integrated forms do not yield $S(t)$ as a function of $W(t)$ explicitly.

Unfortunately, if we try to get round this difficulty with an insouciant application of elementary methods, we run into deep problems. Treating $S(t)$ and $W(t)$ as though they were not random, the elementary integral of (9) is

(13) $$S(t) = S(0)\exp(\beta W(t)).$$

But it is an easy exercise for you to show that for the diffusion process $S(t)$ defined by (13), the instantaneous mean is given by $a(t, S(t)) = \frac{1}{2}\beta^2 S(t)$. This is in direct contradiction to the first representation in (9), where the instantaneous mean of $S(t)$ is zero. Clearly our tentative assumption that the integrals in (10) and (12) could be treated as though they were not random is invalid in this case.

We can gain useful insight into reasons why these problems arise by considering the reversed procedure. Let us consider a process $X(t)$ defined as a function of $W(t)$, and seek to represent it in the form of (1), using the methods introduced in Section 5.4. Thus we assume $X(t) = g(W(t))$, where g has a Taylor expansion, denote $W(t+dt) - W(t) = dW$, and write (neglecting terms of higher order than the second)

$$
\begin{aligned}
dX(t) = X(t+dt) - X(t) &= g(W(t+dt)) - g(W(t)) \\
&= g(W(t) + dW) - g(W(t)) \\
&= dW g'(W(t)) + \tfrac{1}{2}(dW)^2 g''(W(t)) + \cdots.
\end{aligned}
$$
(14)

Taking the expected value conditional on $X(t)$ and $W(t)$ gives the instantaneous parameters as $a = \tfrac{1}{2} g''(W(t))$, and $b = \{g'(W(t))\}^2$, which may be written in standard form if we invert $g(W(t)) = X(t)$. Note that we have glossed over some technical details here; for example, Taylor's expansion is the sum of random variables and its value, and the accuracy of partial sums involve questions of convergence. For random variables there are several types of convergence, so one must decide which is appropriate here. It turns out that mean square convergence is the best framework for our purposes.

The key point in what we have done above is that the usual elementary chain rule for differentials would imply that

$$dX(t) = dg(W(t)) = dW(t) g'(W(t)),$$

which does not include the second term on the right in (14). We can now see that this is the reason why we went astray on the road from (9) to (13). We can also see that the argument was lucky enough to work for the Ornstein–Uhlenbeck process, essentially because the final representation (7) is linear in $W(t)$. The extra term involves the second derivative and vanishes in this case.

Clearly, if we are to reconcile all these results we must develop a satisfactory account of integrals of the form

(15)
$$\int_0^t f(t, W(t)) \, dW(t)$$

that have occurred above. We tackle this in the next section.

Exercises
[$W(t)$ is always the Wiener process.]

(a)* Show that $U(t)$ defined in (7) defines an Ornstein–Uhlenbeck process.

(b) Show that the diffusion defined by $S(t) = \exp(\beta W(t))$ has instantaneous parameters $a(t, S(t)) = \tfrac{1}{2} \beta^2 S(t)$ and $b(t, S(t)) = \beta^2 (S(t))^2$.

(c) Find the instantaneous parameters of the diffusion defined by
$V(t) = \exp\{\mu t + \beta W(t)\}$.

(d) Show that the process $V(t)$ defined in (c) is a martingale if $\mu = -\frac{1}{2}\beta^2$.

5.7 The stochastic integral

It is apparent from our discussion in the previous section (5.6) that if we are to establish a consistent and meaningful theory for stochastic differential equations (SDEs) such as (5.6.1), then this can only be done if we can define what we mean by a stochastic integral of the form

(1) $$I = \int_0^t g(W(s))\,dW(s),$$

where $W(t)$, as always, is the Wiener process, and $g(\cdot)$ is a suitable function.

The presence of the final differential dW may make this integral appear a little unfamiliar, but if $W(t)$ were a non-random function $w(t)$, there is a well-established and straightforward approach to such integrals.

(2) **Stieltjes integrals.** For suitable non-random functions $g(t)$ and $w(t)$, we wish to define $I = \int_0^t g(t)\,dw(t)$. The interval $[0, t]$ is first divided into n disjoint intervals at the points $0 = t_0 < t_1 < \cdots < t_n = t$. Then we form the finite sum

(3) $$I_n = \sum_{r=0}^{n-1} g(t_r)\{w(t_{r+1}) - w(t_r)\}.$$

The term $g(t_r)$ 'represents' the function $g(t)$ in the interval (t_r, t_{r+1}), and the term $w(t_{r+1}) - w(t_r)$ is a discrete version of $dw(t)$. Now we let $n \to \infty$ in such a way that $s_n = \sup_r |t_{r+1} - t_r| \to 0$.

If I_n converges to a limit independent of the way the interval $[0, t]$ was divided up, then we call this limit the Riemann–Stieltjes integral of g with respect to w, and denote it by $I = \int_0^t g(t)\,dw(t)$.

For this to work and be useful, g and w must satisfy some conditions of 'niceness'. For example, $g(t)$ must be smooth enough that $g(t_r)$ does indeed fairly represent $g(t)$ in (t_r, t_{r+1}); we would not want to get a different answer if we replaced $g(t_r)$ by $g(t_{r+1})$ in (3) defining I_n. And second $w(t)$ must be smooth enough that

(4) $$v_n = \sum_{r=0}^{n} |w(t_{r+1}) - w(t_r)|,$$

stays bounded as $n \to \infty$, no matter how the interval $[0, t]$ is divided up on the way. (The limit of v_n as $n \to \infty$ is called the *variation* of $w(t)$ over $[0, t]$.)

So far so good, and this is the obvious method to use in tackling (1). But the properties of the Wiener process place two big obstacles in our path. The first is obvious; $W(t)$ is a random process and so the limit of I_n, as $n \to \infty$, concerns the convergence of a sequence of random variables. As we have discussed above, random sequences may converge in several different ways. Which is appropriate here?

In fact this question is answered for us by the second difficulty, which is far from obvious. It can be shown (we omit the details) that the paths of $W(t)$ are almost surely of unbounded variation. That is, if we replace $w(t)$ by $W(t)$, the limit of v_n in (4) is infinite. We cannot therefore seek to show that $I_n \xrightarrow{\text{a.s.}} I$; the best we can hope for is that $I_n \xrightarrow{P} I$. Of course one way to show this is to prove that $I_n \xrightarrow{\text{m.s.}} I$, and it is therefore extremely important that the process $W(t)$ does have finite so-called *quadratic variation*; here is the proof.

(5) **Quadratic variation of $W(t)$.** Divide the interval $(0, t]$ into n disjoint subintervals $(rh, (r + 1)h]$, $0 \le r < n$, where $nh = t$. Denote the increments thus

$$W[(r + 1)h] - W(rh) = \Delta_r W$$

and note that $E\{\Delta_r W\}^4 = 3h^2$, because it is normally distributed. We shall show that, as $n \to \infty$,

(6) $$E\left\{\sum_{r=0}^{n-1} (\Delta_r W)^2 - t\right\}^2 \to 0,$$

which is to say that for the quadratic variation

(7) $$V_n = \sum_{r=0}^{n-1} (\Delta_r W)^2 \xrightarrow{\text{m.s.}} t.$$

First we note that $E(\{\Delta_r h\}^2) = h = t/n$. Hence, because the increments $\Delta_r W$ are independent, (6) becomes

$$\sum_{r=0}^{n-1} E(\{\Delta_r W\}^2 - h)^2 = \sum_{r=0}^{n-1}(3h^2 - 2h^2 + h^2)$$

(8)
$$= 2nh^2 = 2t^2/n$$

$$\to 0, \quad \text{as } n \to \infty.$$

Note that it can be proved that this limit is independent of the way in which the interval $[0, t]$ is actually divided up, so we have naturally chosen the easiest equipartition. ○

Here is a simple but important example to show how the ideas above do now enable us to define the stochastic integral.

(9) **Example.** What is $\int_0^t W(s)\,dW(s)$?

Solution. Divide the interval $(0, t]$ into the n disjoint subintervals $(rh, (r+1)h]$, $0 \le r < n$, where $nh = t$. Following the procedure outlined above, we form the sum

$$I_n = \sum_{r=0}^{n-1} W(rh)\{W((r+1)h) - W(rh)\}$$

(10)
$$= \frac{1}{2}\sum_{r=0}^{n-1}\{W((r+1)h) - W(rh)\}\{W((r+1)h) + W(rh)\}$$

$$- (W((r+1)h) - W(rh))$$

$$= \frac{1}{2}\sum_{r=0}^{n-1}\{W((r+1)h)^2 - W(rh)^2\} - \sum_{r=0}^{n-1}\{W((r+1)h) - W(rh)\}^2$$

$$= \frac{1}{2}\{W(t)^2 - W(0)^2 - V_n\},$$

where V_n is the quadratic variation defined in (7). Since $V_n \xrightarrow{\text{m.s.}} t$, and $W(0) = 0$, we have shown that

(11)
$$I_n \xrightarrow{\text{m.s.}} I = \tfrac{1}{2}(W(t)^2 - t).$$

This is an important result, and we pause to make a number of remarks and observations.

First, we see the appearance of an extra term in the integral $\int_0^t W(s)\,dW(s) = \tfrac{1}{2}(W(t)^2 - t)$. The ordinary integral of a non-random function $w(t)$ of course yields $\int_0^t w(s)\,dw(s) = \tfrac{1}{2}w(t)^2 - \tfrac{1}{2}w(0)^2$. The ordinary rules of elementary calculus do not all apply to this type of stochastic integral, and this is why our naive attempt to integrate $dX(t) = \beta X(t)\,dW(t)$ failed.

Second, we remark that the integral $\frac{1}{2}(W(t)^2 - t)$ is a martingale; recall (5.5.5). This is a characteristic and important property of this class of integrals.

Third, we stress that the result depends critically on starting with the finite sum

$$I_n = \sum_r W(rh)\{W((r+1)h) - W(rh)\} = \sum_r W(rh)\Delta_r W$$

in which $W(rh)$ is chosen to represent the value of $W(t)$ in $[rh, (r+1)h]$. Had we chosen a different value, we would obtain a different result. It is an exercise for you to show that, for example,

(12) $$J_n = \sum_{r=0}^{n-1} W\{(r+1)h\}\Delta_r W \xrightarrow{\text{m.s.}} \frac{1}{2}(W(t)^2 + t), \quad \text{as } n \to \infty.$$

Other choices will yield different results. It is a natural and important question to wonder why we regard (11) as pre-eminent as the integral of choice. The answer lies in the context of our models, for which t is a time parameter. Consider, for example, the model for stock prices noted above.

The key point is that $S(t)$ is a model for a process that is evolving unpredictably in time. At time t, we know the past, $\{S(u); u \leq t\}$, and we are seeking a value of S to represent the process during the future interval $(t, t + dt)$. At the time t the only value available to us is indeed $S(t)$; choosing another value such as $S(t + dt)$ (or $S(t + \frac{1}{2}dt)$) would amount to looking into the future. We cannot do this in real life, (we can only *predict* the future, which is quite a different thing from looking into it), and our modelling assumptions must reflect reality.

In addition, as a bonus, if we choose as we did in (11), our integrals will be martingales as (11) is (under weak conditions on the integrand). The integral in (12) is of course *not* a martingale, so we cannot apply useful results like the optional stopping theorem to it.

After these preliminaries, we now give a brief sketch of the procedures for defining an integral of the form $I = \int_0^t f(s, W(s))\,dW(s)$; details should be found in more advanced texts. First of all we discuss the conditions that $f(s, W)$ must satisfy for our integral to work. They are twofold;

the first is:

(13) **Mean square integrability.** We require that for all $t \geq 0$,

$$\int_0^t E(f^2) \, dt < \infty.$$

This condition becomes natural when you recall that the integral $\int_0^t W(s) \, dW(s)$ that we considered in (9) was evaluated as a limit in mean square of the approximating sums. The more general integral is defined in the same way, and this condition ensures its existence. (In more advanced work this can be weakened in various ways, but that will not concern us here. For example, one can relax (13) to $P(\int_0^t f^2 \, dt < \infty) = 1$, at the expense of later complications.)

The second condition is this:

(14) **The integrand is non-anticipating.** We require that, for all $t \geq 0$, the value of the function $f(t, W)$ is determined by values of $W(u)$ only for $0 \leq u \leq t$. More formally, we say that a process $X(t)$ is non-anticipating for $W(t)$ if for any $u \leq t < s$, $X(u)$ is independent of the increment $W(s) - W(t)$. Now $f(t, W)$ is required to be non-anticipating for $W(t)$. To put this in a different way, if you know $W(u)$ for $0 \leq u \leq t$, then no further information about $W(s) - W(t)$ for $s > t$ will change your value of f at t. (Of course, this will usually be trivially the case, because f will simply be some elementary function of $W(t)$.)

The condition is seen to be natural when you look at our discussion following equation (12). On modelling grounds, the integrand must not depend on values of $W(t)$ in $(t, t + dt)$ (or any later values), and (14) ensures this.

In more advanced work, f is said to be progressively measurable, or predictable in terms of $\{W(u); 0 \leq u \leq t\}$; and the concept is defined in the context of measure theory. Informally, you can think of this as a requirement that knowledge of the future is not permitted; which seems reasonable.

Finally, we need one more definition.

(15) **Step function process.** A *step function* $f_n(t)$ is simply constant on a set of disjoint intervals in $[0, t]$, and jumps between its values. Such processes are also called *simple* random functions. Formally, if $0 = a_0 < a_1 < \cdots < a_n = t$, we have

(16) $$f_n(t) = C_j, \quad \text{if } a_j < t \leq a_{j+1}$$

and $f_n(t)$ is zero elsewhere. Each random variable C_j is determined by the values of $W(u)$ in $0 \leq u \leq a_j$, and is thus constant on (a_j, a_{j+1}), and non-anticipating for $W(t)$. △

The definition of the stochastic integral of $f_n(t)$ is now rather obvious and easy; we write

$$\text{(17)} \qquad \int_0^t f_n(u)\,dW(u) = \sum_{j=0}^{n-1} C_j \{W(a_{j+1}) - W(a_j)\}$$

and let $s_n = \sup_j |a_{j+1} - a_j|$ be the largest length of a step.

Such simple step functions are useful because of two key theorems, which we state without proof.

(18) **Approximation theorem.** For a function f satisfying (13) and (14), there is a sequence of non-anticipating, that is, predictable, step functions f_n such that, as $n \to \infty$ and $s_n \to 0$,

$$\int_0^\infty E(f_n - f)^2\,dt \to 0.$$

That is to say f_n supplies an arbitrarily good approximation to f, in mean square, as n increases.

(19) **Isometry theorem.** For the integral I_n of the step function f_n

$$EI_n^2 = \int_0^\infty E(f_n^2)\,dt.$$

Combining these two, it follows that there exists I such that, as $n \to \infty$ and $s_n \to 0$,

$$E(I_n - I)^2 \to 0$$

and for any other $\hat{I}$ such that $E(I_n - \hat{I})^2 \to 0$ we have $P(I = \hat{I}) = 1$. We omit the details, but clearly it is natural to designate I as the (Itô) stochastic integral of f with respect to W.

Finally, it is straightforward to show that the integral I has the following most important properties. In each case, the technique is to verify the result for the step functions f_n, and then to confirm that it is preserved in the limit for f as $n \to \infty$. We omit the details, but remark that these results, and many of those above, are considerably easier to prove if we assume that f is continuous. In fact it almost always is, for functions that we consider here.

(20) **Properties of I.** Let $f(t, W)$ satisfy (13) and (14), with Itô integral $I(t) = \int_0^t f\,dW$. Then

$$\text{(21)} \qquad EI(t) = 0,$$

$$\text{(22)} \qquad E(I(t)^2) = \int_0^t (f)^2\,du$$

and $I(t)$ is a martingale with respect to $W(t)$, which is to say that for $0 < s < t$

(23) $\quad \mathrm{E}\left\{\int_0^t f(u, W(u))\,dW(u) \mid W(v), 0 \le v \le s\right\} = \int_0^s f(u, W(u))\,dW(u).$

It is beyond our scope to consider any of the numerous possible generalizations of these results.

Exercises

As usual, $W(t)$ is the Wiener process.

(a)* Verify by evaluating each side that

$$\mathrm{E}\left\{\left(\int_0^t W(u)\,dW(u)\right)^2\right\} = \mathrm{E}\int_0^t W(u)^2\,du.$$

(b)* **Brownian Bridge.** You are given that the process $X(t)$ is continuous in $[0, 1]$, where

$$X(t) = \begin{cases} (1-t)\int_0^t \dfrac{1}{1-s}\,dW(s), & 0 \le t < 1, \\ 0, & t = 1. \end{cases}$$

Show that $\mathrm{cov}(X(s), X(t)) = s \wedge t - st$, and deduce that $X(t)$ is the Brownian Bridge.

(c) By a direct calculation of the limit as in (9), verify that
  (i) $\int_0^t s\,dW(s) = tW(t) - \int_0^t W(s)\,ds.$
  (ii) $\int_0^t (W(s))^2\,dW(s) = \tfrac{1}{3}W(t)^3 - \int_0^t W(s)\,ds.$

[Hint: $x^2(y-x) + x(y-x)^2 = \tfrac{1}{3}\{y^3 - x^3 - (y-x)^3\}$.]

5.8 Itô's formula

After the discussions in the previous section, we now see that the SDE

(1) $\qquad dX = \mu(t, X)\,dt + \sigma(t, X)\,dW,$

which we first introduced as a notational device in (5.1.22), is conveniently interpreted to mean the integral equation

$$(2) \qquad X(t) - X(0) = \int_0^t \mu(s, X(s)) \, ds + \int_0^t \sigma(s, X(s)) \, dW(s),$$

where the final integral in (2) is now well-defined. As in the non-random case, such an equation may have no solutions, or it may have one or more. In fact, fairly mild conditions on μ and σ are sufficient to ensure that (1) has a unique solution (in a suitable sense) that is a continuous Markov process. We omit these results.

As we have remarked above, in the non-random case we may also define derivatives and obtain ordinary differential equations. In this non-random framework, the fundamental theorem of calculus connects the two approaches, and we may switch from one to the other as the mood takes us, or necessity requires.

In stochastic calculus this is not possible; the useful range of techniques is practically restricted to those that deal with integral equations. Of these, by far the most important is that known as Itô's formula, which may be seen as a stochastic chain rule.

Let us recall some elementary non-random chain rules; as usual primes may denote differentiation.

(3) **One-variable chain rule.** If $y(t) = f(g(t))$, then

$$y'(t) = \frac{dy}{dt} = f'(g(t))g'(t),$$

assuming the derivatives f' and g' exist. We may express this in differential notation as

$$dy = f'(g)g'(t) \, dt = f'(g) \, dg.$$

(4) **Two-variable chain rule.** If $y(t) = f(u(t), w(t))$, then

$$\frac{dy}{dt} = \frac{\partial f}{\partial u} \frac{du}{dt} + \frac{\partial f}{\partial w} \frac{dw}{dt},$$

which may be written using differentials as

$$dy = \frac{\partial f}{\partial u} \, du + \frac{\partial f}{\partial w} \, dw = f_u \, du + f_w \, dw,$$

where differentiation may be denoted by suffices in an obvious way. In particular, if $u = t$, we obtain, for $y = f(t, w(t))$,

$$(5) \qquad dy = f_t \, dt + f_w \, dw.$$

Now let us turn to functions of the Wiener process $W(t)$. Our first result, which we state without proof, is

(6) Itô's simple formula. Let $Y = f(t, W)$, where $W(t)$ is the Wiener process, and f is differentiable as required in both arguments. Then we have

$$(7) \qquad dY(t) = \left(f_t + \tfrac{1}{2} f_{ww}\right) dt + f_w \, dW.$$

Notice the second-order term f_{ww}, which has no counterpart in any of the non-random relationships above. (Here f_w and f_{ww} denote the derivatives of f with respect to its second argument, evaluated at $w = W$.) Of course this extra term appears for essentially the same reasons as we have seen before in (5.4); the Taylor expansion of $f(t, W)$ includes a term in $\tfrac{1}{2} f_{ww} (dW)^2$, which has conditional expectation $\tfrac{1}{2} f_{ww} \, dt$, given $W(t)$.

More generally, we may consider a diffusion process $X(t)$, defined by (1), $dX = \mu(t, X) dt + \sigma(t, X) dW$, and define a new process $Y(t)$ by

$$(8) \qquad Y(t) = f(t, X(t)).$$

What is the chain rule for Y, if any? That is to say, what are α and β such that $dY = \alpha \, dt + \beta \, dW$? The answer is this, which we state without proof.

(9) Itô's formula. For $X(t)$ and $Y(t)$ given by (1) and (8),

$$(10) \qquad dY(t) = \left\{\mu(t, X) f_x + f_t + \tfrac{1}{2} \sigma^2(t, X) f_{xx}\right\} dt + \sigma(t, X) f_x \, dW.$$

Thus, $\alpha = \mu f_x + f_t + \tfrac{1}{2} \sigma^2 f_{xx}$, and $\beta^2 = \sigma^2 f_x^2$ are the instantaneous parameters of $Y(t)$. Here f_x denotes the derivative of $f(t, x)$ with respect to its second argument, evaluated at $x = X$; and similarly for f_{xx}, and so on.

Itô's formulas are extremely useful in many ways, particularly in evaluating stochastic integrals. This comes as no surprise, because the differential forms (8) and (10) are (as we have often noted) simply shorthand for integral relationships. We therefore apply these formulas immediately to see how we may solve an important SDE.

Of course, most SDEs do not have nice solutions expressible as a closed form in terms of elementary functions (just like ordinary differential equations, in fact). But when they do, Itô's formula may help to find it.

(11) Exponential process. Solve the SDE

$$(12) \qquad dX = \mu X \, dt + \sigma X \, dW, \quad t \geq 0,$$

where μ and σ are constants.

Solution. We seek a solution of the form

(13) $$Y(t) = Y(0)\exp\{at + bW(t)\} = f(t, W).$$

Preparatory to applying Itô's formula, we calculate $\partial f/\partial t = aY(t)$, $\partial f/\partial w = bY(t)$, $\partial^2 f/\partial w^2 = b^2 Y(t)$, and discover that Itô's formula gives

(14) $$dY = (a + \tfrac{1}{2}b^2)Y(t)\,dt + bY(t)\,dW.$$

This is identical to (12) if we choose $b = \sigma$ and $a = \mu - \tfrac{1}{2}\sigma^2$. Therefore, the required solution is

(15) $$X(t) = X(0)\exp\{(\mu - \tfrac{1}{2}\sigma^2)t + \sigma W(t)\}.$$

Notice that from (12)

$$\int_0^t \frac{dX}{X} = \int_0^t \mu\,dt + \int_0^t \sigma\,dW$$
$$= \mu t + \sigma W(t)$$

and this is *not* equal to

$$\log\left(\frac{X(t)}{X(0)}\right) = (\mu - \tfrac{1}{2}\sigma^2)t + \sigma W(t), \text{ from (15)}$$

This is another way of viewing the reasons why our naive attempt to integrate (5.6.9) by elementary methods failed.

Conversely, given a suitable process defined as a function of $W(t)$, we can use Itô's formula to find a SDE of which it should be the solution. This is a simple routine in principle, but caution may be required.

(16) **A dishonest process.** Let $X(t) = [1 - W(t)]^{-1}$. Then we have $f(t, w) = (1 - w)^{-1}$, and easy calculations give $f_t = 0$, with

$$f_w(t, W) = (1 - W)^{-2} = X^2$$
$$f_{ww}(t, W) = 2(1 - W)^{-3} = 2X^3.$$

Applying Itô's formula formally yields

(17) $$dX = X^3\,dt + X^2\,dW$$

so one might expect that $X(t)$ is the solution of this SDE. The obvious difficulty is that $W(t)$ will take the value $W = 1$ in finite time with probability 1, so $X(t)$ is a dishonest process in the sense defined in Chapter 4. This problem can in fact be resolved, but it is beyond our scope here. ○

Here is a more complicated application of Itô's formula.

(18) **Product rule.** It is often necessary to consider the joint behaviour of random processes $X(t)$ and $Y(t)$ satisfying simultaneous SDEs of the form

$$dX = a(t, X)\,dt + b(t, X)\,dW,$$
$$dY = g(t, Y)\,dt + h(t, Y)\,dW,$$

where a, b, g, h are suitable functions, and each equation is driven by the same Wiener process $W(t)$. Then it is an easy consequence of the general version of Itô's formula (10) that

$$d(X^2) = b^2\,dt + 2X\,dX,$$
$$d(Y^2) = h^2\,dt + 2Y\,dY,$$
$$d\{(X+Y)^2\} = (b+h)^2\,dt + 2(X+Y)(dX+dY).$$

Using the identity $XY = \frac{1}{2}\{(X+Y)^2 - X^2 - Y^2\}$ immediately gives a stochastic product rule for $d(XY)$ in this form:

(19) $$d(XY) = X\,dY + Y\,dX + b(t, X)h(t, Y)\,dt.$$

This is in effect a formula for integration by parts; note the extra term $b\,h\,dt$ appearing in (19), that we do not see in the non-random case. It does vanish if either b or h is zero; see Exercise (5.8(c)) for an example of this. ○

We can also use Itô's formulas to calculate moments. For example, if $X(t) = W(t)^n$, then the simple formula (7) yields

(20) $$dX = d(W^n) = nW^{n-1}\,dW + \tfrac{1}{2}n(n-1)W^{n-2}dt.$$

Integrating and taking expectations, we see that the first term on the right-hand side is zero, by (5.7.22). Hence,

(21) $$EW^n(t) = \frac{1}{2}n(n-1)\int_0^t E(W^{n-2}(s))\,ds,$$

which supplies a simple recurrence relation for EW^n. Thus, starting with $EW^0(t) = 1$, we obtain any required higher moments easily.

Of course we know all these already because $W(t)$ is $N(0, t)$, but the technique can be useful in less trivial circumstances; see the problems.

Our last application of Itô's formula is even more ingenious. In previous sections we have seen many connections between diffusion processes and various differential equations. We conclude this section with yet another remarkable and important example of this.

(22) Feynman–Kac theorem. Let $X(t)$ be the solution of the SDE

$$dX = \mu(t, X)\, dt + \sigma(t, X)\, dW, \quad 0 \le t \le T$$

and suppose that $F(t, x)$ satisfies the partial differential equation

(23) $$\frac{\partial F}{\partial t} + \mu(t, x)\frac{\partial F}{\partial x} + \frac{1}{2}\sigma^2(t, x)\frac{\partial^2 F}{\partial x^2} = 0, \quad 0 \le t \le T$$

with the boundary condition $F(T, x) = K(x)$.

We consider the process $Y(s) = F(s, X)$ in the time interval $t \le s \le T$, so that by Itô's formula (10), (in its integrated form),

(24)
$$\begin{aligned}
F(T, X(T)) - F(t, X(t)) &= \int_t^T \left(\frac{\partial F}{\partial s} + \mu(s, X)\frac{\partial F}{\partial x} + \frac{1}{2}\sigma^2(s, X)\frac{\partial^2 F}{\partial x^2} \right) ds \\
&\quad + \int_t^T \sigma(s, X)\frac{\partial F}{\partial x}\, dW(s) \\
&= \int_t^T \sigma(s, X)\frac{\partial F}{\partial x}\, dW(s),
\end{aligned}$$

because F satisfies (23) in $[0, T]$. Now, provided only that $E(\sigma(\partial F/\partial x))^2 < \infty$, the final remaining integral in (24) is a zero-mean martingale, and therefore we can take expectations conditional on $X(t) = x$ to give, using (5.7.23),

$$E(F(T, X(T)) \mid X(t) = x) = F(t, x).$$

Thus, recalling the boundary condition $F(T, x) = K(x)$, we have that the solution $F(t, x)$ of (23) has the Feynman–Kac representation

(25) $$F(t, x) = E(K(X(T)) \mid X(t) = x).$$

This result has many theoretical implications that we do not pursue; one important practical corollary is that it offers a route to a numerical solution of equation (23) by simulating W and X for relevant values of t and x.

Exercises
(a) By applying Itô's formula to $Y = W(t)^2$, show that Y is a solution of the SDE
$$dY = dt + 2W\,dW$$
and that $\int_0^t W(s)\,dW(s) = \frac{1}{2}(W(t)^2 - t)$.

(b) By applying Itô's formula to $Y = tW(t)$, show that Y is a solution of
$$dY = W\,dt + t\,dW$$
and that $\int_0^t s\,dW(s) = tW(t) - \int_0^t W(s)\,ds$.

(c)* **Ornstein–Uhlenbeck revisited.** Let $U(t)$ satisfy the SDE
$$dU = -\beta U\,dt + dW, \quad U(0) = 0.$$
Use the product rule (18) to show that
$$U(t) = e^{-\beta t} \int_0^t e^{\beta s}\,dW(s)$$
$$= W(t) - \beta \int_0^t e^{-\beta(t-s)} W(s)\,ds.$$

(d) By applying Itô's formula to $W(t)^3$, show that
$$\int_0^t W(s)^2\,dW(s) = \frac{1}{3}W(t)^3 - \int_0^t W(s)\,ds.$$

5.9 Processes in space

Very often in the real world we are presented with several random processes that is necessary to consider jointly. For example, a practical model of a stock market must allow you to possess a portfolio of different stocks, whose prices are unlikely to be independent. The construction of such a model will begin by defining a vector $\mathbf{W}(t)$ of independent Wiener processes

(1) $$\mathbf{W}(t) = (W_1(t), \ldots, W_m(t)).$$

Then the vector $\mathbf{S}(t)$ of individual stock prices $\{S_i(t); 1 \leq i \leq m\}$ may be defined in terms of $\mathbf{W}(t)$ by means of a set of simultaneous SDEs

$$dS_j(t) = \mu_j(t, S_j) \, dt + \sum_k \sigma_{jk}(t, S_j(t)) \, dW_k(t).$$

Of course, a function of this portfolio, denoted by $Y(t) = f(t, \mathbf{S}(t))$, will then require a multidimensional form of Itô's formula to describe its evolution. It is far beyond our scope to explore this territory; here we confine ourselves to examining some entertaining aspects of the behaviour of the Wiener process $\mathbf{W}(t)$ in m dimensions, as defined in (1). Even this simple process is of considerable interest, and is linked in a remarkable way with some famous results of classical mathematical physics.

Typically, $\mathbf{W}(t)$ will denote the position of a particle that is diffusing randomly in $\mathbb{R}^m$, but other spaces are occasionally of interest: for example, one may consider a particle performing a Wiener process on the perimeter of a circle, or on the surface of a sphere. We assume that $\mathbf{W}(0) = \mathbf{w}$, and that all the components of $\mathbf{W}$ have variance parameter unity. For the particle diffusing freely in $\mathbb{R}^m$, the joint density of its position $\mathbf{W}(t)$ is

$$f(\mathbf{x}) = (2\pi t)^{-m/2} \exp\left(-\frac{1}{2t} \sum_{r=1}^m (x_r - w_r)^2\right)$$

(2)
$$= (2\pi t)^{-m/2} \exp\left(-\frac{1}{2t} |\mathbf{x} - \mathbf{w}|^2\right), \quad \mathbf{x} \in \mathbb{R}^m.$$

However, as we have remarked for the one-dimensional case in (5.1), such processes often evolve in the presence of barriers and boundaries, and it is natural to ask about where and when such boundaries are first hit.

We set the problems up formally as follows: the Wiener process $\mathbf{W}(t)$ starts at $\mathbf{W}(0) = \mathbf{w} \in D$, where D is some subset of $\mathbb{R}^m$ with boundary A. (We always assume that D is connected.) If the boundary is divided into disjoint subsets H and M, where $H \cup M = A$, then we may ask for the probability $p(\mathbf{w})$ that the process $\mathbf{W}(t)$ hits H before it hits M. Even more precisely, we may seek the distribution of the point L at which it first hits A, yielding the hitting density

(3) $$h_{\mathbf{w}}(\mathbf{r}), \quad \mathbf{r} \in A.$$

Another quantity of natural interest is the time to hit either of H or M, namely

(4) $$T_A = \inf\{t \geq 0 : \mathbf{W}(t) \in A\}$$

with the usual convention that the infimum of the empty set is ∞. This case may arise if, for example, D is unbounded. The hitting times of H and M are defined similarly, and denoted by T_H and T_M.

If D is bounded, or such that $P(T_A < \infty) = 1$, then we may further ask for the mean hitting time

$$m(\mathbf{w}) = E(T_A \mid \mathbf{W}(0) = \mathbf{w}). \tag{5}$$

Note that we are assuming that all these sets (D, A, H, M) are nice, in the sense that they do not give rise to extraneous technical problems; for example, the kind of difficulties that one might expect if A were a fractal, or included sharp non-differentiable spikes, or were missing isolated points. We evade the details here, but note that smooth sets such as circles, balls, cubes, and other Platonic solids are nice.

After all this preamble, we now need to recall only one bit of notation before stating the key results: we shall make extensive use of the Laplacian $\nabla^2 f$, where in m dimensions, and in Cartesian coordinates,

$$\nabla^2 f(x_1, \ldots, x_m) = \frac{\partial^2 f}{\partial x_1^2} + \cdots + \frac{\partial^2 f}{\partial x_m^2}. \tag{6}$$

Then we have these two remarkable theorems:

(7) **Hitting probability.** Let $p(\mathbf{w})$ be the probability that the process $\mathbf{W}(t)$, started from $\mathbf{W}(0) = \mathbf{w}$, hits H before M; that is to say $p(\mathbf{w}) = P(T_H < T_M \mid \mathbf{W}(0) = \mathbf{w})$. Then,

$$\frac{\partial^2 p}{\partial w_1^2} + \cdots + \frac{\partial^2 p}{\partial w_m^2} = \nabla^2 p(\mathbf{w}) = 0, \quad \mathbf{w} \notin (H \cup M) \tag{8}$$

with the boundary conditions $p(\mathbf{w}) = 1$ for $\mathbf{w} \in H$, and $p(\mathbf{w}) = 0$ for $\mathbf{w} \in M$.

(9) **Hitting time.** Let $m(\mathbf{w})$ be the mean of the hitting time T_A of A, defined in (5). Then

$$\nabla^2 m(\mathbf{w}) = -2, \quad \mathbf{w} \notin A \tag{10}$$

with the boundary condition $m(\mathbf{w}) = 0$ for $\mathbf{w} \in A$.

A brief glance at (5.4.29) and (5.4.32), and a few moments' thought, reveals that these are the natural extensions of those results in one dimension. Indeed, as in one dimension, there are more general results involving more complicated functionals. For example, it can be shown that the function

(11) $$v(\mathbf{w}) = E\left\{\int_0^{T_A} g(\mathbf{W}(s))\,ds \mid \mathbf{W}(0) = \mathbf{w}\right\},$$

satisfies $\nabla^2 v(\mathbf{w}) = -2g(\mathbf{w})$ for $\mathbf{w} \notin A$, with $v(\mathbf{w}) = 0$ for $\mathbf{w} \in A$. Setting $g \equiv 1$ yields (10), of course.

We shall not prove (10) or (11), but note that the easiest way to see why they should be true is to follow through the same argument that we used in one dimension. (See (5.4.27)–(5.4.33) for the basic ideas.) Pursuing the same approach enables us to derive (8); we first assume that $p(\mathbf{w})$ has a Taylor expansion in all its variables. Now we consider the process after a small time-interval h, at which time it is at $\mathbf{w} + d\mathbf{W}$, where $d\mathbf{W} = (dW_1, \ldots, dW_m)$ is the initial increment over $[0, h]$. Then by conditional expectation, and expanding $p(\mathbf{w} + d\mathbf{W})$ about the point $\mathbf{w}$, and using the fact that $\mathbf{W}(t)$ has independent increments, we obtain

(12) $p(\mathbf{w}) = E(p(\mathbf{w} + d\mathbf{W}))$

$$= E\left\{p(\mathbf{w}) + \sum_j dW_j \frac{\partial p}{\partial w_j} + \frac{1}{2}\sum_i \sum_j dW_i\, dW_j \frac{\partial^2 p}{\partial w_i \partial w_j} + \cdots\right\}$$

$$= p(\mathbf{w}) + \frac{1}{2}\sum_j h \frac{\partial^2 p(\mathbf{w})}{\partial w_j^2} + o(h).$$

In making this step we have used the fact that W_i and W_j are independent for $i \neq j$, so that $E(dW_i\,dW_j) = 0$ in this case. Of course $E\{(dW_j)^2\} = h$, and the higher-order terms are all $o(h)$. Now rearranging (12), and letting $h \to 0$, gives (8).

A similar argument applied to $m(\mathbf{w})$ gives

(13) $$m(\mathbf{w}) = m(\mathbf{w}) + h + \tfrac{1}{2}h\nabla^2 m(\mathbf{w}) + o(h)$$

from which (10) follows immediately.

These arguments can be made rigorous, but we omit the details.

We turn aside from these generalities to consider the solutions of (8) and (10) in interesting special cases. These frequently have circular or spherical symmetry, and it is therefore useful to know that in polar coordinates (r, θ) in two dimensions, the Laplacian (6) takes the form

(14) $$\nabla^2 f = \frac{1}{r}\frac{\partial}{\partial r}\left(r\frac{\partial f}{\partial r}\right) + \frac{1}{r^2}\frac{\partial^2 f}{\partial \theta^2},$$

whereas for spherical polar coordinates (r, θ, ϕ) in three dimensions we must use

(15) $$r^2 \nabla^2 f = \frac{\partial}{\partial r}\left(r^2 \frac{\partial f}{\partial r}\right) + \frac{1}{\sin \theta}\frac{\partial}{\partial \theta}\left(\sin \theta \frac{\partial f}{\partial \theta}\right) + \frac{1}{\sin^2 \phi}\frac{\partial^2 f}{\partial \phi^2}.$$

We begin by looking at some elementary problems that illustrate the essential ideas.

(16) **Wedge.** Let $\mathbf{W}(t)$ be a Wiener process in the plane, started at $\mathbf{W}(0) = \mathbf{w}$. We employ the usual plane-polar coordinates (r, θ), and suppose that $\mathbf{W}(t)$ lies in the wedge D such that $-\pi < -\alpha \leq \theta \leq \alpha < \pi$. If H is the line $\theta = \alpha$, and M is the line $\theta = -\alpha$, we may ask for the probability $p(\mathbf{w})$ that $\mathbf{W}(t)$ hits H before M.

By what we have said above, we therefore seek a solution of $\nabla^2 p = 0$, where $\nabla^2 f$ is taken in the form of (14), with the boundary conditions $p = 1$ on $\theta = \alpha$, and $p = 0$ on $\theta = -\alpha$. By inspection, the function

$$p = \frac{\theta + \alpha}{2\alpha}, \quad -\alpha \leq \theta \leq \alpha$$

is the required solution, and so

(17) $$p(\mathbf{w}) = p(w_1, w_2) = \frac{1}{2\alpha}\left(\alpha + \tan^{-1}\frac{w_2}{w_1}\right).$$

○

Now let us consider $m(\mathbf{w})$.

(18) **Wedge again.** In the context of the preceding example, assume that $\alpha < \pi/4$, and let T be the time to hit the boundary of the wedge. What is $m(\mathbf{w}) = E(T \mid \mathbf{W}(0) = (w_1, w_2))$?

Solution. We need a function $m(w_1, w_2)$ satisfying $\nabla^2 m = -2$ in the wedge, and such that $m = 0$ on the boundary. Consider the candidate

(19) $$m(\mathbf{w}) = m(w_1, w_2) = \frac{w_1^2 \sec^2 \alpha - (w_1^2 + w_2^2)}{1 - \tan^2 \alpha}.$$

It is easy to check that in the interior of the wedge

$$\nabla^2 m = \frac{2(\sec^2 \alpha - 1) - 2}{1 - \tan^2 \alpha} = -2$$

and that on the boundary of the wedge, where $w_2 = w_1 \tan \alpha$, we have

$$m(\mathbf{w})(1 - \tan^2 \alpha) = \sec^2 \alpha - 1 - \tan^2 \alpha = 0.$$

Thus $m(\mathbf{w})$ in (19) provides the expected time to hit the boundary from any point (w_1, w_2) in the wedge. ○

Many more special cases can be solved by such elementary means; see the exercises and problems. However, it is natural to ask if there is a more general method of attack applicable in principle to any problem of interest. Remarkably, there is indeed a wonderful apparatus constructed for exactly this purpose in 1828 by George Green. The reason for his interest was the fact that many problems in electrostatics, gravitation, and other branches of mathematical physics can be reduced to finding the solution ϕ of the equations $\nabla^2 \phi = 0$, or $\nabla^2 \phi = g(\mathbf{x})$, in some region D. In these contexts ϕ is called a *potential function*, the name given to it by Green.

Reduced to its bare minimum, the problem and Green's solutions may be sketched like this.

(20) **Green's theorems.** Suppose that ϕ is a solution of Laplace's equation, $\nabla^2 \phi = 0$, in the region D, with boundary A, and satisfies the boundary condition

$$\phi(\mathbf{x}) = \phi_A(\mathbf{x}), \quad \text{for } \mathbf{x} \in A.$$

Of Green's many theorems, an important group shows how to write the solution ϕ in terms of integrals or integral equations. Here are some examples:

(21) **Free space averaging.** Let B be a ball with surface S having area $|S|$, and centre $\mathbf{w}$, which lies entirely in D. Then

(22) $$\phi(\mathbf{w}) = \int_{\mathbf{x} \in S} |S|^{-1} \phi(\mathbf{x}) \, dS.$$

That is to say, $\phi(\mathbf{w})$ is the average of ϕ over the surface S of B. It is interesting that we can prove this directly for the probability $p(\mathbf{w})$ defined in (7); see the problems.

Note that in two dimensions 'a ball B with surface A' will be interpreted as a disk with a circle as boundary.

Here is an even more important application of Green's theorems:

(23) **The Green function.** Let ϕ be a solution of Laplace's equation, $\nabla^2 \phi = 0$, in D, and suppose that on the boundary A, ϕ obeys the boundary condition $\phi = \phi_A$. Then there exists a function $G(\mathbf{w},\mathbf{r})$ such that for any $\mathbf{w} \in D$

$$(24) \qquad \phi(\mathbf{w}) = -\int_{\mathbf{r} \in A} \phi_A(\mathbf{r}) \frac{\partial G}{\partial n} \, dS,$$

which yields the solution ϕ in terms of the Green function G and the boundary values of ϕ on A. Note that $\partial G/\partial n$ denotes the derivative of G in the direction of the outward-facing normal to A at $\mathbf{r}$.

Furthermore, remarkably, G is symmetric in its arguments

$$(25) \qquad G(\mathbf{w},\mathbf{r}) = G(\mathbf{r},\mathbf{w}), \quad \mathbf{r}, \mathbf{w} \in D.$$

Before turning to a detailed discussion of $G(\mathbf{w},\mathbf{r})$, let us turn aside to note a remarkable corollary.

(26) **Hitting density.** For the hitting probability of H from $\mathbf{w}$, denoted by $p(\mathbf{w})$, the boundary condition takes the form

$$\phi_A(\mathbf{r}) = \chi_H(\mathbf{r}) = \begin{cases} 1, & \text{if } \mathbf{r} \in H, \\ 0, & \text{otherwise.} \end{cases}$$

(χ_H is called the indicator, or characteristic, of H.) Hence,

$$(27) \qquad p(\mathbf{w}) = -\int_{\mathbf{r} \in A} \chi_H(\mathbf{r}) \frac{\partial G}{\partial n} \, dS.$$

It follows from our definition of probability density function that $-(\partial G/\partial n)$ is the hitting density of the surface A by a Wiener process started at $\mathbf{w}$; informally, the probability that $\mathbf{W}(t)$ first hits A in the small element of surface dS is $-(\partial G/\partial n) \, dS$.

We shall return to calculate some explicit examples soon; first we make some remarks about the Green function.

The above results are exceptionally interesting, but they will be of equal utility only if we can say what the Green function is. Indeed we can, but the explicit form of the function depends on the number of dimensions, so we must now be specific.

(28) **The Green function in three dimensions.** Green's theorems tell us that in three dimensions, for $\mathbf{w} \in D$ and $\mathbf{r} \in D$, we have

$$(29) \qquad 4\pi G(\mathbf{w},\mathbf{r}) = H(\mathbf{w},\mathbf{r}) + |\mathbf{w} - \mathbf{r}|^{-1},$$

where $H(\mathbf{w},\mathbf{r})$ is a function of $\mathbf{r}$ that satisfies Laplace's equation in D, with the boundary condition

(30) $$4\pi G(\mathbf{w},\mathbf{r}) = H(\mathbf{w},\mathbf{r}) + |\mathbf{w} - \mathbf{r}|^{-1} = 0, \quad \mathbf{r} \in A.$$

At this point it is necessary to insert yet another parenthetical remark about names and notation. We refer to $G(\mathbf{w},\mathbf{r})$ as 'the Green function', or 'Green's function' (*not*, we stress, 'the Green's function'). However, because these results are applicable in such a wide range of areas in mathematical physics and engineering, there is a lack of uniformity in this. Some writers call H the Green function. Furthermore, some writers will change the signs of H and G. Beware!

Having issued this warning, we now return to the practical point at issue.

(31) **Green's function in two dimensions.** In this case Green's theorems tell us that, for $\mathbf{w} \in D$ and $\mathbf{r} \in D$, we have

(32) $$2\pi G(\mathbf{w},\mathbf{r}) = H(\mathbf{w},\mathbf{r}) - \log|\mathbf{w} - \mathbf{r}|,$$

where $H(\mathbf{w},\mathbf{r})$ is a function of $\mathbf{r}$ that satisfies Laplace's equation in D, with the boundary condition

(33) $$2\pi G(\mathbf{w},\mathbf{r}) = H(\mathbf{w},\mathbf{r}) - \log|\mathbf{w} - \mathbf{r}| = 0, \quad \mathbf{r} \in A.$$

Of course, the important thing about Green's method is that it is often relatively easy to find H in situations of practical interest. Here are some examples; many more can be found in books dealing with several branches of mathematical physics, and also in engineering.

(34) **Two-dimensional Wiener process.** Let $\mathbf{W}(t)$ be a two-dimensional Wiener process started at $(w,0) = \mathbf{w}$ where $w > 0$. Let A be the y-axis, $x = 0$. We seek $H(\mathbf{w},\mathbf{r})$ such that $\nabla^2 H = 0$ for $x > 0$, and

$$H - \log|\mathbf{r} - \mathbf{w}| = 0, \quad x = 0.$$

By inspection we see immediately that a suitable H is

$$H = \log|\mathbf{r} + \mathbf{w}|.$$

To check the details, note that on $x = 0$

$$H - \log|\mathbf{r} - \mathbf{w}| = \log(y^2 + w^2)^{1/2} - \log(y^2 + w^2)^{1/2} = 0$$

and for $x > 0$

$$\nabla^2 H = \frac{1}{2}\frac{\partial^2}{\partial x^2}\log[(x+w)^2 + y^2] + \frac{1}{2}\frac{\partial^2}{\partial y^2}\log[(x+w)^2 + y^2]$$

$$= \frac{\partial}{\partial x}\frac{(x+w)}{(x+w)^2 + y^2} + \frac{\partial}{\partial y}\frac{y}{(x+w)^2 + y^2} = 0.$$

Hence, the Green function is

(35) $$G(\mathbf{w}, \mathbf{r}) = \frac{1}{2\pi}\log\left|\frac{\mathbf{w}+\mathbf{r}}{\mathbf{w}-\mathbf{r}}\right|.$$

Now, using the result in (26), we can calculate the density of the position of the point at which $\mathbf{W}(t)$ hits the y-axis (the hitting density), as

$$-\frac{\partial G}{\partial n} = \frac{\partial G}{\partial x}\bigg|_{x=0} = \frac{1}{2\pi}\left\{\frac{\partial H}{\partial x} - \frac{1}{2}\frac{2(x-w)}{(x-w)^2 + y^2}\right\}\bigg|_{x=0}$$

(36) $$= \frac{1}{\pi}\frac{w}{y^2 + w^2}, \quad -\infty < y < \infty.$$

Recall that you obtained this by another method in Exercise (5.5.6). ○

It is important and instructive to notice that the function H simply reproduces at $-\mathbf{w}$ the singularity $\log|\mathbf{r} - \mathbf{w}|$ seen at $\mathbf{r} = \mathbf{w}$; that is to say it appears as the reflexion in the boundary $x = 0$. This observation can be made the basis of a systematic technique for finding Green functions, called the *method of images*. We do not elaborate here.

(37) **Three-dimensional Wiener process.** Let $\mathbf{W}(t)$ be a three-dimensional Wiener process started at $\mathbf{W} = (w, 0, 0)$ where $w > 0$. Let A be the (y, z)-plane, $x = 0$. As prescribed by Green, we seek a function H such that $\nabla^2 H = 0$ for $x > 0$, and

$$H + |\mathbf{r} - \mathbf{w}|^{-1} = 0, \quad \text{for } x = 0.$$

Guided by the preceding example, we consider the function

$$H(\mathbf{w}, \mathbf{r}) = -|\mathbf{r} + \mathbf{w}|^{-1}.$$

Then it is easy to see that

$$H(\mathbf{w}, \mathbf{r}) + |\mathbf{r} - \mathbf{w}|^{-1} = 0, \quad \text{for } x = 0$$

and some calculation by you verifies that $\nabla^2 H = 0$ in $x \geq 0$. Hence, the Green function is

$$(38) \qquad G(\mathbf{w}, \mathbf{r}) = \frac{1}{4\pi} \left\{ \frac{1}{|\mathbf{r} - \mathbf{w}|} - \frac{1}{|\mathbf{r} + \mathbf{w}|} \right\}$$

and, by calculating $\partial G / \partial x$ at $x = 0$, it is routine to show that the density of the position of the point at which $\mathbf{W}(t)$ hits the (y, z)-plane, A, is

$$(39) \qquad h_A(y, z) = \frac{w}{2\pi} (w^2 + y^2 + z^2)^{-3/2}, \quad (y, z) \in \mathbb{R}^2.$$

Hence, for example, the probability that $\mathbf{W}(t)$ hits A inside the disk $y^2 + z^2 \leq c^2$ is

$$\iint_{y^2 + z^2 \leq c^2} h_A(y, z) \, dy \, dz = 1 - \frac{w}{(c^2 + w^2)^{1/2}}.$$

○

For our final application of (25), which is a classic example, here is:

(40) **Hitting density for the Wiener process in a sphere.** Here A is the surface of the sphere B with centre at 0, and radius a. Suppose the Wiener process begins at $\mathbf{W}(0) = \mathbf{w}$, where $|\mathbf{w}| < a$. By Green's theorems we seek a function $H(\mathbf{w}, \mathbf{r})$ such that $H - |\mathbf{w} - \mathbf{r}|^{-1} = 0$ on A, and $\nabla^2 H = 0$ in B.

At this point we recall a little classical geometry. The inverse point $\mathbf{w}'$ of $\mathbf{w}$ lies at $\mathbf{w}' = a^2 \mathbf{w} / |\mathbf{w}|^2$, and for any point $\mathbf{r} \in A$ it is a routine exercise (using similar triangles) to show that

$$\frac{|\mathbf{r} - \mathbf{w}|}{|\mathbf{r} - \mathbf{w}'|} = \frac{|\mathbf{w}|}{a}.$$

It follows that we can set

$$H = -|\mathbf{w}' - \mathbf{r}|^{-1} \frac{a}{|\mathbf{w}|}, \quad \mathbf{r} \in B,$$

because H is a solution of Laplace's equation in B, and is easily seen to satisfy the required condition on $|\mathbf{r}| = a$. It follows that we can write

$$(41) \qquad 4\pi G(\mathbf{w}, \mathbf{r}) = |\mathbf{w} - \mathbf{r}|^{-1} - \frac{a}{|\mathbf{w}|} \left| \frac{a^2}{|\mathbf{w}|^2} \mathbf{w} - \mathbf{r} \right|^{-1}, \quad \mathbf{r} \in B.$$

Of course in this case $-(\partial G / \partial n) = -(\partial G / \partial r)$ on A. Therefore (after a tedious differentiation, which is left to you to verify), we obtain the hitting density of the sphere for $\mathbf{r} \in A$ as

(42) $$h_A(\mathbf{r}) = \frac{a^{-1}}{4\pi} \frac{a^2 - |\mathbf{w}|^2}{|\mathbf{w} - \mathbf{r}|^3}.$$

It is notable that this procedure can be repeated for the general ball in m dimensions, for which it can be shown that the density of the position of the point at which the Wiener process started at $\mathbf{w}$ hits A with radius a is

$$h_A(\mathbf{r}) = \left\{\Gamma\left(\frac{1}{2}m\right) \Big/ (2\pi^{m/2})\right\} \frac{a^2 - |\mathbf{w}|^2}{a|\mathbf{w} - \mathbf{r}|^m}, \quad m \geq 1.$$

The case $m = 1$ is an exercise for you, and $m = 2$ is Problem (23). (Recall that $\Gamma(\cdot)$ denotes the gamma function.) ○

Here is an application of this result.

(43) **Wiener process hitting a polar cap.** For ease of reference and analysis, we employ spherical polar coordinates (r, θ, ϕ) in this example. Let $\mathbf{W}(0)$ lie at distance w from 0 on the axis of symmetry, where $|w| < a$. Here B is the sphere of radius a, and A is divided into the polar cap H, where $|\theta| < \alpha < \pi$, $0 \leq \phi < 2\pi$, and the complement M, where $|\theta| > \alpha$, $0 \leq \phi < 2\pi$. Then the probability that the Wiener process started at $\mathbf{W}(0)$ hits H (the polar cap of semiangle α), before it hits $M = H^c$, is the integral over the cap of the hitting density given above in (42). That is:

(44) $$\int_0^\alpha \frac{a^2 - w^2}{(a^2 + w^2 - 2aw\cos\theta)^{3/2}} \frac{a^2 \sin\theta}{4\pi a} 2\pi \, d\theta$$

$$= \frac{a^2 - w^2}{2w} \left\{\frac{1}{a - w} - \frac{1}{(a^2 + w^2 - 2aw\cos\alpha)^{1/2}}\right\}.$$

In order to appreciate the beauty of Green's method to the full, you should spend some time trying to obtain these results by any other method. ○

Now we return from this excursion to reconsider the other basic problem, the solution of Poisson's equation in the form that we discussed after (11), that is to solve the following in D, with boundary A:

(45) $$\nabla^2 \phi = -2g(\mathbf{w}), \quad \mathbf{w} \in D$$

with the boundary condition $\phi(\mathbf{w}) = 0$ for $\mathbf{w} \in A$. This equation arises in (for example) gravitation and electrostatics, where $g(\mathbf{w}) > 0$ indicates the presence of matter or electric charge, respectively. In our context, it arises as the differential equation satisfied by the functional ϕ of the Wiener process given by

$$(46) \qquad \phi(\mathbf{w}) = \mathrm{E}\left\{\int_0^{T_A} g(W(s))\,ds \,\bigg|\, W(0) = \mathbf{w}\right\},$$

where T_A is the first-passage time to A. Of course when $g(\mathbf{x}) = 1$, then $\phi(\mathbf{w}) = \mathrm{E}(T_A \mid W(0) = \mathbf{w}) = m(\mathbf{w})$, and we have

$$(47) \qquad \nabla^2 m = -2, \quad \mathbf{w} \in D.$$

Now Green's theorems give the solution $\phi(\mathbf{w})$ of (45) in terms of the Green function G as

$$(48) \qquad \phi(\mathbf{w}) = \int_{\mathbf{r}\in D} 2g(\mathbf{r})G(\mathbf{w},\mathbf{r})\,dv.$$

(Here, dv is thought of as a small element of volume, parameterized by $\mathbf{r}$; to stress the parameterization, we may denote it by $dv(\mathbf{r})$.) This result has several interesting corollaries:

(49) **Expected visiting time.** Let $g(W(s))$ be the indicator of the event that $W(s) \in C \subseteq B$, where C is any subset of B. Formally, we may denote this by

$$g(\mathbf{x}) = I(\mathbf{x} \in C).$$

Then $\phi(\mathbf{w})$ as defined in (46) is V_C, the expected total time spent visiting C before T_A. But (48) tells us that

$$V_C = \int_{\mathbf{r}\in D} 2I(\mathbf{r}\in C)G(\mathbf{w},\mathbf{r})\,dv.$$

Thus the function $\Delta(\mathbf{r}) = 2G(\mathbf{w},\mathbf{r})\,dv(\mathbf{r})$ has a physical interpretation as the expected total time that $W(t)$ spends in the small volume $dv(\mathbf{r})$, before arriving at A.

But of course we have another expression for this; recall our discussions of the visiting times of Markov processes in (4.4.7) and elsewhere. To see it, let the transition density function of $W(t)$ started at $\mathbf{w}$ and absorbed when it hits A be $f(t,\mathbf{r}\mid \mathbf{w})$. Then also

$$\Delta(\mathbf{r}) = \int_0^\infty \mathrm{E}(I(W(s)\in dv(\mathbf{r})))\,ds$$
$$= \int_0^\infty f(s,\mathbf{r}\mid \mathbf{w})\,dv(\mathbf{r})\,ds.$$

Hence the Green function satisfies

(50) $$2G(\mathbf{w},\mathbf{r}) = \int_0^\infty f(t,\mathbf{r}\mid\mathbf{w})\,dt.$$

This remarkable identity is hard to verify explicitly except in the simplest cases; here is one of them.

(51) Wiener process in $\mathbb{R}^3$. Let $\mathbf{W}(t)$ be the unconstrained Wiener process in three dimensions started at $\mathbf{W}(0) = 0$. The Green function takes the trivial form

$$G = \frac{1}{4\pi|\mathbf{r}|}$$

and the transition density function of $\mathbf{W}(t)$ is

$$f(t,\mathbf{r}\mid 0) = \frac{1}{(2\pi t)^{3/2}} \exp\left\{-\frac{1}{2t}(x^2+y^2+z^2)\right\}$$

(52) $$= (2\pi t)^{-3/2} \exp\left\{\frac{-|\mathbf{r}|^2}{2t}\right\}.$$

We recognize (perhaps with some surprise) that as a function of t this is proportional to the first-passage density (5.3.7). Since we know that the integral of (5.3.7) is 1, we have trivially by inspection that

(53) $$\int_0^\infty f(t,\mathbf{r}\mid 0)\,dt = \frac{1}{2\pi|\mathbf{r}|} = 2G.$$

Thus the Wiener process in three dimensions is transient, and $2G$, (roughly speaking), tells us how much time it spends at $\mathbf{r}$ in total. ○

This result can be extended to find that explicitly in m dimensions, $m \geq 3$,

$$f(t,\mathbf{r}\mid\mathbf{w}) = (2\pi t)^{-m/2} \exp\left\{-\frac{1}{2t}|\mathbf{r}-\mathbf{w}|^2\right\}$$

and

(54) $$\int_0^\infty f(t,\mathbf{r}\mid\mathbf{w})\,dt = \Gamma\left(\frac{m}{2}-1\right)\frac{1}{2}\pi^{-m/2}|\mathbf{r}-\mathbf{w}|^{2-m} = 2G(\mathbf{w},\mathbf{r}).$$

Of course it does not supply anything interesting in lower dimensions, because for $m = 1$ and $m = 2$ the Wiener process is recurrent; the expected amount of time spent in any subset C of the state space is infinite. (You may care to check this by showing that the appropriate integrals do diverge.)

These low-dimensional processes are rather special. The case $m = 1$ we have considered extensively above, and therefore finally in this section we turn our attention once more to the Wiener process in the plane. By a happy chance, there is one extra technique available to help us tackle such problems in two dimensions. We identify the point (x, y) in $\mathbb{R}^2$ with the point $z = x + iy$ in the complex plane. Then we can use the theory of functions of a complex variable z; in particular we use these facts (note that while the complex numbers z and w are vectors, they are usually not set in boldface type):

(55) **Laplace's equation and the complex plane.** Let

$$f(z) = f(x + iy) = u(x, y) + iv(x, y),$$

be a differentiable function of the complex variable z, in an open region D, with real and imaginary parts u and v. It is easily shown that u and v satisfy the so-called Cauchy–Riemann equations

$$\frac{\partial u}{\partial x} = \frac{\partial v}{\partial y}, \quad \frac{\partial u}{\partial y} = -\frac{\partial v}{\partial x}$$

and hence that u and v both satisfy Laplace's equation

$$\nabla^2 u = \nabla^2 v = 0.$$

More importantly, there is a converse: suppose that the function $u(x, y)$ satisfies $\nabla^2 u = 0$ in some region R. Then there exists a function $f(z)$ such that f is differentiable and $u = \Re e f$, (the real part of f), in R.

Furthermore, suppose that $g(z) = u + iv$ is a differentiable function that maps D to R, which is to say that the image of D under $g(\cdot)$ is the region R. If $\phi(u, v)$ satisfies Laplace's equation in R, then it is straightforward to show that

$$\psi(x, y) = \phi(u(x, y), v(x, y)),$$

satisfies Laplace's equation in D. ○

All these facts, (whose proofs can be found in any text on functions of a complex variable), can be combined to yield a useful routine procedure for solving Laplace's equation with given boundary conditions. The program may be sketched thus:

(56) **Method of maps.** We seek ϕ such that $\nabla^2 \phi = 0$ in D and $\phi = \phi_A$ on the boundary A of D. Let $g(z) = u + iv$ be a function that maps D with boundary A to R with boundary B; for technical reasons we assume

that g is differentiable, invertible, and $g'(z) \neq 0, z \in D$. Now seek a solution ψ in R of Laplace's equation $\nabla^2 \psi = 0$, with the boundary condition

(57) $\quad\quad \psi_B(w) = \phi_A(g^{-1}(w)), \quad \text{for } w \in B, \; g^{-1}(w) \in A.$

From our remarks above, we have that $\phi = \psi(u(x,y) + iv(x,y)) = \psi(g(z))$ satisfies Laplace's equation in D, and (57) ensures that on the boundary A of D

$$\phi(z) = \psi(g(z)), \quad z \in A$$
$$= \phi_A(g^{-1}(g(z)))$$
$$= \phi_A(z), \quad \text{as required.}$$

The key idea is to choose R such that it is easy to find the solution ψ; the method is best grasped by seeing some examples. Note that we often use the polar representation $z = r \exp(i\theta)$ for points in the complex plane.

As an easy introduction, we tackle a familiar problem.

(58) **Wiener process in a wedge.** Let $\mathbf{W}(t)$ be a Wiener process started at $\mathbf{W}(0) = \mathbf{w}$ in the interior of the wedge D where $-\alpha < \theta < \alpha$. What is the probability that the process hits $H = \{\theta = \alpha\}$ before $M = \{\theta = -\alpha\}$? We seek to solve $\nabla^2 \phi = 0$, with the boundary conditions $\phi = 1$ on H and $\phi = 0$ on M. By inspection, the function $g(z) = \log z = \log r e^{i\theta} = \log r + i\theta = u + iv$ takes the wedge to the region

$$R = \{(u, v) : u \in \mathbb{R}, -\alpha < v < \alpha\}.$$

In that region (a strip) we seek a solution ψ of Laplace's equation with the boundary conditions $\psi = 1$ when $v = \alpha$, and $\psi = 0$ when $v = -\alpha$. Trivially, we have then

$$\psi = \frac{v + \alpha}{2\alpha}, \quad \text{in } R,$$
$$\phi = \frac{\theta + \alpha}{2\alpha}, \quad \text{in } D.$$

This we have already seen in (16) by another method, but (using this new technique) we can now obtain more details by using a different map.

(59) **Wiener process in quadrant.** Suppose that $\mathbf{W}(t)$ is a Wiener process started at $\mathbf{W}(t) = \mathbf{w}$ in the positive quadrant Q of the (x, y)-plane. The boundary A comprises the positive parts of the x and y axes; for hitting probabilities the problem of interest thus reduces to solving $\nabla^2 p(\mathbf{w}) = 0$ in Q, with $p(\mathbf{w}) = 1$ for $\mathbf{w} \in H$, and $p(\mathbf{w}) = 0$ for $\mathbf{w} \in M$.

It is easy to see that the function $s(z) = z^2$ maps the quadrant Q into the upper half plane U. Likewise the function $c(z) = (z-i)/(z+i)$ maps the upper half plane U into the unit disk R. We define $g(z)$ as the composition of the functions $c(\cdot)$ and $s(\cdot)$, thus

$$g(z) = \frac{z^2 - i}{z^2 + i},$$

which maps Q to R. The sets H and M are mapped to disjoint subsets $\hat{H}$ and $\hat{M}$ of the perimeter B of the unit disk R. Now, if we write $w = w_1 + iw_2$, where $\mathbf{w} = (w_1, w_2)$, then the original problem is thus essentially transformed to finding the probability that a Wiener process started at

$$\hat{w} = \frac{w^2 - i}{w^2 + i} \in R,$$

hits $\hat{H}$ before it hits $\hat{M}$. This can be done using more sophisticated methods from the theory of functions of a complex variable. But this is a problem to which we may obtain the solution using Green's function; see Problem (23). Indeed that supplies the hitting density $h_B(\mathbf{r})$ of B, which in turn will yield the hitting density of the boundary A of Q.

From (42) with $m = 2$, we know that if $v = \xi + i\zeta \in B$

$$h_B(v) = \frac{1 - |\hat{w}|^2}{2\pi |\hat{w} - v|^2}.$$

The hitting density of the x-axis (say) from $w \in Q$ then follows when we identify $v = (x^2 - i)/(x^2 + i)$, and naturally obtain

$$h_A(x) = h_B(v(x)) \frac{dv}{dx}.$$

The probability $p(w)$ of hitting any given interval $[a, b]$ of the real axis from the point $w \in Q$ is then easily calculated in principle (using plane polar coordinates) as

$$p = \int_a^b h_A(x)\, dx = \int_\alpha^\beta h_B(r(\theta))\, d\theta = \int_\alpha^\beta \frac{1 - |\hat{w}|^2}{2\pi |\hat{w} - e^{i\theta}|^2}\, d\theta,$$

where $\alpha = \arg\{(a^2 - i)/(a^2 + i)\}$ and $\beta = \arg\{(b^2 - i)/(b^2 + i)\}$.

As a special case (and a check on the arithmetic so far), consider $\mathbf{W}(0) = w = e^{i\pi/4}$. Then easy calculations show that $\hat{w} = 0$, and we have trivially

$$P(\text{hit}[a,b]) = \frac{\beta - \alpha}{2\pi}.$$

Then, for example, if $a = 0$ we see $\alpha = \pi$, and as $b \to \infty$, $\beta \to 2\pi$. Therefore, we confirm (what is also obvious by symmetry) that

$$P(\text{hit the } x\text{-axis before the } y\text{-axis from } e^{i\pi/4}) = \tfrac{1}{2}.$$

For another special value, you can show similarly that

$$P(\text{hit } [0,1] \text{ first from } e^{i\pi/4}) = P(\text{hit } [1,\infty) \text{ first from } e^{i\pi/4}) = \tfrac{1}{4}.$$

○

There are numerous further amusing applications of these ideas to the Wiener process in various regions; we confine ourselves to a brief look at two of them.

(60) Wiener process in a semicircle. The Wiener process $\mathbf{W}(t)$ starts at $\mathbf{W}(0) = \mathbf{w} = (w_1, w_2)$, inside the upper half D of the disk $|z| \leq 1$, where $z = x + iy$. Find the probability p that it leaves D via the diameter $H = (-1, 1)$.

Solution. The map $v(z) = ((z+1)/(z-1))^2$ takes D into the lower half plane. Therefore, the map $g(z) = \{v(z) - v(w)\}/\{v(z) - \overline{v(w)}\}$, where $\bar{v}$ is the complex conjugate of v, takes D to the interior of the unit disk and, in particular, $g(w) = 0$, where $w = w_1 + iw_2$. Hence, arguing as in the previous example, we observe that $g(1) = 1$ and $g(-1) = v(w)/\overline{v(w)}$, and so

$$1 - p = \frac{1}{2\pi} \arg \frac{v(w)}{\overline{v(w)}} = \frac{1}{\pi} \arg \left(\frac{w+1}{w-1} \right)^2$$

$$= \frac{2}{\pi} \arg \frac{(\bar{w}-1)(w+1)}{|w-1|^2} = \frac{2}{\pi} \arg(w_1^2 + w_2^2 - 1 - 2iw_2)$$

$$= \frac{2}{\pi} \tan^{-1} \frac{2w_2}{1 - |w|^2}.$$

It is an exercise for you to check that this does indeed satisfy Laplace's equation.

○

Finally, here is a problem in three dimensions.

(61) Wiener process in a square pipe. Let P be a semi-infinite hollow pipe of square cross-section. Suppose $\mathbf{W}(t)$ starts at a point on the axis of symmetry of P, at a distance d from the open end, and inside the pipe. What is the probability p that $\mathbf{W}(t)$ escapes through the open end of the pipe before hitting any of the four sides?

Solution. Without loss of generality, let the sides have width π. Now consider the planar problem of a Wiener process in a semi-infinite canal D in the complex plane, viz:

$$D \equiv \{(x, y) : 0 \leq x \leq \pi; y \geq 0\}$$

with

$$\mathbf{W}(0) = w = \frac{\pi}{2} + id, \quad d > 0.$$

Now observe that the map

$$g(z) = \frac{\cos z - \cos w}{\cos z - \overline{\cos w}},$$

takes D into the interior of the unit disk, and in particular $g(w) = 0$. Also

$$g(\pi) = \frac{1 + i \sinh d}{1 - i \sinh d} = \overline{g(0)}$$

and using the by now familiar argument we see that the probability that this process hits the sides of the canal before hitting the end $[0, \pi]$ is

$$\frac{1}{2\pi} 2 \arg g(\pi) = \frac{2}{\pi} \arg(1 + i \sinh d)$$

$$= \frac{2}{\pi} \tan^{-1}(\sinh d).$$

Hence, by the independence of the components of the original $\mathbf{W}(t)$ in three dimensions,

$$p = \left\{ 1 - \frac{2}{\pi} \tan^{-1}(\sinh d) \right\}^2.$$

○

Exercises

(a) Let $\mathbf{W}(t)$ be a Wiener process in the plane, started at $\mathbf{W}(0) = w = (w_1, w_2)$, where $w_1 > 0$. Show that with probability 1, $\mathbf{W}(t)$ visits the negative x-axis in finite time. Does it visit any finite nonzero interval of the x-axis in finite time?

(b)* Let $\mathbf{W}(t)$ be a Wiener process in the plane started at w, which is outside the circle centre 0 and radius ϵ, and inside the circle centre 0 and radius b. Find the probability $p(w)$ that $\mathbf{W}(t)$ hits the smaller circle first, and show that as $b \to \infty$, $p(w) \to 1$. What do you deduce about the recurrence or transience of $\mathbf{W}(t)$?

(c)* Following the idea of (b), let $\mathbf{W}(t)$ be a Wiener process in three dimensions started at w between two concentric spheres of radii ϵ

and b, respectively. Show that $p(\mathbf{w}) = \epsilon/b$ in this case. What do you deduce about the recurrence or transience of $\mathbf{W}(t)$?

(d)* **Hitting a conic section.** Let A be the boundary of the ellipse $(x^2/a^2) + (y^2/b^2) = 1$, and let $\mathbf{W}(t)$ be a Wiener process in the plane starting at $\mathbf{w}$ inside A. Find the expected time to hit A from $\mathbf{w}$. Repeat the problem when A is a hyperbola, and then again when A is a parabola, being careful to say when and where your answers are finite.

(e) **One-dimensional Green function.** Let $W(t)$ be a Wiener process in one dimension where $W(0) = w$, $b < w < a$. As usual, let $h_a(w)$ be the probability of hitting a before b, $h_b(w)$ the probability of hitting b before a, and $m(w)$ the expected time to hit either. By analogy with $W(t)$ in higher dimensions, if there is a Green function $G(w, x)$ it should satisfy

$$h_a(w) = -\frac{\partial G}{\partial x}(w, z),$$

$$h_b(w) = \frac{\partial G}{\partial x}(w, b),$$

$$m(w) = \int_b^a 2G(w, x)\,dx.$$

Verify that the function below does so.

$$G(w, x) = \begin{cases} \dfrac{(a-w)(x-b)}{(a-b)}, & b \le x \le w \le a, \\ \dfrac{(w-b)(a-x)}{(a-b)}, & b \le w \le x \le a. \end{cases}$$

Problems

(1)* Let $W_1(t), W_2(t), W_3(t)$ be independent Wiener processes. Show that both $S_2(t) = W_1^2 + W_2^2$ and $S_3(t) = W_1^2 + W_2^2 + W_3^2$ are Markov processes.

(2)* Find the equilibrium distribution for the drifting Wiener process between reflecting barriers at 0 and $a > 0$.

(3)* Find the equilibrium distribution for the Ornstein–Uhlenbeck process between reflecting barriers at 0 and $a > 0$.

(4)* Let $D(t)$ be the drifting Wiener process given by $D(t) = \mu t + W(t) + x$, where $D(0) = x$ and $b < x < a$. Let $r(x)$ be the probability that $D(t)$

hits a before b. Show that $\frac{1}{2}r''(x) + \mu r'(x) = 0$, and deduce that

$$r(x) = \frac{e^{-2\mu b} - e^{-2\mu x}}{e^{-2\mu b} - e^{-2\mu a}}.$$

What is $r(x)$ when $\mu = 0$?

(5) Let $X(t)$ have instantaneous parameters $a(x)$ and $b(x)$, and let T be the first-passage time of X to a or b, where $b < x = X(0) < a$. Show that the mgf

$$M(s, x) = \mathrm{E}(e^{-sT} \mid X(0) = x)$$

satisfies the equation

$$\frac{1}{2}b(x)\frac{\partial^2 M}{\partial x^2} + a(x)\frac{\partial M}{\partial x} = sM.$$

Hence calculate M when $X(t)$ is $W(t)$, the Wiener process, and $b \to -\infty$.

(6) With the same notation as the previous question (5), let

$$g(x) = \mathrm{E}\left\{\int_0^T k(X(u))\,\mathrm{d}u \mid X(0) = x\right\}.$$

Show that $g(x)$ satisfies the equation

$$\tfrac{1}{2}b(x)g''(x) + a(x)g'(x) = -k(x).$$

What are the boundary conditions at a and b?

(7) Let $W(t)$ be the Wiener process, and let $M_1(t), M_2(t), M_3(t), M_4(t)$ be the martingales defined in (5.5.5), (5.5.8), and (5.5.9). Let T_a and T_b be the first-passage times to a and b, respectively, where $b < 0 = W(0) < a$. Define $T = \min\{T_a, T_b\}$, the indicator $I_a = I\{T_a < T_b\}$, and denote $p_a = \mathrm{E}I_a = \mathrm{P}(T_a < T_b)$. Prove that

$$0 = \mathrm{E}M_3(T) = a^3 p_a + b^3(1 - p_a) - 3a\mathrm{E}(T I_a) - 3b\mathrm{E}(T(1 - I_a))$$

and

$$0 = \mathrm{E}M_4(T) = a^4 p_a + b^4(1 - p_a) - 6a^2\mathrm{E}(T I_a) - 6b^2\mathrm{E}(T(1 - I_a)) + 3\mathrm{E}(T^2).$$

Deduce that $\mathrm{var}\,T = -\tfrac{1}{3}ab(a^2 + b^2)$.

(8)* Let $D(t)$ be the drifting Wiener process given by $D(t) = \mu t + W(t) + x$, where $b < x = D(0) < a$. Show that $M(t) = \exp(-2\mu D(t))$ is a martingale, and hence deduce the result of Problem (4) above. In particular, show that, if $a = -b = 1 = \mu$, then $r(0) = e^2/(1+e^2)$, where $r(x)$ is defined in Problem (4).

(9)* (a) **Wiener process on a circle.** Let $W(t)$ be the Wiener process and set $Z(t) = \exp(iW(t))$. Show that $Z(t) = X(t) + iY(t)$ is a process on the unit circle satisfying

$$dX = -\tfrac{1}{2}X\,dt - Y\,dW, \qquad dY = -\tfrac{1}{2}Y\,dt + X\,dW.$$

(b) **Wiener process on a sphere.** Let $S(t)$ be a Wiener process on the surface of the unit sphere and let A be the equator of the sphere. Find the expected time until $S(t)$ hits A, and show in particular that the expected time to hit A starting from either pole is $4\log\sqrt{2}$. [Hint: Recall Laplace's equation in the form (5.9.15), and note that for this problem there is no dependence on r or ϕ.]

(10)* By applying Itô's formula to $Y = f(t)W(t)$, where $f(t)$ is a non-random differentiable function, show that

$$\int_s^t f(u)\,dW(u) = f(t)W(t) - f(s)W(s) - \int_s^t W(u)f'(u)\,du.$$

Deduce that $\int_0^t f(u)\,dW(u)$ has a normal distribution, and is a martingale. Prove also that $Z(t)$ is a martingale, where

$$Z(t) = \left(\int_0^t f(u)\,dW(u)\right)^2 - \int_0^t (f(u))^2\,du$$

provided that all the integrals exist.

(11)* Show that the Ornstein–Uhlenbeck process has the representation $U(t) = e^{-\beta t}\int_0^t e^{\beta s}\,dW(s)$.

(12)* Let $X(t)$ be the exponential Wiener process such that

$$dX = \mu X\,dt + \sigma X\,dW.$$

Write down Itô's formula for $Y = X^n$, and deduce that

$$E[X(t)^n] = EX(0)^n + \int_0^t \left\{n\mu + \tfrac{1}{2}n(n-1)\sigma^2\right\} E(X(s)^n)\,ds.$$

Hence find $E(X(t)^n)$, $t \geq 0$.

(13)* Let $U(t)$ be the Ornstein–Uhlenbeck process for which

$$dU = -\beta U\, dt + dW$$

and let $E(U(t)^n) = m_n(t)$. By writing down Itô's formula for $Y = U^n$, show that

$$m_n(t) = m_n(0) + \int_0^t \left\{ -\beta n m_n(s) + \frac{1}{2}n(n-1)m_{n-2}(s) \right\} ds.$$

(14) **Squared Ornstein–Uhlenbeck process.** If $Y = U^2$, where $dU = -\beta U\, dt + dW$, show that

$$dY = (1 - 2\beta Y)\, dt + \sqrt{Y}\, dW.$$

(This is a special case of a so-called 'square root process', which may be used to model interest rates.)

(15) Show that the solution of the SDE

$$dX = -\tfrac{1}{2}X\, dt + \sqrt{1 - X^2}\, dW$$

is $X = \cos W$.

(16) **Doob's inequality.** If $W(t)$ is the standard Wiener process, show that

$$E\left(\max_{0 \leq s \leq t} W(s)^2 \right) \leq 4E(W(t)^2).$$

[Hint: Recall the martingale maximal inequalities given in Chapter 2.]

(17) **Quotient rule.** Use the product rule (5.8.18), and Itô's formula, to show that if $Q = X/Y$, where

$$dX = \mu\, dt + \sigma\, dW, \qquad dY = \alpha Y\, dt + \beta Y\, dW$$

then

$$Y\, dQ = \{\mu - \sigma\beta + (\beta^2 - \alpha)X\}\, dt + (\sigma - \beta X)\, dW.$$

(18)* Let $W(t)$ be the standard Wiener process, and $M(t)$ its maximum in $[0, t]$:

$$M(t) = \sup_{0 \leq s \leq t} W(s).$$

Use the reflection principle to show that for $x > 0$,

$$P(W(t) \leq c - x, M(t) \geq c) = P(W(t) \geq c + x).$$

Deduce that $W(t)$ and $M(t)$ have joint density

$$f_{W,M}(w, m) = \sqrt{\frac{2}{\pi}} \left\{ \frac{2m - w}{t^{3/2}} \right\} \exp\left\{ -\frac{(2m - w)^2}{2t} \right\}, \quad w < m, \; m > 0.$$

(19) Suppose that $\mathbf{W}(t)$ is a Wiener process in three dimensions. Let $\mathbf{w} \in D$, and define, as usual,

$$p(\mathbf{w}) = P(T_H < T_M \mid \mathbf{W}(0) = \mathbf{w}).$$

Let B be the ball of radius a with surface S, having area $4\pi a^2$ in three dimensions, where $B \subset D$. Let T_S be the first-passage time of the Wiener process $\mathbf{W}(t)$ from $\mathbf{w}$ to S. Show that $P(T_S < \infty) = 1$, and then use the strong Markov property at T_S to deduce that

$$4\pi a^2 p(\mathbf{w}) = \int_{\mathbf{x} \in S} p(\mathbf{x}) \, dS.$$

Repeat the problem in two dimensions.

(20) Let A be a right circular cone of semiangle α, with the x-axis as its axis of symmetry. Let $\mathbf{W}(t)$ be a Wiener process started at $\mathbf{W}(0) = \mathbf{w} = (w_1, w_2, w_3)$ in the interior of A, that is, where $\tan^{-1}\{w_2^2 + w_3^2)^{1/2}/w_1\} < \alpha$. Show that the expected time for $\mathbf{W}(t)$ to hit the boundary of the cone is

$$m(\mathbf{w}) = \frac{w_1^2 \sec^2 \alpha - |\mathbf{w}|^2}{2 - \tan^2 \alpha}, \quad |\alpha| < \frac{\pi}{4}.$$

(21) **Hitting a quadric.** Suppose a Wiener process $\mathbf{W}(t)$ in three dimensions is started at $\mathbf{W}(0) = \mathbf{w}$ in the interior of the ellipsoid E given by

$$\frac{x^2}{a^2} + \frac{y^2}{b^2} + \frac{z^2}{c^2} = 1.$$

Find the expected time until $\mathbf{W}(t)$ hits E. Discuss the equivalent problem for the hyperboloids $(x^2/a^2) + (y^2/b^2) - (z^2/c^2) = 1$, and $(x^2/a^2) - (y^2/b^2) - (z^2/c^2) = 1$.

(22) **Hitting a triangle.** Suppose that a Wiener process in two dimensions is started at $\mathbf{W}(0) = (x, y)$ where (x, y) is in the interior of the triangle D

determined by

$$t(x, y) = (a - x)(x - \sqrt{3}y + 2a)(x + \sqrt{3}y + 2a) = 0.$$

Show that the expected time until $\mathbf{W}(t)$ hits the triangle is $t(x,y)/6a$.

(23) **Hitting density for the circle.** Show that Green's function for the interior of the circle D with centre 0 and radius a is

$$2\pi G(\mathbf{w}, \mathbf{r}) = \log\left\{\frac{|\mathbf{w}|}{a}\frac{|\mathbf{r} - \mathbf{w}'|}{|\mathbf{r} - \mathbf{w}|}\right\}, \quad \mathbf{r}, \mathbf{w} \in D,$$

where $\mathbf{w}' = (a^2/|\mathbf{w}|^2)\mathbf{w}$.

Deduce that the hitting density of the perimeter A of D, for a Wiener process started at $\mathbf{W}(0) = \mathbf{w} \in D$, is

$$h_A(\mathbf{r}) = \frac{a^2 - |\mathbf{w}|^2}{2\pi a |\mathbf{r} - \mathbf{w}|^2}, \quad \mathbf{r} \in A.$$

(24) **Wiener process between coaxal circles.** Let $\mathbf{W}(t)$ be a Wiener process in the (x, y)-plane. Write $z = x + iy$, and suppose that two circles A and B are determined, respectively, by the equations

$$\left|\frac{z - \alpha}{z - \beta}\right| = \lambda_a, \quad \left|\frac{z - \alpha}{z - \beta}\right| = \lambda_a.$$

Suppose that $\mathbf{W}(0) = w = w_1 + iw_2$, where w lies between A and B. Show that the probability that $\mathbf{W}(t)$ hits A before B is

$$\frac{1}{\log \lambda_a - \log \lambda_b}\left\{\log\left|\frac{w - \alpha}{w - \beta}\right| - \log \lambda_a\right\}.$$

6 Hints and solutions for starred exercises and problems

Chapter 1

Exercises

1.1(a)

(i) A and A^c are disjoint with union Ω; thus
$$P(A) + P(A^c) = P(\Omega) = 1,$$
and (7) follows.

(ii) By the above, $P(\phi) = P(\Omega^c) = 1 - P(\Omega) = 0$.

(iii) $A = (A \cap B^c) \cup (A \cap B)$, and $B = (B \cap A^c) \cup (A \cap B)$, $A \cup B = (A \cap B^c) \cup (A \cap B) \cup (B \cap A^c)$. Now use (5).

(iv) If $A \subseteq B$ then $B = A \cup (B \cap A^c)$. Use (5).

(v) Use induction on n, remembering that you have established the case $n = 2$ above in (iii).

(vi) If (B_n) is an increasing sequence of events, $B_1 \subseteq B_2 \subseteq \cdots \subseteq B = \lim_{n \to \infty} B_n$, then by (5),
$$P(B) = P(B_1) + \lim_{n \to \infty} \sum_{r=1}^{n-1} [P(B_{r+1}) - P(B_r)] = \lim_{n \to \infty} P(B_n).$$

The same is true for a decreasing sequence (D_n).
Let $D_n = \bigcup_{m=n}^{\infty} A_m$, and $B_n = \bigcap_{m=n}^{\infty} A_m$. Then, D_n and B_n are decreasing and increasing, respectively, with limit A. The result follows.

1.1(c) Use induction and (1.1.9).

1.2(a) $P(A \cap B) = 6^{-2} = P(A)P(B)$, and so on. But
$$6^{-2} = P(A \cap B \cap C) \neq P(A)P(B)P(C) = 6^{-3}.$$

1.3(c) $c = \pi^{-1}$.

1.4(b) No, because $(\partial^2 F)/(\partial x \partial y) = (1-xy)\exp(-xy) < 0$ for $xy > 1$.

1.5(b) The Jacobian of this transformation is $|w|$. Thus, W and Z have the joint density $(2\pi)^{-1}|w|\exp\{-\frac{1}{2}w^2(1+z^2)\}$. Integrate with respect to w to obtain the required density of Z. The last part follows by symmetry.

1.6(b) Using 1.6(a), $\mathrm{cov}(M, X_r - M) = \mathrm{cov}(M, X_r) - \mathrm{var} M = (1/n)\mathrm{var} X_r - (1/n)\mathrm{var} X_1 = 0$.

1.7(b) By (1.5.12), $X + Y$ is Poisson $(\lambda + \mu)$. Therefore, $P(X = k \mid X + Y = n) = P(X = k, Y = n - k)/P(X + Y = n) = p^k(1-p)^{n-k}\binom{n}{k}$, where $p = \lambda/(\lambda + \mu)$. We recognize this as Binomial, $B(n, p)$, and the result follows.

1.8(c) Using 1.8(b) and conditional expectation gives

$$E(s^X t^Y) = E(E(s^X t^Y \mid N)) = E(E(s^X t^{N-X} \mid N))$$
$$= E\left(t^N\left(1 - p + p\frac{s}{t}\right)^N\right) = \exp\{\lambda[(1-p)t + ps - 1]\},$$

using (1.8.11) and (1.8.12). Since this factorizes as $\exp[\lambda p(s-1)]\exp[\lambda(1-p)(t-1)]$, we have that X and Y are independent Poisson with parameters λp and $\lambda(1-p)$, respectively.

1.8(d) Use (1.8.13) to find that S is Exponential $(p\lambda)$.

1.9(b) Use (1.9.5) and (1.9.6).

Problems

1. By (1.6.7), $EX^n = \int_0^\infty P(X^n > y)\,dy$. Now set $y = x^n$ in the integral. Likewise $Eg(X) = \int_0^\infty P(g(X) > y)\,dy$. Set $y = g(x)$.

2. By (1.5.6), $P(U \leq u) = 1 - \exp(-(\lambda+\mu)u)$; and by (1.7.19), $P(X < Y) = \lambda/(\lambda + \mu)$. Hence,

$$P(U \leq u, V \geq v, I = 1) = P(X \leq u, Y - X \geq v)$$
$$= \int_0^u \lambda\exp\{-\lambda x - \mu(v + x)\}\,dx = \frac{\lambda}{\lambda + \mu}e^{-\mu v}(1 - e^{-(\mu+\lambda)u}).$$

(i) The factorization yields the required independence, and $P(V \geq v) = (\lambda/(\lambda+\mu))e^{-\mu v} + (\mu/(\lambda+\mu))e^{-\lambda v}$.

(ii) $P(W = 0) = P(X < Y) = \lambda/(\lambda+\mu)$, and $P(W > w) = \mu/(\lambda+\mu)e^{-\lambda w}$.

4. Mathematical induction is a straightforward approach. Alternatively, show that
$$\mathrm{E}e^{-\theta T_n} = \left(\frac{\lambda}{\lambda+\theta}\right)^n = \int_0^\infty e^{-\theta x} f(x)\,dx$$
and appeal to the uniqueness theorem for moment-generating functions (mgfs).

5. If $\mathrm{E}Y^2 = 0$, then $P(|Y| > 0) = 0$. Now consider $Y = X - \mathrm{E}X$.

7. If either of $\mathrm{E}X^2$ or $\mathrm{E}Y^2$ is zero, the result is immediate, by Problem 5. Otherwise, the condition for $\mathrm{E}[(sX - tY)^2] \geq 0$, viewed as a quadratic in t, is $\mathrm{E}[(XY)^2] - \mathrm{E}(X^2)\mathrm{E}(Y^2) \leq 0$. It has exactly one root if and only if equality holds, and then, for some s, t, $\mathrm{E}[(sX - tY)^2] = 0$. Now use Problem 5 again.

For Schlömilch's inequality consider the special case when $X = 1$ and $Y = |Z - \mathrm{E}Z|$.

8. Simply take the expected value of each side of the inequality, which is preserved.

9. $dM(r)/dr = r^{-2}[\mathrm{E}(X^r)]^{(1/r)-1}\{\mathrm{E}(X^r \log X^r) - \log(\mathrm{E}(X^r))\mathrm{E}(X^r)\} \geq 0$, because $g(x) = x \log x$ is a convex function and we can use the result of Problem 8, Jensen's inequality. Thus $M(r)$ increases in r.

10(a) Suppose the biased die shows Z. The other is fair, so
$$P(X + Y = r, \text{modulo } 6) = \sum_{k=1}^{6} f_Z(k)\frac{1}{6} = \frac{1}{6}, \quad 0 \leq r \leq 5.$$

(b) Any 3 of the dice includes at least one fair die. Roll these and calculate the sum of any two, modulo 6, then add the third, modulo 6. By the argument in (a), the result is uniform on $\{1, 2, 3, 4, 5, 0\}$. Do this twice to obtain the net effect of two fair dice.

11. Like Problem 10 with an integral replacing the sum.

12(a) The probability-generating function (pgf) of X_1 is $\sum_{k=1}^{\infty} s^k p(1-p)^{k-1} = ps/(1-(1-p)s) = G_X(s)$. By (1.8.5), the pgf of Y is $\{G_X(s)\}^n$, and the coefficient of s^k is the required probability.

(b) The intervals between successive heads are independent Geometric (p), so by construction the required distribution is that of Y.

13(a) Use integration by parts to evaluate $\mathrm{E}g'(X) = \int g'(x)(1/\sqrt{2\pi}\sigma) \exp\{-\frac{1}{2}(x-\mu)^2/\sigma^2\}dx = \mathrm{E}\{g(X)(X-\mu)/\sigma^2\}$.

(b) Same procedure as (a), using the binormal density (1.9.3).

14. The probability that any given ordering of $X_1, \ldots, X_n$, denoted by $X_{\pi(1)}, \ldots, X_{\pi(n)}$, satisfies

$$x_{(1)} < X_{\pi(1)} < x_{(1)} + dx_{(1)}, \ldots, x_{(n)} < X_{\pi(n)} < x_{(n)} + dx_{(n)},$$

where $0 \leq x_{(1)} < x_{(2)} < \cdots < x_{(n)} \leq t$ is $\prod_{r=1}^{n} dx_{(r)} t^{-n}$. Since there are $n!$ such orderings, which are distinct, the result follows.

16(a) (i) $f(\theta) = \frac{1}{2} \sin \theta$, $0 \leq \theta \leq \pi$. (ii) Now draw a diagram.

(b) A and B cannot communicate if $(\pi/2) < \theta < \pi$. By conditional probability, using the first part, the probability required is

$$\frac{1}{4\pi} \int_{\pi/2}^{\pi} (\pi - \theta) \sin \theta \, d\theta = (4\pi)^{-1}.$$

Chapter 2

Exercises

2.1(a)

(i) Consider the first n of the Xs, and let V be the number of inspections needed to prove the conjecture wrong. Now $P(V > n)$ is the probability that the first is the largest, and by symmetry this is n^{-1}. Hence $P(V > n) \to 0$, but $EV = \sum_n P(V > n) = \infty$.

(ii) $EX_N = E(X_1 \mid X_1 > a) = \int_a^\infty x f(x) \, dx / P(X_1 > a)$.

(iii) N is Geometric with parameter $P(X_1 > a)$. Therefore, $E(N - 1) = P(X_1 \leq a)/P(X_1 > a)$. Also, as above, $E(X \mid X \leq a) = \int_0^a x f(x) \, dx / P(X_1 \leq a)$. Finally, by conditioning on N, $ES_N = E(N - 1)E(X_1 \mid X_1 < a) + E(X_1 \mid X_1 > a) = EX_1/P(X_1 > a)$.

2.1(b)

(i) $E\Sigma_r X_r = \Sigma \lambda_r^{-1}$; if the sum converges then the expectation exists, and we must have $P(\Sigma_r X_r < \infty) = 1$.

(ii) $E(\exp -\Sigma_r X_r) = \lim_{n\to\infty} \prod_{r=1}^n E(\exp -X_r) = \prod_{r=1}^\infty (1 + (1/\lambda_r))^{-1} \to 0$, if $\sum_{r=1}^\infty \lambda_r^{-1} = \infty$. Hence, $P(\Sigma_r X_r = \infty) = 1$.

2.2(a)

(i) Consider successive disjoint sequences of consecutive trials, where each sequence is of length $|a| + |b|$. Any such sequence comprises entirely zeros with probability $p^{|a|+|b|}$. Let W be the number of sequences until one is indeed all zeros. Then $EW = p^{-|a|-|b|}$, because W is Geometric ($p^{|a|+|b|}$). But by construction $N \leq (|a| + |b|)W$, and the result follows.

(ii) T is less than or equal to the number of flips required to obtain $|a| + |b|$ successive heads. Now use the first part.

2.3(a)

(i) For any $a > 0$, $P(X \geq \epsilon) = P(X + a \geq a + \epsilon) \leq E(X + a)^2/(a + \epsilon)^2$. The right-hand side is smallest when $a = (\text{var} X)/\epsilon$, yielding

$$P(X \geq \epsilon) \leq \{\text{var} X + ((\text{var} X)^2/\epsilon^2)\}/\{(\text{var} X/\epsilon) + \epsilon\}^2$$
$$= \text{var} X/(\text{var} X + \epsilon^2).$$

(ii) If m is a median of X,

$$\frac{1}{2} = P(X \geq m) = P(X - \mu \geq m - \mu) \leq \frac{\sigma^2}{\sigma^2 + (m - \mu)^2}.$$

2.4(a)

(i) Follows from (ii).
(ii) Follows from 2.4.2(a).
(iii) First, $E|(q/p)^{S_n}| = ((q/p)p + (p/q)q)^n = 1 < \infty$. Likewise $E((q/p)^{S_{n+1}}|S_0, \ldots, S_n) = (q/p)^{S_n}$.
(iv) It is easy to check that

$$E|S_n^2 - 2\mu n S_n + (n\mu)^2 - n\sigma^2| \leq n^2 + 2\mu n^2 + n^2\mu^2 + n\sigma^2 < \infty.$$

Now

$$E(S_{n+1}^2 - 2\mu(n+1)S_{n+1} - (n+1)\sigma^2 + (n+1)^2\mu^2 \mid S_0, \ldots, S_n)$$
$$= S_n^2 + 2\mu S_n + 1 - 2\mu(n+1)(S_n + \mu) - n\sigma^2 - \sigma^2$$
$$+ n^2\mu^2 + 2n\mu^2 + \mu^2$$
$$= S_n^2 - 2\mu n S_n + (n\mu)^2 - n\sigma^2$$

as required.

2.4(b) By 2.1(a), $ET < \infty$ and $\operatorname{var} T < \infty$. Assume $p \neq q$ (you can do the case $p = q$).

(i) $(q/p)^{S_n \wedge T}$ is uniformly bounded, so by optional stopping,

$$1 = E\left(\frac{q}{p}\right)^{S_T} = P(A)\left(\frac{q}{p}\right)^a + (1 - P(A))\left(\frac{q}{p}\right)^b,$$

which gives $P(A)$ and $P(B)$.

(ii) Since $ET < \infty$, we can use dominated convergence for the martingale $S_n - n(p - q)$, which gives

$$0 = ES_T - (p - q)ET = aP(A) + b(1 - P(A)) - (p - q)ET.$$

Solve this to obtain ET.

(iii) Likewise, using the final martingale in Exercise (a),

$$a^2 P(A) + b^2(1 - P(A)) - 2\mu E(T S_T) + (p - q)^2 ET^2 - \sigma^2 ET = 0.$$

As $a \to \infty$, $a^2 P(A) \to 0$; $P(B) \to 1$; $ET \to -b/(q - p)$; $E(T S_T) \to -b^2/(q - p)$. After some work, $\operatorname{var} T = -4pqb(q - p)^{-3}$.

2.5(a)

(i) Binomial $B(n, \sigma)$.
(ii) Use Exercise 1.8(c).
(iii) By the independence, Geometric (σ).

2.6(b) Condition on X_1. If $X_1 = x < a$, then $b(a)$ is x plus the expected further time required, where this further time has the same distribution as the original. If $X_1 = x > a$, then $b(a) = a$. Hence, $b(a) = \int_0^a \{x + b(a)\} f_X(x)\, dx + \int_a^\infty a f_X(x)\, dx$, and the result follows. For the last part, set $f_X(x) = \lambda e^{-\lambda x}$.

2.7(a) Let $g(x) = \mathcal{G}(x) - x$, so that $g(\eta) = 0$ and $g(1) = 0$. First, we dispose of the trivial case when $\mathcal{G}(x)$ is linear, $\mathcal{G}(x) = q + px$. Here $\mu < 1$ and $\eta = 1$. Otherwise $g''(x) = E\{Z_1(Z_1 - 1)x^{Z_1 - 2}\} > 0$, so $g'(x)$ is strictly increasing in x.

A well-known theorem (Rolle) tells us that between any two zeros of a differentiable function $g(x)$, its derivative $g'(x)$ has a zero. Therefore,

$g(x)$ has at most two zeros in $[0, 1]$, which is to say that it has at most one zero in $[0, 1)$. Consider the two cases:

(I) If $g(x)$ has no zero in $[0, 1)$, then $g(x) > 0$ in $[0, 1)$, which in turn yields $g(1) - g(x) < 0$. Hence $g'(1) = \lim_{x \uparrow 1}(g(1) - g(x))/(1-x) \leq 0$. Thus, $\mu \leq 1$.

(II) If $g(x)$ has a zero η in $[0, 1)$, then $g'(x)$ has a zero z in $[\eta, 1)$. But $g'(x)$ is increasing, so $g'(1) > g'(z) = 0$. Thus, $\mu > 1$.

2.8(a) Let X_i have parameter q_i, and let T_i be a collection of independent Bernoulli random variables such that $ET_i = p_i/q_i$, and set $Z_i = X_i T_i \leq X_i$. Then $EZ_i = p_i$, and

$$R(\mathbf{p}) = E\phi(\mathbf{Z}) \leq E\phi(\mathbf{X}), \quad \text{because } \phi \text{ is monotone,}$$
$$= R(\mathbf{q}).$$

Problems

2. In calculating ES_n^4, use the fact that terms like $E(X_1^3 X_2)$, $E(X_1^2 X_2 X_3)$, and $E(X_1 X_2 X_3 X_4)$ all have zero mean. There are n terms like EX_1^4, and $\binom{4}{2}\binom{n}{2}$ terms like $E(X_1^2 X_2^2)$; to see this last, note that there are $\binom{n}{2}$ ways to choose the suffices, and $\binom{4}{2}$ ways to choose the pairs when the suffices are fixed. By Markov's inequality, with $S_n' = \sum_1^n X_r' = \sum_1^n (X_r - EX_r)$,

$$\sum_n P(|(S_n'/n| > \epsilon) \leq \sum_n ES_n'^4/(\epsilon n)^4 = A \sum_n n^{-3} + B \sum_n n^{-2}$$

for some constants A and B. Since the sums converge, the result follows using the Borel–Cantelli lemma.

3. For any plate S, let R and L be the plates on the right and left of S. Whichever of R or L is visited first by the fly, the probability that S is last to be visited is N^{-1}. If the fly can move anywhere, the same is true by symmetry.

If the fly has visited k plates, we consider the expected number of steps n_k until it first visits one that it has not visited before, thus yielding an obvious recurrence relation. Under the first regime using (2.2.13), we see that $n_k = k$, and so $EC = \frac{1}{2}N(N+1)$. Under the second regime, the number of flights required to see a new plate is Geometric $((N-k+1)/N)$, so $n_k = N/(N-k+1)$, and

$$EC = 1 + \frac{N}{N-1} + \cdots + \frac{N}{1} \simeq N \log N.$$

4. We use the result of (2.2.13) repeatedly.

(a) The probability that the walk ever visits j from k (before being absorbed at N, since $k > j$) is $(N-k)/(N-j)$. By conditioning on the first step, the probability that the walk ever revisits j is $\frac{1}{2}((N-j-1)/(N-j)) + \frac{1}{2}((j-1)/j) = p$, (say). Let $q = 1 - p$. Then the expected number of revisits is geometric with mean p/q. Thus, finally, the expected number of visits to j from k is

$$\frac{N-k}{N-j}\left(1 + \frac{p}{q}\right) = \frac{N-k}{N-j} \cdot \frac{1}{q} = \frac{2j(N-k)}{N},$$

after a little algebra, yielding $q = N/(2j(N-j))$. (*)

(b) Interchanging the roles of N and 0, when $k < j < N$ the expected number of visits to j is $2k(N-j)/N$.

(c) Let W be the event that the walk stops at N. Let p be the probability of revisiting the start before absorption. Clearly X is geometric with parameter $1 - p$. Furthermore,

$$P(X = x \mid W) = \frac{P(W \cap \{X = x\})}{(k/N)}$$

$$= p^x \frac{1}{2} \frac{1}{N-k} \frac{N}{k} = p^x(1-p),$$

after a glance at (*) in (a). We have shown that $P(X = x \mid W) = P(X = x)$, which yields the independence.

5(a)

(i) $E|Z_n \mu^{-n}| = 1 < \infty$, and

$$E(Z_{n+1}\mu^{-(n+1)} \mid Z_0, \ldots, Z_n) = E\left(\sum_{1}^{Z_n} X(i)\mu^{-(n+1)} \mid Z_0, \ldots, Z_n\right)$$

$$= Z_n \mu^{-n}.$$

(ii) $E|\eta^{Z_n}| \le 1 < \infty$, and
$$E(\eta^{Z_{n+1}} \mid Z_0, \ldots, Z_n) = (\mathcal{G}(\eta))^{Z_n} = \eta^{Z_n}.$$

(b)
$$E|H_n^{Z_n}| = \mathcal{G}_n(H_n(s)) = s < \infty,$$
and
$$E(H_{n+1}^{Z_{n+1}} \mid Z_0, \ldots, Z_n) = \{\mathcal{G}(H_{n+1})\}^{Z_n} = (H_n(s))^{Z_n},$$
where the last equality follows because
$$\mathcal{G}_n(H_n(s)) = s = \mathcal{G}_{n+1}(H_{n+1}(s)) = \mathcal{G}_n(\mathcal{G}(H_{n+1}(s))).$$

7. It is elementary to show that, however many people there are, if they all select a coat at random the expected number of matched coats is 1. It follows that $M_n = X_n + n$ is a martingale. By the optional stopping theorem,
$$EM_T = E(X_T + T) = ET = E(X_0 + 0) = C.$$

8. Condition on the event A_N that N events of the process occur in $[0, t]$. They are independently and uniformly distributed on $[0, t]$ by (2.5.15). Therefore, if we consider the sum $S = \sum_{r=1}^{N} g(S_r)$ conditional on A_N, we find that
$$E(S \mid A_N) = E\left(\sum_{1}^{N} g(S_r) \mid A_N\right) = N \frac{1}{t} \int_0^t g(u) \, du,$$
$$E(S^2 \mid A_N) = N \frac{1}{t} \int_0^t g(u) \, du + N(N-1) \frac{1}{t^2} \left\{\int_0^t g(u) \, du\right\}^2.$$

But $EN = \lambda t$, $EN(N-1) = \lambda^2 t^2$, so the result follows when we remove the conditioning and let $t \to \infty$.

13. Suppose you are rewarded at unit rate while the machine is working. Your expected reward during a cycle is therefore EZ_1, and the expected length of a cycle is $EZ_1 + EY_1$. By the renewal–reward theorem, with an obvious indicator I,
$$\lim_{t \to \infty} \frac{1}{t} \int_0^t I(\text{the machine is working at } u) \, du = \frac{EZ_1}{EZ_1 + EY_1}.$$

But this is what you are asked to prove.

14. The result of Problem 13 holds in discrete time also. The expected time to carry the umbrella is $(1-p)^{-1}$, and the expected time to be reunited with it if separated is $n-1$, in both cases. By the renewal–reward theorem, the long-run proportion of umbrella-free trips is $(n-1)/\{n-1+(1-p)^{-1}\}$. The result follows by the independence of the rain.

15. $P(A_n \cap A_m) = 1/mn$ by symmetry. But $P(A_n) = 1/n$, so A_n and A_m are independent. Since $\sum_n 1/n$ diverges, the second Borel–Cantelli lemma gives that A_n occurs infinitely often.

16. Let $X_m(a) = X_m \wedge a$, where $a < \infty$. By the strong law,

$$\frac{1}{n}\sum_{m=1}^n X_m \geq \frac{1}{n}\sum_{m=1}^n X_m(a) \to E(X_m \wedge a) < \infty, \quad \text{as } n \to \infty,$$

almost surely. Now let $a \to \infty$ and use monotone convergence.

Chapter 3

Exercises

3.1(d) By symmetry, the probability that the walk visits 0 at all before returning to $j \neq 0$ is also θ. Therefore, the number of visits to j on an excursion from 0, denoted by V, has distribution $P(V=0) = 1-\theta$, and $P(V=k) = \theta(1-\theta)^{k-1}\theta$, $k \geq 1$. It is easy to calculate $P(V \geq k) = \theta(1-\theta)^{k-1}$, $k \geq 1$, and $EV = \sum_{k=1}^{\infty} P(V \geq k) = 1$.

3.2(d) Because of the symmetry between plates (mutually) and moves to and from the tureen, the question of interest can be answered by considering the chain with two states which are: $0 \equiv$ tureen; $1 \equiv$ not the tureen. For this chain, $p_{00} = 0$, $p_{01} = 1$, $p_{10} = K^{-1}$, and $p_{11} = 1 - K^{-1}$. Hence, from (3.2.13), the probability of being at the tureen after n steps is $1/(K+1)\{1-(-(1/K))^{n-1}\}$.

3.3(b) Since X is irreducible, for any state s there is a finite sequence $i_1, \ldots, i_{n(s)}$ such that $p(s, i_1) \ldots p(i_{n(s)}, k) > p_s > 0$. Let $0 < p < \min_s p_s$, and $m = \max[n(s) + 1]$. Then in any sequence of m steps the probability of not visiting k is less than $1-p$. So the probability of not hitting k from j in mr steps is less than $(1-p)^r$. Both results now follow.

3.4(b) For $m \geq 1$, $Y_m = X(T_{jA}^{(m+1)})$, where $T_{jA}^{(m+1)}$ is the time of the $(m+1)$th visit of X to A. This is easily seen to be a stopping time for X, so by the strong Markov property at these times we see that Y is a Markov chain, where

$$P(Y_{m+1} = k \mid Y_m = j) = P(X \text{ next hits } A \text{ at } k,$$
$$\text{starting from } j \in A)$$
$$= a_{jk}(\text{say}),$$

where the a_{jk} are obtained by finding the minimal (positive) solution of $a_{jk} = p_{jk} + \sum_{i \notin A} p_{ji} a_{ik}$, $j, k \in S$.

(c) Let $R_j(k)$ be the time of first return to j after a visit to k. We have

$$\min\{n > 0 : X_n = j \mid X_0 = j\}$$
$$\leq \min\{n > 0 : X_n = j \text{ after a visit to } k \mid X_0 = j\}$$

so $ET_j \leq ER_j(k) = ET_{jk} + ET_{kj}$, applying the strong Markov property at T_{jk}.

3.5(a) $\qquad$ P(j occurs infinitely often $\mid X_0 = j$)

$$= P\Big(j \text{ occurs i.o.} \bigcap T_{jk} < \infty \mid X_0 = j\Big)$$
$$+ P\Big(j \text{ occurs i.o.} \bigcap T_{jk} = \infty \mid X_0 = j\Big).$$

The first term on the right is 0 because $k \not\to j$, and the second term on the right is less than $P(T_{jk} = \infty \mid X_0 = j) < 1$, since $j \to k$. Hence j is transient.

(b) By Exercise (a), we must have $k \to j$. Now use Theorem (1).

(c) Let θ_{jk} be the probability of visiting k before returning to j, when the chain starts at j; θ_{kj} is defined likewise, and both are nonzero because the chain is irreducible. Let V_j^k be the number of visits to j before returning to k. If $\theta_{jk} = 1$, then $EV_j^k = \theta_{kj}$. Otherwise, by the strong Markov property, $P(V_j^k = r) = \theta_{kj}(1 - \theta_{jk})^{r-1}\theta_{jk}$, and the result follows.

3.6(b) In general $x_k = (1 - K)(p/q)^k + K$. Minimality entails $K = 0$

3.7(b) We seek a solution to $\pi_j(1 - (j/m)) = ((j+1)/m)\pi_{j+1}$, i.e.

$$\frac{\pi_j}{\pi_0} = \frac{(m-j-1)(m-j)\cdots m}{j(j-1)\cdots 1} = \binom{m}{j}.$$

Imposing the requirement that $\sum_j \pi_j = 1$ gives the result.

3.8(b) Let X have transition probabilities $p_{01} = 1$, $p_{10} = a$, $p_{11} = 1-a$, with state space $\{0, 1\} = S$. Provided $0 < a < 1$, this is reversible and ergodic, with $\pi_0 = a/(1 + a)$, $\pi_1 = 1/(1 + a)$. But if the iteration stops at T, one of the two simulations must have been in state 0 at time $T - 1$ and the other in state 1. (For otherwise they would have coupled before T.) But from 0, the chain goes to 1 with probability 1, so $P(W = 1) = 1 \neq \pi_1$.

3.9(b) There are two overlaps, T and THT. So by the argument of (2), the answer is $m_H + \mu_{THT} + \mu_{THTHT} = 2 + 8 + 32 = 42$.

Problems

1.(a) Easy to check that there is a stationary distribution uniform on S.

(b) Likewise, the invariant measure is constant.

3. $P(Y_{n+1} = (k, \ell) \mid Y_n = (i, j)) = \begin{cases} p_{k\ell}, & \text{if } j = k, \\ 0, & \text{if } j \neq k. \end{cases}$

The stationary distribution of Y is $(\pi_j p_{jk}; j, k \in S)$.

4. Conditioning on the first step, and using the strong Markov property as usual, shows that e_0 is a root of $e_0 = G(e_0)$. By (3.3.12) it is the smallest root. But this is now equivalent to the result of Exercise 2.7. For the example given, $e_0 = 1$ if $p \leq \frac{1}{3}$. If $p > \frac{1}{3}$ then $e_0 = \{q + (4q - 3q^2)^{1/2}\}/(2p)$.

7. By symmetry, the problem reduces to a type of simple random walk on $(0, 1, \ldots, \ell)$. Elementary calculations give the answers (a) $2\ell^2 - \ell$, (b) $2\ell^2 + \ell$.

10. The chain is reversible in equilibrium. Use the detailed balance equations to give $\pi_j = \binom{m}{j}^2 / \binom{2m}{m}$.

11. Proceed as in (10) to find that $\pi_j = c (\beta/\alpha)^j \binom{m}{j}^2$, where c is chosen so that $\sum \pi_j = 1$.

12. The two states $\{3, 4\}$ are a closed irreducible subchain with unique stationary distribution $\{\frac{1}{2}, \frac{1}{2}\}$, so if X arrives in $\{3, 4\}$ the only possible value for c is $\frac{1}{2}$. This requires that the initial distribution in $\{3, 4\}$ should also be $\frac{1}{2}$, or what is in effect the same, the parity of the arrival time in $\{3, 4\}$ must be equally likely to be even or odd. Considering the first two steps of X shows that this happens if and only if $p_{13}(1) = p_{13}(2)$, which is to say $1 - a = ab$, so that $a = (1 + b)^{-1}$, and then $c = \frac{1}{2}$ if $b > 0$. If $b = 0$, so that $a = 1$, then the chain never enters $\{3, 4\}$ and $c = 0$.

13. Use Problem 12.

Chapter 4

Exercises

4.1(b)

(i) Verify the first part by substitution, and the deduction follows by induction.

(ii) Integrate by parts, and use (1.8.5) to write the mgf of J_k as the desired product.

(iii) Use the uniqueness theorem (1.8.4).

(c) By inspection of (4.1.26), you see that, in general (any λ), $X(t)$ has the Geometric ($e^{-\lambda t}$) distribution, when the initial population is $X(0) = 1$. If $X(0) = j$, then the population at t is the sum of j independent Yule processes having the Geometric ($e^{-\lambda t}$) distribution. By Problem (1.12), the required distribution is the negative binomial distribution as displayed.

4.2(a) In general, when $X(0) = 0$, $p'_{0n}(t) = -\lambda p_{0n}(t) + \lambda p_{0,n-1}(t)$, $n \geq 1$; and in particular $p'_{00}(t) = -\lambda p_{00}(t)$. Solve these iteratively, or define $G = \sum_{n=0}^{\infty} p_{0n}(t) z^n$, and discover that $G(z, t) = \exp\{\lambda t(z - 1)\}$. Hence, $p_{0n}(t) = (\lambda t)^n e^{-\lambda t}/n!$ and, likewise, even more generally, $p_{jk}(t) = (\lambda t)^{k-j} e^{-\lambda t}/(k - j)!$, $k \geq j \geq 0$.

4.2(e) By (1.8.13), Y has mgf $G_N(M_X(\theta)) = p\lambda/(\lambda - \theta - (1-p)\lambda) = \lambda p/(\lambda p - \theta)$. Therefore, Y is Exponential (λp).

4.3(c) Suppose that $q_j < b < \infty$ for all j, so that $\sup q_j < b$. For any holding time U in state j, we know from Exercise 1.3(f) that $V = q_j U$ is Exponential (1). Hence, with an obvious notation,

$$bJ_\infty = \begin{cases} \infty, & \text{if any } q_j = 0, \\ \sum_{n=0}^{\infty} bU_n \geq \sum_{n=0}^{\infty} q_{Z_n} U_n = \sum_{n=0}^{\infty} V_n, & \text{if } q_j \neq 0 \text{ for any } j. \end{cases}$$

Here, each V_n is Exponential (1), so by Exercise 2.1(b), $J_\infty = \infty$ and there is no explosion.

Obviously, $\sup q_j < b < \infty$ if the state space is finite, so we turn to the final condition. Again, if $q_j = 0$, the result is trivially true. If $q_j > 0$ and j is persistent, then $X(t)$ revisits j infinitely often with probability 1, and is held there each time for an Exponential (q_j) interval. The result follows by using Exercise 2.1(b) again.

4.4(a)

(i) If $X(t)$ spends only a finite amount of time visiting j, then $\sum_n P(X(nh) = j) < \infty$, and j is transient for X_n.

(ii) The event that $X_n = j$ includes the possibility that the process $X(t)$ arrives in j during $((n-1)h, nh)$ and remains there until time nh. Hence,

$$P(X_n = j \mid X_0 = j) \geq p_{jj}(t) e^{-q_j h}, \quad (n-1)h \leq t \leq nh.$$

Hence, since all terms are positive,

$$\int_0^\infty p_{jj}(t)\, dt \leq h \exp(hq_j) \sum_{n=1}^\infty P(X_n = j \mid X_0 = j),$$

and the result follows.

4.5(b) The distribution of V is a mixture of an exponential density and an atom of probability at zero. To be specific, let θ be the probability that the walk visits r before returning to 0, and ϕ the probability that the walk, started at r, revisits r ever without a visit to 0. Then the number of visits to r before any visit to 0 is geometric with parameter $(1-\phi)$, and by Exercise 4.2(e) the total time spent in r is Exponential $\{(\lambda+\mu)(1-\phi)\}$, because each holding time is Exponential $(\lambda+\mu)$. This outcome occurs at all with probability θ; otherwise $P(V=0) = 1-\theta$.

Elementary calculations from Section (2.2) yield

$$\theta = \frac{q-p}{(q/p)^r - 1} + p \wedge q$$

and

$$\phi = q\{(q/p)^{r-1} - 1\}/\{(q/p)^r - 1\} + p \wedge q.$$

where $p = \lambda/(\lambda+\mu)$, $q = \mu/(\lambda+\mu)$, $p \wedge q = \min\{p,q\}$, and $p \neq q$. The case $p = q = \frac{1}{2}$ is left for you.

4.6(a) We find the stationary distribution of $X(t)$ by solving $\pi Q = 0$, which yields the set of equations

$$(j+1)\mu \pi_{j+1} - j(\lambda+\mu)\pi_j + (j-1)\lambda \pi_{j-1} = 0, \quad j \geq 2,$$

with $2\mu\pi_2 - (\lambda + \mu)\pi_1 + \pi_0 = 0 = \pi_1\mu - \lambda\pi_0$, and $\Sigma\pi_j = 1$. It is easy to show that

$$\pi_j = \frac{\pi_0 \lambda^j}{j\mu^j}, \quad j \geq 1,$$

which yields a stationary distribution if $\sum_{j=1}^{\infty} (\lambda/\mu)^j (1/j) < \infty$, that is, if $\lambda < \mu$.

The jump chain is simply a random walk reflected at the origin, and it is easy to find that when $\lambda < \mu$ there is a stationary distribution of the form

$$\hat{\pi}_j = \frac{\hat{\pi}_0 (\lambda + \mu)\lambda^{j-1}}{\mu^j}, \quad j \geq 1$$

whence

$$\hat{\pi}_j = \frac{\mu^2 - \lambda^2}{2\mu^2} \left(\frac{\lambda}{\mu}\right)^{j-1}, \quad j \geq 1, \qquad \hat{\pi}_0 = \frac{\mu - \lambda}{2\mu}.$$

The difference arises naturally because the holding times of $X(t)$ depend on the state of the process. From (4.6.6) we know

$$\hat{\pi}_j \propto \pi_j q_j = \left(\frac{\lambda}{\mu}\right)^j (\lambda + \mu)\pi_0, \quad j \geq 1,$$

and

$$\hat{\pi}_0 \propto \pi_0 q_0 = \pi_0 \lambda.$$

Now we set $\Sigma\hat{\pi}_j = 1$, yielding $\hat{\pi}_0 = (\mu - \lambda)/2\mu$ as above. Incidentally, $\pi_0^{-1} = 1 - \log(1 - (\lambda/\mu))$.

4.7(a) The equations $\pi Q = 0$ take the form

$$\pi_{j+1}\mu_{j+1} - \pi_j \lambda_j = \pi_j \mu_j - \pi_{j-1}\lambda_{j-1}, \quad j \geq 1$$

with $\pi_1 \mu_1 = \pi_0 \lambda_0$. Therefore, $\pi_j \mu_j = \pi_{j-1}\lambda_{j-1}$, and π exists, so $X(t)$ is reversible.

(c) If $q_{01} = \alpha$ and $q_{10} = \beta$, it is easy to check that $\pi_0 = \beta/(\alpha + \beta)$ and $\pi_1 = \alpha/(\alpha + \beta)$. So $\pi_0 q_{01} = \pi_1 q_{10}$, and the result follows.

(d) Because the detailed balance equations hold for X and Y, and these are independent, it is straightforward to see that the detailed balance equations for (X, Y) are satisfied. In an obvious notation

$$\pi_X(i)\pi_Y(m)q_X(i,j)q_Y(m,n) = \pi_X(j)\pi_Y(n)q_X(j,i)q_Y(n,m).$$

4.8(a) Let W_1 and W_2 be the respective waits, and let X_1 and X_2 be the respective queue lengths. Then

$$P(W_1 = 0, W_2 = 0)$$
$$> P(\text{nobody in either queue when the customer arrives})$$
$$= P(X_1 = 0, X_2 = 0) = P(X_1 = 0)P(X_2 = 0)$$
$$= P(W_1 = 0)P(W_2 = 0).$$

(c) By the independence proved in (4.8.1) we need only consider one typical M/M/1 queue. If the length of the queue is X, the total time spent in it by an arriving customer if $\sum_0^X S_r$, where $(S_r; r \geq 0)$ are independent Exponential (μ). By (4.6.13) and Exercise 4.2(e) the result follows.

4.9(a) From (4.9.4), differentiating twice and setting $s = 1$, we have

$$s(t) = G_{tss}|_{s=1} = \frac{d}{dt}E(X(t)^2 - X(t))$$
$$= 2(\lambda - \mu)G_{ss}|_{s=1} + 2(\lambda + \nu)G_s|_{s=1}$$
$$= 2(\lambda - \mu)s(t) + 2(\lambda + \nu)m(t).$$

(b) First we solve (4.9.4) with $\nu = 0$, and λ, μ both constant, to find that, for $G(z, 0) = z$,

$$G = 1 + \frac{(z-1)(\lambda - \mu)e^{(\lambda - \mu)t}}{\lambda z - \mu - \lambda(z-1)e^{(\lambda - \mu)t}}, \quad \lambda \neq \mu.$$

Now do the integral in (4.9.12) to obtain the given solution.
If $\lambda = \mu$, then the corresponding solution to (4.9.4) with $\nu = 0$ is

$$G = 1 + \frac{z-1}{\lambda t(1-z) + 1}.$$

Hence the required result is

$$\exp\left(\nu \int_0^t \frac{z-1}{\lambda s(1-z) + 1} ds\right) = \{1 + \lambda t(z-1)\}^{-\nu/\lambda}.$$

Hints and solutions 313

Problems

1. This follows by using the binomial theorem.

3. $p_{00}(t) = p_{11}(t) = P(N(t) \text{ is even}) = \frac{1}{2}(1 + e^{-2\lambda t})$; $p_{01}(t) = 1 - p_{00}(t)$, etc. Substitution verifies the claim.

4. Using reversibility $\pi_{n+1}\mu = \pi_n \lambda b_n$; whence $\pi_{n+1} = (\lambda/\mu)^n \prod_{r=0}^{n} b_r \pi_0$. The stationary distribution exists if $\sum_n (\lambda/\mu)^n \prod_{r=0}^{n} b_r < \infty$, which certainly holds if any $b_n = 0$. It also holds if, as $n \to \infty$, $(\lambda/\mu)b_n \to a < 1$.

5. Either solve $\partial G/\partial t = \mu(1-s)(\partial G/\partial s)$, or use (4.9.7) with μ constant; or use the fact that lifetimes are independent Exponential (μ), to give

$$p_{jk}(t) = \binom{j}{k}(1 - e^{-\mu t})^k (e^{-\mu t})^{j-k}, \quad 0 \le k \le j.$$

6. Either use the result of Exercise 4.9(b) (with $\mu = 0$), or use (4.9.6) to find that the mean of a Yule process is $e^{\lambda t}$, and hence, by conditioning on the time of the first birth, obtain an integral equation with the given solution. Or, by conditioning on events in $[0, h]$, obtain the backward equation $(d/dt)EX(t) = \nu e^{\lambda t}$.

8(a) With the convention that $p_{0,-1}(t) = 0$, the forward equations are

$$\frac{d}{dt}p_{0n}(t) = -\lambda(t)p_{0n}(t) + \lambda(t)p_{0,n-1}(t), \quad n \ge 0.$$

Defining $G = Es^{N(t)}$, you obtain $\partial G/\partial t = -\lambda(t)(1-s)G$.

(b) Let $L(t) = \int_0^t \lambda(u)du$. We follow the same argument as that used in Theorem (2.5.15). Conditional on the event that $N(t) = n$, the probability that these events occur in the intervals $(t_1, t_1+dt_1), \ldots, (t_n, t_n+dt_n)$, where $0 \le t_1 \le t_1 + dt_1 \le t_2 \le \cdots \le t_n + dt_n \le t$, is

$$(e^{-L(t_1)}\lambda(t_1)dt_1 \, e^{-[L(t_2)-L(t_1)]}\lambda(t_2)dt_2 \ldots e^{-[L(t)-L(t_n)]} + \cdots)$$

$$\times \frac{n!}{[L(t)]^n e^{-L(t)}} = n![L(t)]^{-n} \prod_{r=1}^{n} (\lambda(t_r)dt_r) + \cdots,$$

where we have neglected higher-order terms in dt_r, $1 \le r \le n$.

Now by essentially the same argument as that used in Problem (1.14), this is seen to give the joint density of the order statistics of n independent random variables with density $\lambda(x)/L(t)$, $0 \le x \le t$.

(c) Now the same argument as in (4.9.10) gives

$$Ez^{X(t)} = E(Ez^{X(t)} \mid N(t))$$

$$= G_N \left\{ \frac{1}{L(t)} \int_0^t G(z, t-u)\lambda(u)du \right\}$$

$$= \exp\left\{ \int_0^t (G(z, t-u) - 1)\lambda(u)du \right\}$$

on recalling that $N(t)$ is Poisson $(L(t))$.

(d) Let the time of the most recent disaster before t be $C(t)$. Suppose the disasters occur at rate $d(t)$, and $N(t)$ has rate $\lambda(t)$. Let $L(t) = \int_0^t \lambda(u)du$, and $D(t) = \int_0^t d(u)du$. $C(t) \geq x$ if and only if there are no disasters in $(t-x, t)$ which has probability $e^{-[D(t)-D(t-x)]}$. Hence $C(t)$ has density $d(t-x)e^{-[D(t)-D(t-x)]}, 0 \leq x < t$, and $P[C(t) = t] = e^{-D(t)}$. Hence the expected population at t is, conditioning on $C(t)$,

$$EX(t) = E\{L(C(t))\}$$

$$= L(t)e^{-D(t)} + \int_0^t L(x)d(t-x)e^{-D(t)+D(t-x)}dx.$$

10. **Yule process.** If the first split is at $u > t$, with probability $e^{\lambda t}$, then $G = z$. If the first split is at $u < t$, then $X(t) = X_1(t-u) + X_2(t-u)$ where $X_1(t)$ and $X_2(t)$ are independent, having the same distribution as $X(t)$. The equation follows by conditional expectation. To solve it, multiply by $e^{\lambda t}$, and change the variable in the integral to give $e^{\lambda t}G = z + \int_0^t \lambda e^{\lambda u} G^2 du$. Differentiating for t yields $\partial G/\partial t = \lambda G(G-1)$, which integrates as

$$\lambda t = \int \lambda dt = \int \frac{dG}{G(G-1)} = \int \frac{dG}{G-1} - \int \frac{dG}{G} = \log \frac{G-1}{G} + \text{constant}.$$

Inserting the initial condition $G(z, 0) = z$ yields $(G-1)/G = e^{\lambda t}(z-1)/z$, whence

$$G = \frac{ze^{-\lambda t}}{(1 - z + ze^{-\lambda t})}.$$

Finally, recall the result of Exercise 4.1(c), and write $e^{-\lambda t} = p(t)$, to give

$$P(X(s) = j \mid X(t) = n) = P(X(t) = n \mid X(s) = j)P(X(s) = j)/P(X(t) = n)$$

$$= \binom{n-1}{j-1} p(t-s)^j (1 - p(t-s))^{n-j} (1 - p(s))^{j-1} p(s) \times$$

$$\times \{(1 - p(t))^{n-1} p(t)\}^{-1}$$

$$= \binom{n-1}{j-1} \{\frac{1 - p(t-s)}{1 - p(t)}\}^{n-j} \{\frac{p(t-s) - p(t)}{1 - p(t)}\}^{j-1},$$

which is Binomial $(n-1, (1 - p(t-s))/(1 - p(t)))$.

11. **M/M/1.** If the number completing service is Z, then

$$P(Z = r) = \frac{\lambda}{\lambda + \mu}\left(\frac{\mu}{\lambda + \mu}\right)^r, \quad 0 \le r \le k-1$$

and

$$P(Z = k) = \left(\frac{\mu}{\lambda + \mu}\right)^k.$$

13. **Busy period.** Let customers arriving during the service time of any customer C_i be called scions of C_i. The number of scions of different customers are independent and identically distributed because arrivals are a Poisson process. The busy period is the service time of C plus all the busy periods that arise from successive servicing of his scions. But this is the required equation.

Conditional on S, Z is Poisson (λS), and

$$E\left\{\exp\left(\theta \sum_{r=1}^{Z} B_r \Big| S\right)\right\} = \exp\{\lambda S\{M(\theta) - 1\}\}.$$

Therefore, as S is Exponential (μ), using conditional expectation

$$M(\theta) = E\left(\exp\left\{\theta\left(S + \sum_{1}^{Z} B_r\right)\right\}\right)$$

$$= \frac{\mu}{\mu - (\theta + \lambda M(\theta) - \lambda)}.$$

Solving $\lambda M^2 - (\lambda + \mu - \theta)M + \mu = 0$ gives the result. Finally,

$$EB = \frac{1}{\mu - \lambda}, \quad \text{var } B = \frac{\mu + \lambda}{(\mu - \lambda)^3}.$$

17. This is a Markov process, and $X(t)$ is easily seen to be the immigration–death process of Section (4.9). One approach is to solve $\partial G/\partial t = \mu(1-s)\partial G/\partial s + \lambda(s-1)G$ to give the pgf of $X(t)$ as

$$G = \{1 + (s-1)e^{-\mu t}\}^{X(0)} \exp\left\{\frac{\lambda}{\mu}(s-1)(1 - e^{-\mu t})\right\}.$$

Thus, $X(t)$ has a distribution which is that of a sum of a Binomial and a Poisson random variable. As $t \to \infty$, the limit is Poisson (λ/μ) with mean $\lambda/\mu = \lambda ES$, where S is a service time.

Alternatively, use the method of (4.9.10). If λ is a function of t, then integrating the equation for G is tedious. It is easier to use (4.9.10) and the result of Problem (8). If $X(0) = 0$, then $X(t)$ is Poisson with parameter $\int_0^t \lambda(u)\{1 - e^{-\mu(t-u)}\}du$. One can investigate this as $t \to \infty$, but it is easier to use Little's theorem in which $m = L = \lambda w = c/\mu$, so in the limit the number of busy servers is Poisson (c/μ).

18. **M/G/∞.** Now $X(t)$ is not a Markov process. But we can still use the alternative method, conditioning on the Poisson process of arrivals. In the first case $X(t)$ is Poisson with parameter

$$\int_0^t \lambda\{1 - F(t-u)\}\,du = \int_0^t \lambda(1 - F(u))\,du \to \lambda ES, \quad \text{as } t \to \infty.$$

In the non-homogeneous case $X(t)$ is Poisson with parameter $\int_0^t \lambda(u) \times (1 - F(t-u))\,du$. Using Little's theorem as above the limit m is found to be cES.

Chapter 5

Exercises

5.1(b)

(i) $-W(0) = 0$, $-W(t)$ is continuous; the increments remain independent, and normally distributed because $N(0, t)$ is an even function.

(ii) The process starts at 0, and is continuous. The increment $W(a + t + s) - W(a + s)$ is still $N(0, t)$, and increments over disjoint intervals are independent because those of $W(t)$ are.

(c) It is easy to check that the proposed density satisfies the forward equations; and further that it satisfies the required boundary condition at a, because $f(t, a) - f(t, -a) = \phi_t(a) - \phi_t(-a) = 0$, because the Normal density $\phi_t(a)$ is an even function. For another way of deriving this result, see the passage on the reflection principle.

5.2(a) Clearly $\hat{B}(t)$ has multinormal joint distributions. Set $t = 1 - (1/u)$, and use the fact that $\lim_{u\to\infty} W(u)/u = 0$, to find that $\hat{B}(t)$ is continuous in $[0, 1]$. Finally, calculate $\text{cov}(\hat{B}(s), \hat{B}(t)) = (1-s)(1-t)\min\{s/(1-s), t/(1-t)\} = s \wedge t - st$, as required.

(d) Conditioning on $U(t)$ is effectively conditioning on $W(e^{2t})$. Hence, using the properties of the Wiener process we have

$$E(U(t+h) - U(t) \mid U(t)) \qquad (*)$$
$$= E(e^{-(t+h)}\{W(e^{2(t+h)}) - W(e^{2t})\} + (e^{-(t+h)} - e^{-t})$$
$$\times W(e^{2t}) \mid W(t))$$
$$= -h e^{-t} W(e^{2t}) + o(h),$$

as required. Similarly, evaluating the square of the first argument in $(*)$ conditional on $W(t)$ yields $e^{-2(t+h)}(e^{2(t+h)} - e^{2t}) + o(h) = 2h + o(h)$, as required.

5.3(a) Simply change the variable in the $N(0, m^{-2})$ density with the substitution $t = x^{-2}$.

(b) (i) Follows immediately from (5.3.12) by conditional probability, when we notice that the event {no zeros in (s, u)} is included in the event {no zeros in (s, t)}.

(ii) Let $s \to 0$ in part (i); and then relabel t and u. (You need to remember that $\sin \sqrt{s/t} = \sqrt{s/t} + o(\sqrt{s/t})$, as $s \to 0$.)

(c) Using 5.3(a), above, $P(T_m < \infty) = P(X^2 \neq 0) = 1$.

5.4(a) We need to solve (5.4.29), which reduces to $p''(x) = 0$, giving the required answer. Since we divided by $b(x)$, the method may fail if $b(x) = 0$ for any x in (β, α), or if $b(x)$ is unbounded in (β, α).

5.5(a) Consider $W(t)$ over unit time intervals $(n, n+1)$, $n \geq 0$. Let X be the first n such that $|W(n+1) - W(n)| > |a| + |b|$. Then, X is a geometric random variable with finite mean, and $T < X$.

(b) By conditional probability, $f_V(v) = \int_0^\infty \phi_t(v) f_T(t) dt$ where $f_T(t)$ is the first-passage density of (5.3.7), and $\phi_t(v)$ is the standard normal $N(0, t)$ density. The integral is evaluated easily; set $t = y^{-1}$.

(c) The first equality is immediate because $V(T)$ is normally distributed, given T; the second equality follows using (5.5.22).

5.6(a) Without loss of generality let $U(0) = 0$. Then $EU(t) = 0$, and after extensive integration you may calculate the covariance

$$\text{cov}(U(t), U(t+s)) \propto e^{-s} - e^{-(2t+s)}.$$

Now the integral in (7) is the limit of a sum of independent normally distributed random variables, and it follows that $U(t)$ has multinormal joint distributions. Finally, a glance at (5.2.16) and the following remarks shows that (7) defines the same process as the Ornstein–Uhlenbeck process:

$$U_1(t) = e^{-\beta t} W\left\{\frac{1}{2\beta}(e^{2\beta t} - 1)\right\}.$$

(Remember that multivariate normal distributions are determined by their first and second moments.)

5.7(a)

$$\text{LHS} = E\{\tfrac{1}{4}(W(t)^2 - t)^2\} = \tfrac{1}{4}\{3t^2 - 2t^2 + t^2\} = \tfrac{1}{2}t^2$$

$$\text{RHS} = \int_0^t EW(u)^2 du = \int_0^t u\, du = \frac{1}{2}t^2.$$

(b) $EX(t) = 0$. Assume $s < t$, and then

$$\text{cov}(X(s), X(t))$$

$$= E\left\{(1-s)\int_0^s \frac{1}{1-u}dW(u)\left((1-t)\int_0^s \frac{1}{1-u}dW(u)\right.\right.$$

$$\left.\left.+(1-t)\int_s^t \frac{1}{(1-u)}dW(u)\right)\right\}$$

$$= (1-s)(1-t)\int_0^s \frac{1}{(1-u)^2}du, \quad \text{by (5.7.23)}$$

$$= s(1-t).$$

It is clear that $X(t)$ has multivariate normal joint distributions, and is continuous, so the result follows.

5.8(c) Apply the product rule (18) to $d(U(t)e^{\beta t})$ and to $d(W(t)e^{\beta t})$.

(b) Solve Laplace's equation in plane polar coordinates with the appropriate boundary conditions, to find that, for $\epsilon < r < b$ and $W(0) = (r, \theta)$, we have $p(r) = \log(r/b)/\log(\epsilon/b) \to 1$, as $b \to \infty$. We deduce that the Wiener process in the plane is recurrent, in the sense that with probability 1 it approaches arbitrarily close to any point.

(c) Solve Laplace's equation using spherical polar coordinates to yield, for $\epsilon < r < b$,

$$p(r) = ((1/r) - (1/b))/((1/\epsilon) - (1/b)) \to \epsilon/r,$$

as $b \to \infty$. Since this is less than 1, we deduce that the Wiener process in three dimensions is transient.

(d) For the ellipse, when the process starts inside A,

$$m(x, y) = \left(1 - \frac{x^2}{a^2} - \frac{y^2}{b^2}\right)\left(\frac{1}{a^2} + \frac{1}{b^2}\right)^{-1}.$$

The expected hitting time from a point outside A is infinite. The parabola and the hyperbola also divide the plane (in general) into a region in which the expected hitting time is given by the obvious quadratic form, and a disjoint region in which it is infinite. A glance at (5.9.18) tells you which is which.

Problems

1. Consider the three-dimensional problem. Using the independence of increments and Pythagoras' theorem, the result follows if the distribution of $S_3(t + s)$, conditional on $S_3(s)$, depends only on $S_3(s) = \sum_{i=1}^{3} W_i^2(s)$. Since the W_i are normally distributed, this is equivalent to the claim that the sum of the squares of independent normal random variables $N(\lambda, 1)$, $N(\mu, 1)$, $N(\nu, 1)$ depends on λ, μ, and ν, only through $\lambda^2 + \mu^2 + \nu^2$. Moving the origin, this is equivalent to the assertion that the integral of $\exp\{-\frac{1}{2}(x^2 + y^2 + z^2)\}$ over a sphere centred at $(0, 0, a)$, depends on a only through a^2.

Taking spherical polar coordinates centred at a, with the same axis of symmetry $(0, 0, z)$, this integral is (using the cosine formula)

$$((*)) \qquad 2\pi \int_0^c \int_0^\pi e^{-(1/2)(a^2 + 2ar\cos\theta + r^2)} r^2 \sin\theta \, d\theta \, dr.$$

Now

$$\int_0^\pi e^{-ar\cos\theta} \sin\theta \, d\theta = \frac{1}{a}(e^{-ar} - e^{+ar}) = \sum_{k=0}^\infty c_{2k} a^{2k},$$

where c_{2k} are constants. The integral (∗) thus depends on a only through a^2, and we are done. Likewise a similar integral in the plane verifies that S_2 is Markov.

2. If the drift is d, you need to integrate $f_{xx}(x) = 2df_x(x)$, with the conditions $f_x(0) = 2df(0)$ and $f_x(a) = 2df(a)$.

3. The equation to be solved is of the form $d^2 f/dx^2 = -2c(d/dx)(xf)$, with the conditions $f_x(0) = 0$ and $f_x(a) = -2acf(a)$.

4. When $\mu = 0$, $r(x) = (x-b)/(x-a)$.

8. From (5.5.5), for any real λ

$$e^{\lambda W - (1/2)\lambda^2 t} = e^{\lambda D - (1/2)\lambda^3 t - \lambda \mu t}$$

is a martingale. Set $\lambda = -2\mu$ to get $M(t)$. Now use the optional stopping theorem. For the last part,

$$r(0) = \frac{e^2 - 1}{e^2 - e^{-2}} = \frac{e^2}{1 + e^2}.$$

9. (a) Because $X(t) = \cos W(t)$ and $Y(t) = \sin W(t)$, we use Itô's formula to get the result.

   (b) Consider using spherical polar coordinates; there is no r-dependence, and by symmetry there is no ϕ-dependence. The only terms surviving from (5.9.15) give this equation for the expected hitting time $m(\theta)$:

$$\sin\theta m''(\theta) + \cos\theta m'(\theta) + 2\sin\theta = 0.$$

   A routine integration gives the result,

$$m(\theta) = 4\left(\log\left|\cos\frac{\theta}{2}\right| + \log\sqrt{2}\right).$$

10. Using (5.9.6) with $f(t, W) = f(t)W$, we see $f_t = f'(t)W$, $f_w = f(t)$, $f_{ww} = 0$. Hence,

$$dY = d(f(t)W(t)) = f'(t)W(t)dt + f(t)dW$$

and the first result follows. The rest of the question follows using linearity and the normal distribution of the independent increments of $W(t)$.

11. One neat approach uses (5.9.8)–(5.9.10) with $Y = e^{-\beta t}X(t)$ and $dX = e^{\beta t}dW(t)$.

12. If $dX = \mu dt + \sigma dW$, and $Y = X^n$, then Itô tells us that

$$dY = d(X^n) = \left(\mu n X^{n-1} + \tfrac{1}{2}\sigma^2 n(n-1) X^{n-1}\right) dt + \sigma n X^{n-1} dW.$$

Integrating and taking expectations then gives the result.

13. See Problem 12.

18. Draw essentially the same picture as Figure 5.1. Now it is clear that

$$P\left(W(t) \leq c - x \bigcap M(t) \geq c\right) = P\left(W \leq c - x \bigcap T_c \leq t\right)$$
$$= P\left(2c - W \leq c - x \bigcap T_c \leq t\right), \quad \text{by the reflection principle}$$
$$= P\left(W \geq c + x \bigcap T_c \leq t\right)$$
$$= P(W(t) \geq c + x), \quad \text{because this event is in } \{T_c \leq t\}.$$

Set $c = m$, and $m - x = w$, and differentiate $1 - \Phi_t(2m - w)$ to get the final expression.

Further reading

(**a**) For anyone with an interest in probability and random processes, the following book is essential reading:

Feller, W. (1968). *An Introduction to Probability and its Applications*, Vol. 1 (3rd edition). Wiley, New York.

Feller defined the nomenclature, notation, and style, for large parts of probability and random processes. All books on the subject since then acknowledge this fact, either overtly or implicitly.

(**b**) The following books contain much material that is useful at our level:

Cox, D. R. and Miller, H. D. (1965). *The Theory of Stochastic Processes*. Chapman and Hall, London.

Grimmett, G. R. and Stirzaker, D. R. (2001). *Probability and Random Processes* (3rd edition). Oxford University Press, Oxford.

—— and —— (2001). *One Thousand Exercises in Probability*. Oxford University Press, Oxford.

Karlin, S. and Taylor, H. M. (1975). *A First Course in Stochastic Processes* (2nd edition). Academic Press, New York.

Moran, P. A. P. (1968). *An Introduction to Probability Theory*. Oxford University Press, Oxford.

Ross, S. (1997). *Introduction to Probability Models* (6th edition). Academic Press, New York.

(**c**) For some reading that goes more deeply into topics that we have merely sketched, you can explore the harder sections of the books cited above, and also these:

Feller, W. (1971). *An Introduction to Probability Theory and its Applications*, Vol. 2 (2nd edition). Wiley, New York.

Karlin, S. and Taylor, H. M. (1975). *A Second Course in Stochastic Processes*. Academic Press, New York.

Norris, J. R. (1997). *Markov Chains*. Cambridge University Press, Cambridge.

Ross, S. (1996). *Stochastic Processes* (2nd edition). Wiley, New York.

Williams, D. (1991). *Probability with Martingales*. Cambridge University Press, Cambridge.

Index

Note: Page numbers in *italics* refer to definitions of terms.

absorbing barriers 226
absorbing states, Markov chains *118*
acceptance matrix *152*
accessibility, Markov chains *117*, 191, 192
age (renewal processes) *85*
algebra of events *see* event space
almost sure convergence 58, 59
aperiodic state, Markov chains *118*
approximation theorem 266
arcsine laws
    for maximum of Wiener process 241
    for zeros of Wiener process 238–9

backward equations 180–1
    in birth processes 187–91
    Brownian motion 222
    diffusion equations 221
    diffusion processes 225
    Yule process 183
balking 214
Barker's method 156
barriers 226–8
Bayes' rule 5
Bernoulli random variables *8*
    sum of 20
Bessel process 245, 250
    first-passage time 248
beta density 12
biased coin-flipping 160
binary system, indicator of functioning system 96
binomial distribution 7–8
    negative 12, 42
    pgf 36
binormal density 37–9
birth-death chain 195
birth-death processes 215
    non-homogeneous 210–13
    reversibility 149
birth-immigration processes 214, 216–17
birth processes 186–91, 214
birth rates *187*
Black–Scholes option-pricing theorem 245
Boole's inequality 4
Borel–Cantelli lemma, second 58
boundaries 226

branching processes 88–90, 94
    geometric branching limit 91–3
    independence of past 107
    Markov chains 120
    martingales 91
    refreshed branching 93
    total population 90–1
bridge systems, reliability 99
Brownian Bridge 233–5, 241, 267
    instantaneous parameters 242
Brownian meander 239–40
Brownian motion 219–20
    infinitesimal parameters 222–4

Campbell–Hardy theorem 104
candidate variable *152*
Cauchy density 12, 21
    characteristic function 36
Cauchy–Riemann equations 286
Cauchy–Schwartz inequality 42
censored state space 204
central limit theorem 60–1
certain event 2
chain rules 268
Chapman–Kolmogorov equations 116–17, 171
    for transition density 221
characteristic function (cf) *36*
Chebyshov inequality 57, 63
Chernov inequality 63
circles, Wiener process 293, 296
classes of Markov chains 118, 132–3
classical laws of motion 257
closed migration processes 206–7
closed states, Markov chains *118*
coaxial circles, Wiener process between 296
coin-flipping, expectations for patterns 157–61, 164
communication, in Markov chains *117*
complex plane, Laplace's equation 286
conditional density *28*
conditional expectation 27–32, *101–2*
conditional independence 5, 32
conditional mass function *27*
conditional probability *4*
conditional variance 31

conditioning of Wiener process 248–50
cone, Wiener process in 295
conic sections, Wiener process 291
continuity property of P(.) 3
continuity theorem 34, 59–60
continuous random variables 8–9
  transformations 17–19
continuous time, Markov chains 169–76
convergence 57, 58, 62–3
  central limit theorem 60–1
  laws of large numbers 60, 61
  of martingales 69–70
convergence in distribution 59
convergence in mean square 59
convergence in probability 58–9
convergence theorem 141–2
Conway magic formula 161
correlation coefficient 26
countable additivity, axiom of 3
counting processes 71–2
  see also Poisson process; renewal processes
coupling 20, 22
coupling from the past 155–6
covariance 26
  Wiener process 229
covering time 103
current life (renewal processes) 85
cut sets 98
cycles
  in Markov chains 118
  in renewal-reward processes 82

decomposition of Markov chains 133–4
delayed renewal 87, 131
DeMoivre trials 160
density function 9–10
detailed balance equations 148
diagrams, as representation of Markov chains 111–12
die rolls, Markov chains 109, 113–14, 115, 123, 160
differential operator 223
diffusion equations 221
diffusion processes 224–5
  first-passage times 245–8
  martingales 250–6
  transition density function 225
  see also Wiener process
disaster processes 125, 127, 212–13
discrete random variables 7–8
  transformations 19–21
discrete renewal 111, 116
dishonest processes 189, 270
disjoint events 2
distribution function 7
distribution of random vectors 13
DNA substitution model 213–14
dominated convergence 62

Doob's inequality 294
doubly stochastic matrices 108–9
drifting Wiener process 224, 256, 291–2, 293

Ehrenfest model 150
elementary optional stopping 67
elementary renewal theorem 80–1
ellipsoid, Wiener process in 295
equilibrium 114, 135, 137–8
  in continuous time 196–8
equivalence classes in Markov chains 118, 132–3
ergodic reward theorem 144
ergodic theorem 143, 151
  in continuous time 200
Erlang's loss formula 201–2
events 2
event space 2
exact simulations 155
excess life (renewal processes) 85
expectation 22–5
  conditional 27–8
expected duration, diffusion processes 247
expected value (mean)
  of continuous random variable 23
  of discrete random variable 22
  of random sum 30–1
expected visiting times, Wiener process 284–5
experiment 2
exploration time 103
exponential density 9, 12
exponential flip-flop circuit 170
exponential martingales 254–5
exponential mean 23, 24
exponential mgf 35
exponential random variables, lack-of-memory property 32
exponential Wiener process 233, 244–5
extinction, probability of 89, 94
extinct particles 88
extrinsic boundary points 226

factorization of joint mgf 34
Feynman-Kac theorem 272–3
filtrations 102
finite stochastic processes 47–8
first exit time 253
first-passage time
  diffusion processes 235, 241, 245–8
  Markov chains 121, 129–30, 192
first renewal theorem 80
Fisher-Wright model, neutral 250
flight, random 50
flip-flop chain 120, 144–5
  regular 119
flip-flop circuit 110–11
  exponential 170, 178–9

forward equations 178–9
  in birth processes 187–91
  Brownian motion 222
  diffusion equations 221
  diffusion processes 225
  Yule process 182
free space averaging 278
functions
  expected value 24–5
  moment generating function (mgf) 33–5
fundamental matrices 125–6

Gambler's Ruin problem 51–2, 112, 123, 245
gambling models
  martingales 64, 69
  stopping times 65
gambling systems, the martingale 70–1
Gamma density 12, 19
generator of Markov chain 178
genetic drift, Markov chain 109
geometric branching limit 91–3
geometric distribution 8
geometric mean 23, 24
geometric Wiener process 233, 244–5
Gibbs sampler 153
G matrix 178
graph, random walks on 50
  reversibility 149–50
Green function 279
  one-dimensional 291
  two-dimensional 280
  three-dimensional 279–80
Green's theorems 278–80

hard-core model 153–4
hitting density, Wiener processes 274, 279–82
  polar cap 283
  in sphere 282–3
hitting probability
  Markov chains 122
  Wiener processes 275
hitting times 122, 123–6, 128
  in continuous time 194–5
  martingales 253–4
  Wiener processes 275
holding time 172
homogeneous Markov chains 108
honest processes 189–91

immigration-birth processes 214, 216–17
impossible event 2
inclusion-exclusion identity 3
independence 5
  of components 97
  conditional 32
  future events of past 107–8
independent increments, property of 72

independent random variables 15
indicators 8
infinitesimal construction, Markov chains 184
infinite stochastic processes 47–8
inspection paradox 85–7
instantaneous mean 223, 225
instantaneous parameters, diffusion process 225
instantaneous variance 223, 225
intercommunication, in Markov chains 117
intrinsic boundaries 226
invariant measure 137, 138–40
irreducibility, Markov chains 118, 133, 135, 147, 191
isometry theorem 266
Itô's formulas 268, 269, 273
  solution of stochastic differential equations 269–71

Jensen's inequality 42
joint density, sum of normal random variables 37
joint density function 14–15, 18–19
joint distribution function 14
joint distributions
  stochastic processes 47
  Wiener process 229–31
joint mass probability function 12–13
joint mgf 33, 34
jump chain 172
jump processes 219

knight's moves 165
Kolmogorov's criterion 150, 205
Kolmogorov's theorem 48
k-out-of-n systems 98

lack-of-memory property 32
Lancaster's theorem 38
Laplace's equation and complex plane 286
Laplacian polar coordinates 276–7
last in first out queues 216
law of rare events (small numbers) 21
laws of motion, classical 257
left-continuous random walk 165
length biasing 87
line, random walks on 49–50
Little's theorem 208–9

maps, method of 286–7
marginal distributions 13–14
Markov chains 108–15
  applications 157–64
  backward equations 180–1, 183
  for calculation of sums 151–6
  in continuous time 169–76
    birth processes 186–91

Markov chains (cont.)
  convergence 199–200
  hitting and visiting 194–6
  queues 205–9
  recurrence 198–9
  recurrence and transience 191–4
  reversibility 202–5
  stationary distribution 196–8
 decomposition 133–4
 equivalence classes 118, 132–3
 flip-flop chain 144–5
 forward equations 178–9, 182
 fundamental matrices 125–6
 hitting times 122, 123–6
 infinitesimal construction 184
 non-homogeneous birth and death
   processes 210–13
 n-step probabilities 116–17
 passage times 121
 Q matrix 177–8
 and renewal processes 131
 reversible 147–50
 simulation of 151–2
 stationary distribution 135–47, 152–4
 structure 117–21
 survival chain 145–7
 uniformization 184–5
 visits 131–2
Markov inequality 57
Markov property 107–8, 109–10, 114–15
 strong 128–31
martingale, the (gambling system) 70–1
martingales 64–5, 71, 250–1
 in branching processes 91
 convergence 69–70
 inequalities 68–9
 matching 104
 optional stopping 67–8
 orthogonal increments 70
 simple systems theorem 65–6
 Wiener process 251–2
matching martingale 104
Matsumoto-Yor property 43
maximum of Wiener process 237, 240–1
m dimensional space, Wiener processes
  274–5
mean
 of continuous random variable 23
 of discrete random variable 22
 instantaneous 223, 225
meander process 239–40
mean recurrence time, Markov chains 121,
  192
mean square integrability 265
measures 101
median of random variables 11–12
meteorite counting 72–5
method of images 281

method of maps 286–7
Metropolis method 153
M/G/∞ queues 217
minimal cut sets 98, 99
minimal invariant measure 139, 147
minimal path sets 98, 99
minimal solutions 123, 124–5, 190
M/M/1 queue 200–1, 202
 busy period 215–16
 last in first out 216
M/M/∞ queues 217
moment generating function (mgf) 33–5
 of binormal density 38, 39
moments 25
 calculation using Itô's formula 271
 conditional 27
monotone convergence 62
monotonicity of system 97
MOTTO 160–1
moves chain 128–9
multinormal density 39
multiplication rule 4, 5
multiserver queues 203–4
mutually accessible states 191

negative binomial distribution 42
negative binomial e 42
neutral Fisher-Wright model 250
nice sets 275
normal random variables, sum of 37
non-anticipating processes 265
non-measurable sets 101
non-null recurrent states 193
normal density 9–10
 scaling and shifting 11
normal distribution 26, 37
normal mgf 34
n-out-of-n (series) systems 98
null events 101
null-recurrent states 193
null state, Markov chains 122

one-dimensional Green function 291
one-one maps 32
one-out-of-n (parallel) systems 97–8
one-variable chain rule 268
one-way Markov switch 169–70
optional stopping 67–8
 martingales 253
Ornstein–Uhlenbeck process 231–3, 235,
  242–3, 258, 260, 273, 293
 equilibrium distribution 291
 squared 294
orthogonal increments of martingales 70

P(.) see probability function
pairwise independence 5
parallel (one-out-of-n) systems 97–8

particles *87–8*
    Markov chains 120–1
partition lemma 28–30
partition rule 4–5
passage probability, Markov chains 122
passage times, Markov chains 121
path sets 98
periodic simple random walk *113*
period of Markov chains *118*
persistent states, Markov chains *122*, 132–3
P-matrix 108–9, 116–17
Poisson approximation 20–1
Poisson arrivals 211–12
Poisson distribution 8
    transformations 19
Poisson-driven Markov chain 173–4
Poisson mgf 35
Poisson process 72–6, 171, 173
polar cap, hitting density, Wiener process 283
Polya's urn 71
populations *see* birth-death processes; birth processes
potential functions 278
Pratt's lemma 62
predictable functions 265
probability 1–3
    conditional 4
probability function 2–3, 7
probability generating function (pgf) 35
    of binomial distribution 36
probability space 3
    completions 101
product rule, stochastic differential equations 271
progressively measurable functions 265
proposal matrix 152
Propp–Wilson method 155–6, 157
pseudo-random numbers *151*
pull-through property 31

Q-matrix 177–8
quadrant, Wiener process in 287–9
quadratic variation 262–3
quadric, Wiener process in *295*
queues 46–7, 162–4, 200–2
    balking 214
    busy periods 215–16
    closed migration processes 206–7
    last in first out 216
    Little's theorem 208–9
    M/G/∞ 217
    M/M/∞ 217
    multiserver 203–4
    renewal theory 77
    shared waiting room 204
    tandem 205–6, 209

quicksort procedure 94–5
quotient rule 294

randomness 1
random processes *45*
    pitfalls 102
random sum
    expectations of 30–1
    mgf 35
random variables 6
    continuous 8–9
    discrete 7
    independent *15*
    median of 11–12
    transformations 16–22
random vectors 12–16
random walks 46, 49–51
    independence of past 108
    left-continuous 165
    as model for gambling 51–2
    and renewal theory 77
    returns to origin 52–5, 56
    reversibility 149–50
    transition probabilities 109
    see also simple random walks
rare events, law of 21
Rayleigh distribution 240
RCLLs (right continuous processes with limits from left) 102
realizations (sample paths) 47
recurrence 139–40
    in continuous time 198–9
recurrence times 128
recurrent states *122*, 130, 132–3, 134
    in continuous time *192*–4
reflected Wiener process 227–8
reflecting barriers 226–8
reflection principle, Wiener process 235–7
refreshed branching 93
regular boundary points 226
regularity, long-run 56–7
regular Markov chains *118*, 119
reliability function 97
    bounds 99
renewal function 76, 77
renewal processes 76–80, 87, 217
    elementary renewal theorem 80–1
    first renewal theorem 80
    and Markov chains 131, 197–8
renewal-reward processes 81
    delayed 87
renewal-reward theorem 82–5, 134
    inspection paradox 85–7
repulsive model 153–4
reversed Markov chains 148
reversible Markov chains *148*, 149–50
    continuous time 202–5
reward function *82*

Riemann–Stieltjes integral 261
ruin probability 246

sample mean 50
sample paths 47, 102
sample space 2
scaled Wiener process 231
Schlömilch's inequality 42
second Borel–Cantelli lemma 58
self sort procedure 95–6
semicircle, Wiener process in 289
series (n-out-of-n) systems 98
shared waiting room 204
sigma field *see* event space
simple random functions 265
simple random walks 51, 53–5
   invariant measure 137
   Markov chains 112–13, 116, 123, 124–5
   and martingales 67, 71
   returns to origin 52, 56
simple systems theorem 65–6
simulation of Markov chains 151–2
skeletons 199
small numbers, law of 21
sorting
   quicksort procedure 94–5
   self sort procedure 95–6
sphere, Wiener process 282–3, 293
squared Wiener process 243
square pipe, Wiener process in 289–90
standard bivariate normal density 37–8
standard normal density 10
   characteristic function 36
standard Ornstein–Uhlenbeck process 231–3
state space 45, 108, 132
   censored 204
stationary distribution 114, 120, 135–47
   in continuous time 196–8
step functions 265–6
Stieltjes integrals 261–2
Stirling's formula 56
stochastic differential equations (SDEs) 267–8
   calculation of moments 271
   Feynman–Kac theorem 272–3
   product rule 271
   use of Itô's formulas 269–71
stochastic integrals 261, 264, 266–7
   mean square integrability 265
   non-anticipating processes 265
   quadratic variation 262–3
   step functions 265–6
   Stieltjes integrals 261–2
stochastic matrices 108
stochastic processes 45–8
stock portfolio model 273–4
stock price model 258–60

stopping times 65, 127–8, 175, 252–3
   martingales 253
   Wiener process 237
strong law of large numbers 60
   proof 61
strong Markov property 128–31
   in continuous-time 175
   Wiener process 237
structure, of Markov chains 117–21
structure function, indicator of functioning system 97
subcritical processes 92
submartingales 64
sum of continuous random variables 17
summation lemma 58
supermartingales 64
survival chain 145–7
symmetric random walks 51, 55, 56
   and martingales 67, 71
systems, reliability 96–100

tail theorem 23
tandem queues 205–6, 209
Taylor's expansion 260
telegraph process 174, 176, 214
three-dimensional Wiener process 281–2, 285, 290–1
tied down Wiener process 233
   see also Brownian Bridge
time inverted Wiener process 231
time-parameter 45
time of system failures 100
topology of Markov chain 117
tower property 32
transformations of random variables 16–22
transformed diffusion processes 243–4
transient states 122, 130, 134
   in continuous time 192–4
transition density function
   diffusion processes 225
   Wiener process 221
transition probabilities 108, 109
transition rate 177
tree of a particle 90
triangle, Wiener process in 295–6
trinomial distribution 13–14
two-dimensional Wiener process 280–1, 285, 290
two-variable chain rule 268

uniform density 9
uniformization, Markov chains 184–5
uniqueness of mgf 34

variance 25
   instantaneous 223, 225
   of random sum 31

vectors
    conditioning on 32
    of Wiener processes 273–4
visiting times 195, 196

waiting room, shared 204
waiting times 201
Wald's equation 68
weak law of large numbers 60
    proof 61
wedges, Wiener processes 277, 287
Wiener processes 220–1, 228–9
    Bessel process 245
    Brownian Bridge 233–5, 241, 242
    Brownian meander 239–40
    and circles 293, 296
    conditioning of 248–50
    in a cone 295
    and conic sections 291
    covariance 229
    drifting 224, 256, 291–2, 293
    expected visiting times 284–5
    exponential (geometric) 233, 244–5
    hitting density 274, 279–82
        polar cap 283
        in sphere 282–3
    joint distributions 229–31
    martingales 251–2
    maximum of 237, 240–1
    m dimensional spaces 273–6
    Ornstein–Uhlenbeck process 231–3, 242–3
    in a quadrant 287–9
    in a quadric 295
    reflected 227–8
    reflection principle 235–7
    scaled 231
    in a semicircle 289
    on a sphere 282–3, 293
    squared 243
    in a square pipe 289–90
    stopping times 253
    in a strip 253, 255–6
    three-dimensional 281–2, 285, 290–1
    time inverted 231
    in a triangle 295–6
    two-dimensional 280–1, 285, 290
    vectors of 273–4
    in a wedge 277–8, 287
    zeros of 238–9, 242
Wright's formula 228

Yule process 174, 176, 191, 215
    forward and backward equations 182–3
zeros of Wiener process 238–9, 242